T0210884

Electronic Collaboration in the Humanities

Issues and Options

Electronic Collaboration in the Humanities
Issues and Options

Edited by

James A. Inman
University of South Florida

Cheryl Reed
Naval Health Research Center, San Diego, California

Peter Sands
University of Wisconsin at Milwaukee

Routledge
Taylor & Francis Group
New York   London

First published by Lawrence Erlbaum Associates, Inc., Publishers
10 Industrial Avenue
Mahwah, New Jersey 07430

Transferred to digital printing 2010 by Routledge

Routledge

270 Madison Avenue
New York, NY 10016

2 Park Square, Milton Park
Abingdon, Oxon OX14 4RN, UK

Cover design by Kathryn Houghtaling Lacey

Library of Congress Cataloging-in-Publication Data

Electronic collaboration in the humanities : issues and options / edited by James A.
  Inman, Cheryl Reed, Peter Sands
      p.   cm.
  Includes bibliographical references and index.
  ISBN 0-8058-4146-6 (cloth : alk. paper) — ISBN 0-8058-4147-4 (pbk. : alk. paper)
    1. Humanities—Study and teaching (Graduate)—United States.   2. Humanities—Data
processing.   3. Humanities—Technological innovations.   I. Inman, James A.   II. Reed,
Cheryl, 1952–   . III. Sands, Peter.

AZ183.U5 .E44   2002
001.3′071′173—dc21   2002069250
                                                                             CIP

10   9   8   7   6   5   4   3   2   1

Contents

A Word to the Fore

Myka Vielstimmig

You know, "fore" is the word you yell at someone your golf ball is about to hit. Are you warning us about something here? ;)

(What do you mean by "us" and "you"?)

"To the fore," I was thinking. As in: a word to those who run ahead of us, to the fore word thinkers, to the garde avant de nous. But also "a word at the fore," at the fore of the essays whose forward thinking this foreword precedes.

Which word, though? Or words? If you were to forward four words to foreword this volume, which would they be?

Collaboration-humanities-electronic-space?

~~or~~

1. Genre
2/3. Rhetorical situation
4. Identity

~~or~~

a) front b) matter c) audience d) absent?

Or no, here's my question: what does one expect from a foreword, and what from a foreword fronting a volume like this? What does the genre enforce, and what does this volume?

I don't know about forewords: as a genre, they are unappreciated and undertheorized. I'm not sure that they are Carolyn Miller's recurring social situation, and Swales doesn't acknowledge them at all.

I think a foreword has a bona fide social purpose, a la Miller. Not the same as an Introduction, which contributes substantively, an intellectual map to the volume; not a preface, which is a word on "how this came to be." A foreword is written by a non-contributor.

A foreword is written afterward, but is forwarded in the
volume; it's even forward of the Introduction.

And its function is epideictic, right? That's a bona fide social
function.   It celebrates, christens, or commends after the
writing is done. It lands at the fore like the queen's sword on the
bowed shoulder of the book.

More pragmatically, the best one I've read, I think, was the one where
Bob Connors plays in a very informal voice to front Greg Clark's
book, or maybe it was Derek Owens' on the Heilker book on essays,
also playful and inventive. Each was smart, but mostly, they were
fun. And substantive: in purpose, they added something to the *idea*
of the volume, rather than to its specifics. They also provided a frame
of some sort.

So our task here is to add something to the idea of the volume,
to be playful perhaps, but in a tone that commends the volume.
Not to forget: one is honored to be invited to write a foreword.
One's own work is perceived worthy, as well as related to the
focus of the volume.

And if I'm not mistaken, writing a foreword
ushers one officially into Geezer Status.

I'm so glad Geezer is a masculine noun. ;)

(Why would that reassure you? Online, nobody knows you're
a femme.)

Exactly: who's online now? ;)

❦

Sherry Turkle's opening essay in this book reminds us that all
who write in electronic spaces suit ourselves in invented
personae.

Which this invented persona isn't willing to stipulate, should anyone
inquire.

(Right: I'll make a note.)

It's nothing unknown in other discourse creations—the essay,
the novel, the sermon, the news report—but truly, is there, as
Turkle implies, something about life on the screen

Now there's a title. ;)

that especially invites invention?

Yes. And no.

Easy for you to say.

Her suggestion that the concreteness of our experience online has opened newly aesthetic levels of work is, I think, provocative. And didn't we already nominate Turkle for the oxymoron of the 1990s: the Romantic Postmodern?

But I think this claim attempts to peculiarize the online experience, like you wouldn't bring your uninvented self to the screen. Sorry, but when I email my dean tomorrow, she'll not think of postmodern narrators. (More's the pity.) The assumption seems that some "real" chasm—geographical, psychological?—separates rl and virtuality.

According to Pierre Levy, it's **all** virtual.

It's all virtual, of course, so you do agree. One doesn't need to peculiarize the online experience to claim this. There is no uninvented self in any discourse (is there?), no face uncomposed, and thus no chasm. I don't think I've seen a dean online or off who wasn't pulling a face. This is not an insult; it's a feature of discourse, of genre, of dialogue.

Some online genres are new, as Chandler suggests: web pages, for example, are both personal and public in a way other texts don't seem to be. And email has the recipient built in/to it. But on the other hand, it's not like we're multiple personalities—

Though Myka certainly sounds so ;)

one for rl, one for vl. I think there is some overlap, like zones. But not, please, contact zones!

(Oh, never. And never multiple personalities. I mean: people you think you know adopting usernyms, pseudonyms, screen identities? Some of them even have multiple nyms, you know. Who *is* Myka really, and what do they have to hide?)

Of course, what Myka is interested in varies, as per . . .

While early work on computers in the humanities classroom focused on pragmatics

and word processing

and later work on the rhetoric of online spaces, what I find myself most interested in is the dawning understanding that the *poetic* is available too. Each utterance online offers itself in as many dimensions as offline utterances do. We've called this an inexact asterisk (Vielstimmig 2001),

(We did?)

where the rhetorical crosses with the poetic with the cultural with the ethical. These intersecting planes, this asterisk of online discourse, is perhaps no more inherent than it is in print or f2f discourse, but there's something about the early years of a medium that invites experimentation. That's my theory, anyway. Experimentation invites play invites the poetic.

The same play we see in forwards?

Foreword movement. (Can you believe how many people misspell that word?)

The concreteness of the medium as it has developed since, say 1995, has become irresistible to many writers.

What do you mean by the concreteness of the medium? Isn't that an oxymoron for what you are describing?

I mean point-n-click, drag-n-drop, icons, colors, mpegs, jpegs, tiffs, pdfs, fonts, scalable point size, moveable margins, zooming, curving, filtering, mode adjusting, layering, blurring, cropping, boxing. It's all there now, it's concrete in that it's manipulable as a trowel. This is Turkle's point, not mine, but even in the writing of code, it's no longer lines and lines of agonized syntax. It's point -n-click, drag-n-drop . . .

So now, the merest English prof can design her/his very own page on the screen. And suddenly we're messing about with the page as much as with the screen.

So cause and effect? Back and forth?

Yea, verily. Migration. (*Must* I quote you to yourself? *Must* I lob you these softballs?)

Precisely: it's not mere migration. That is one option, to be sure: an online phone book looks a lot like a print phone book, which transparency is part of its charm. Functionality matters. But for most of what's inside *these* covers, the *composing* isn't migratory so much as multi-modal: online, onphone, onpage.

And in Myka, of course, composing *is* composition's aesthetic.

Suddenly writing in the humanistic academy is about design as much as it is about critique.

Grusin and Bolter discuss remediated texts, that is, the tendency of the new to create—or re-create—itself through the technology, the frame and the medium of the old. Web as TV, TV as film, and so on. One night in class we looked at web pages and could see one remediated as print magazine, another remediated as TV. Perhaps this is another way to make your point: remediation in cross-dressing.

The box is floating, and I'm having trouble moving it: I assume that's because you shifted it into RTF. I've left it in that format.

Sorry you're having trouble with the box. It was floating and disappearing for me, too. (Hiding, actually, behind the text.) FYI, I did leave it in MS Word—a terrible program for design.

I know. I don't know how it went to RTF. Your file to me was indeed in Word. Aggghhh.

No. They tried remedial cross-dressing with me. I *still* can't do that clasp in the back.

Another set of metaphors we use to describe online writing is located in the relationship of space to place. Look how often we invoke the virtues of cyberspace, how frequently we talk about it as a new space that can be mapped and inhabited in new ways. Tuan, the humanist geographer, helps us understand this in yet another way, however. He talks about how we transform "space" into "place," about how the transformation isn't quite that but rather partly that, only partly a migration from the old. In other words, suppose we did think of the online as a space that through our habitation we *make into* a place, one that—as you point out above—is indebted to print, but one that is different from print, too. Or: we make the space of online into a place where we write, where we create a new kind of reality—a reality that isn't *merely* virtual—not if everything is already virtual.

Abstract space, lacking significance other than strangeness, becomes concrete place, filled with meaning. Much is learned, but not through formal instruction.
(Tuan ,1977)

If the screen can be a place made by writing, then what sort of habitation does it make for the word? A new kind of reality, you're saying.

I think this is what we are trying to discern and create simultaneously.

A new kind of writing, too, then?

> That's the hope and the claim. There's almost an impulse to will it into being (so), then a kind of (critical) reaction, then a move forward. ;)

> > But don't you think *this* tends to peculiarize the
> > experience of the online?

It's not peculiar to the extent that it replicates what is, which is Bolter's point. It is peculiar in that it desires the new and claims the unique.

❧

Insofar as this volume is preoccupied with collaboration, does it suggest that the online habitation socializes the life on the screen toward collaborative work? You can see it in Inman's cross-temporal collaboration, in Gajjala and Mamidipudi's cross-national collaboration, in many other chapters here. Something about the online dispensation, they might say, invokes not just the virtuality of identity, but the real-world effect of a collaborative.

> In its connectedness, the medium encourages a collective authorial identity, yes.

> Not just composed selves, but concrete P/T rewards in the
> humanities for collaborated work. Not just a new academy,
> > but a new world. Vitanza would
> > > agree.

I am interested in the shaping forces of electricity on literacy that wants to become electracy. On what will have become...

Total Collaboration!

(Vitanza, 2001)

Which in its own way is the same point that Bolter makes, only his field of interest is medium. You can't invent anything without explicit reference to, allusion to even, a prior. Daniel Chandler (1998) talks about the bricolage of the old being bootlegged into the new. And if it's new, then it will require a new language, new metaphors: the language of the old will not suffice.

> Especially if you think that language
> plays a part in the construction of
> > the online.

Everything is a priori.

Another thing that seems clear to me is that it is a new rhetorical situation (and I think it's prompting new genres like the web); the networking capacity itself constitutes a new rhetorical situation, I think.

The linking in fact, is material, and—as Madonna reminds us—we're material girls.

(Online, no one knows you're a "grrl.")

We are very much nonhierarchical collaborators, personalizing in our own writing process the antidote to painful disconnects we saw harming so many ... The natural link to maintain this kind of writing relationship and to meet publishing deadlines was technology.

Reed and Formo (this volume)

In terms of this book, I think the most interesting part is that the response the (networked) situation is evoking, if we want to use that language, is a new authorial identity. That's the claim of the book, and that's the experience of Myka.

Chapters here suggest, however, that new identity (or even successful collaboration) is not guaranteed in the electronic composing media—even if it's true that the rhetorical situation online invites it. When the medium becomes such a powerful partner, exerting its own shaping influence on the writing, it is not surprising that many of these chapters relate narratives of struggle as well as hope. Struggle with the dynamics of collaboration, but struggle as well with the demands of media. This is *also* Myka's experience. (In spades.)

In fact, we jumped to a typesetting program in the middle of writing this foreword because designing the page itself has become an issue, and we couldn't manage design as well in a word-processing program, let alone on e-mail. (This choice, in turn, has created issues for the volume editors, and the publisher's editorial staff, too—other collaborators whose invisible electronic hands have shaped things here.)

To borrow from our volume editors here, "just as collaboration changes the nature of any project, so do information technologies change the nature of collaboration." This seems a fundamental premise to the work of all we who collaborate online,

and the book moves from this point forward.

(Oh, nice segue.)

References

Chandler, D. 1998. Personal Home Pages and the Construction of Identities on the Web. Aberystwyth (Wales) Post-International Group Conference Issues in the Politics of Identity. (September): www.aber.ac.uk/media/Documents/short/webident.html.

Grusin, R., and J. D. Bolter. 2000. *Remediation: Understanding new media.* Boston: MIT Press.

Levy, P. 1998. *Becoming virtual: Reality in the digital age.* Trans. Robert Bononno. New York: Plenum.

Miller, C. 1984. Genre as Social Action. *Quarterly Journal of Speech,* 70:151-67

Reed, C. and D. Formo. 2002. Writers anomalous. *Electronic collaboration in the humanities.* Ed. Inman et al. Mahwah, NJ: Erlbaum.

Swales, J. 1990. *Genre analysis: English in academic and research discourse.* Cambridge: Cambridge University Press.

Tuan, Yi-Fu. 1977. *Space and place: The perspective of experience.* Minneapolis: University of Minnesota Press.

Turkle, S. 1995. *Life on the screen.* NY: Simon and Schuster.

Vielstimmg, M. 2001. A play on texts. In *New worlds, new words.* Ed. Barber and Grigar. NJ: Hampton.

Vitanza, V. 2001. The shaping force of electronic texts and journals on our professional work. *The writing instructor.* Beta 1.0. http://flansburgh.english.purdue.edu/twi/

PREFACE

Issues and Options for Electronic Collaboration in the Humanities: A Framework

James A. Inman
University of South Florida

Cheryl Reed
Naval Health Research Center, San Diego, California

Peter Sands
University of Wisconsin at Milwaukee

Contemporary humanities education is challenging, insecure. Where teacher-scholars once could examine intellectual traditions of the past to remind themselves of academic values and practices, the postmodern, postindustrial era has seen a re-action against such definable concepts, especially in terms of their representation and inclusion of minority and alternate voices. In many respects, the nature of the humanities itself has been threatened with fragmentation beyond reasonable assembly and, at times, erasure from educational memory, as its core values have diffused widely and diversely enough that they can no longer be easily recognized. Although it would be too reactionary to suggest that the survival of the humanities is in doubt, clearly the contemporary academy challenges the humanities in new and conse-quential ways. We do not know what the future will bring.

One of the influences impacting humanities education most prominently and profoundly is the rise of information technologies. As Jean Francois Lyotard de-scribes in *The Postmodern Condition: A Report on Knowledge* (1997), electronic media require a new paradigm of knowledge production and consumption, one with serious implications. Instead of individuals locating and making use of print-only resources for projects, the contemporary knowledge quest may often be met by computer database access, an interface that well reflects shifting values in the humanities because it calls into question the nature of research practices them-selves. But, understanding transformations in knowledge transactions does not re-veal everything. Indeed, information technologies also enable new genres of communication, important opportunities with critical implications. Finally, then, humanities scholars cannot imagine information technologies as invisible conduits for research and communication; instead, the technologies must be critically and thoroughly examined because they impact educational practices.

Electronic Collaboration in the Humanities is one response to the challenges information technologies bring to the humanities. This book's development and composition reflect the following five ideals, each boldfaced for emphasis:

1. Humanities scholars with all levels of access are doing important work with technology. Readers will recognize scholars like Sherry Turkle, whose reputations as scholars have been built on understanding how electronic technologies shape the way we think and communicate and whose work has, in fact, shaped the profession itself. The reader will also recognize other senior scholars, like Anne Ruggles Gere and Stephen Tchudi, who have brought important voices to social and cultural issues in literacy studies. More, readers will encounter the voices of artist-innovators, such as John Craig Freeman and Timothy Allen Jackson, who imagine technological futures amidst contemporary aesthetic media. *Electronic Collaboration* also includes scholars such as Cheryl Reed, Dawn M. Formo, and Alice L. Trupe, who have varying levels of access and support who are working to bring whatever level of access they have to their own work and to their students. Finally, the book includes the voices of scholars such as Randall Bass and Tari Fanderclai, who have significant and consistent access and who have used it to shape their own and their students' negotiations of the academy.

2. Humanities scholars' projects with technology reflect significant diversity, both across and within disciplinary bounds. *Electronic Collaboration*'s chapter authors well reflect contemporary diversity in the humanities—their disciplinary foundations include literary studies, writing studies, media studies, history of science and technology, women's studies, information science, American studies, literacy studies, technical and professional communication, graphic art and design, and communication studies.[1] In assembling this collection, we have worked to provide these scholars with opportunities to demonstrate how their disciplinary expertise and experiences can inform broader issues in the humanities, and in taking this approach, we have greatly benefited from their previous and current pursuits of interdisciplinary projects. Indeed, interdisciplinarity seems to be a hallmark of the contemporary humanities. Readers of *Electronic Collaboration* will encounter different genres of research writing, from formal interrogations of complex subjects to shorter reflective engagements of key issues to montage and other arrangements of appropriate humanities texts, and we believe this diversity is best shared on its own terms.

---

[1]Many readers may know humanities computing as a field of inquiry, as connected to the Humanist discussion list and the work of senior scholars such as Willard McCarty. Although chapters in *Electronic Collaboration* do share scholarship in humanities computing, we believe electronic collaboration in the humanities is best constructed broadly, not as an area of specialization but as an interdisciplinary investment in sound teaching and research practices in the 21st century.

3. Using information technologies in the humanities is a continuous conversation. We do not mean *Electronic Collaboration* as the final word on electronic collaboration in the humanities, though we believe it well represents key issues currently evolving. Instead, we wish to suggest that careful and well-reasoned progress for the humanities must come as a result of consistent dialogic encounters among a diverse range of teacher-scholars. In editing this collection, we have emphasized the need for conversation in several ways:

- First, the section responses authored by senior scholars in the humanities bring forward central elements of individual chapters and articulate a logic for their relation.
- The first chapter is a version of a talk originally presented as a keynote address to the Eighth International Symposium on Electronic Art. It is an example of the kind of conversation that we believe needs to happen between social scientists and arts practitioners.
- Finally, the collection's genesis is in conversations among participants in the Computer Research Section of the 1998 Midwest Modern Language Association Convention.

4. Information technologies offer new options for humanities education. At the outset of this preface, we noted that the rise of information technologies is one of many influences reshaping the contemporary humanities, and in *Electronic Collaboration* we imagine these technologies as not necessarily threatening. Instead, we see them as adding new and important options to humanities teacher-scholars' catalogs of effective pedagogical and research practices. Readers will encounter technology success stories in this book, such as Dagmar Stuehrk Corrigan and Simone M. Gers's cross-institutional collaboration in writing classes across urban and minority populations and Karen L. McComas's professional renewal in collaborative MOO communities. Additionally, chapter authors articulate pedagogical opportunities available only through technologies, including Christina L. Prell's experience with service learning and electronic document authoring and support and Paul J. Morris's forging of connections between the academy and industry. However, *Electronic Collaboration* also presents a strong critical voice, arguing for the sound and reasonable introduction of information technologies into humanities spaces; chapter authors such as Jami Carlacio and Donna N. Sewell articulate and apply critical frameworks for analysis, demonstrating their use in both theoretical and pragmatic dimensions.

5. Just as collaboration changes the nature of any project, so do information technologies change the nature of collaboration—its speed, character, methods, and possible implementations. Collaborative projects have become increasingly prominent in the humanities in recent years, led by scholars such as Lisa Ede and Andrea Lunsford, who have coauthored several highly regarded books, including *Singular Texts/Plural Authors: Perspectives on Collaborative*

Writing (1990), and Kenneth Bruffee, whose *Collaborative Learning: Higher Education, Interdependence, and the Authority of Knowledge* (1993) is a standard read across the contemporary academy. Only now, however, is detailed attention being given to the specific way that information technologies impact collaboration in the humanities. In *Electronic Collaboration,* readers will find essays that expand models of collaboration based on the influence of technology, including Radhika Gajjala and Annapurna Mamidipudi's employment of postcolonial theory in section 1, as well as James A. Inman's linking of the concept of electracy with cross-temporal collaboration. This book project, we hope, serves as a first step by validating a suite of electronic collaborative practices and demonstrating how teacher-researchers bring specific strengths to their work in the contemporary humanities, strengths that enable electronic collaboration in the humanities to be both unique and rich with promise.

Acknowledging and engaging these five ideals is at the heart of *Electronic Collaboration,* because we believe they serve as themes for the future of the humanities.

OVERVIEW

Section 1 opens with an essay by Sherry Turkle that highlights the issues from her well-known *Life on the Screen* that she believes to be most relevant to students of online identity in the context of the psychodynamic, literary, and cultural discourses of the humanities. Turkle considers the nature of online identities and roles and the relationship between the psychoanalytic culture of the past century and the computer culture of the coming century and millennium. In "What's So Democratic About CMC? The Rhetoric of Techno-Literacy in the New Millennium," Jami Carlacio questions the uncritical rhetoric often associated with the rise of computer technologies in the contemporary academy, especially because that rhetoric assumes social and cultural empowerment that may not ever be realized; Carlacio grounds her critique in excerpts of her own classroom practice, making tellingly evident the necessarily dialogic nature of rhetorical critique. Crafting an important link between writing across the curriculum projects and the nature of computer-mediated communication (CMC) in "Computer-Mediated Communication as Reflective Rhetoric-in-Action: Dialogic Interaction, Technology, and Cross-Curricular Thinking," Rebecca J. Rickly explains how information technologies enable new collaborative relationships to be encouraged and constructed. James A. Inman's contribution, "Electracy for the Ages: Collaboration With the Past and Future," explores the possibility of collaborating across the past, present, and future; expanding on Gregory Ulmer's theory of electracy, Inman works through the implications of a temporally expanded concept of collaboration. Finally, in their collaborative chapter, "Collaborating Across Contexts: Rethinking the Local and the Global, Theory and Practice," Radhika Gajjala and Annapurna Mamidipudi experiment with physical form and display both the context and content of their own electronic collaboration across the physical and cultural divide separating mid-America and India. Stephen Tchudi authors this section's response.

In section 2, "Student Collaboration and Electronic Media," writers focus on the social and cognitive collisions electronic collaborations provoke. Nancy Knowles and M. Wendy Hennequin speak of the emergence and the convergence of three different collaborative relationships: student–student, teacher–teacher, and student–teacher in "New Technology, Newer Teachers: Computer Resources and Collaboration in 'Literature and Composition.' " Knowles and Hennequin address a range of technologies, including electronic discussion lists, the World Wide Web, synchronous conferencing, and audiovisual distance learning, to ask how a technology-based collaboration might represent an appropriate blurring of disciplinary boundaries. In "Voices Merged in Collaborated Conversation: The Peer Critiquing Computer Project," Mary E. Fakler and Joan E. Perisse suggest that students' analytical and interpretive skills are sharpened through technologically enhanced peer critique. In a semester-long project using asynchronous conferencing software, Fakler and Perisse found that students developed a heightened awareness of different styles of writing, learned techniques for functioning within group dynamics, and developed an increased awareness of socioeconomic and gender issues in writing. Alice L. Trupe analyzes a classroom ethnographic study of reentry women students at an urban community college in "Reentry Women Students' Online Collaboration Patterns: Synchronous Conferencing in a Basic Writing Class." Analyzing student interactions in an online environment, she discusses factors at work both in narratives of success—the productive use of networked conferencing—and in narratives of failure—resistance and nonparticipation that resulted in unsatisfactory attempts to engage the writing task. In "Using a Virtual Museum for Collaborative Exhibit Design in an Undergraduate Course," Jo B. Paoletti, Mary Corbin Sies, and Virginia Scott Jenkins describe the collaborative formation and construction of the Virtual Greenbelt Museum, an electronic companion to the local Greenbelt Museum; they extend their discussion to demonstrate how the virtual museum has been useful in undergraduate American studies classes, and they reflect on their experiences to outline specific lessons for humanities educators to learn about electronic collaboration. Dagmar Stuehrk Corrigan and Simone M. Gers discuss the impact of students' personal literacies on academic and technological literacies in "Across the Cyber Divide: Connecting Freshman Composition Students to the 21st Century." Pairing English as a second language student populations from a rural community college located in the desert southwest with e-pals from an inner-city, 4-year university located in the south, Corrigan and Gers explore students' abilities to generate meaningful and productive discussions through electronic media. They discuss how students' evolving understandings of audience affected their sense of writing purpose, style, and tone. In "Web Writing and Service Learning: A Call for Training as a Final Deliverable," Christina L. Prell takes a hard look at the sustainability of class projects that require students to write Web pages as a community service. What happens, she asks, when the semester ends, students move on, and the recipients of this service are faced with maintaining a Web presence that may well exceed their own technological expertise? Prell offers productive suggestions on how instructors can create service-learning projects that extend beyond the confines of the current school term. Bill Freidheim authors this section's response.

Section 3, "Faculty Collaboration and Electronic Media," shares scholarly perspectives on collegial and professional collaboration, shared spaces, and ventures too rarely addressed in humanities scholarship. In "Writers Anomalous: Wiring Faculty Research," Cheryl Reed and Dawn M. Formo articulate the dynamics of their collaborative research and writing relationship; they introduce readers to a series of collaborative scenes and reflect on their own practices, demonstrating for readers how collaboration can be foregrounded as a fundamental aspect of contemporary professional practice in the humanities. Donna N. Sewell, in "What's in a Name? Defining Electronic Community," forwards a cross-media perspective on electronic collaboration in the humanities, suggesting that some characteristics of both synchronous and asynchronous forums can and should be mutually informing. Sewell uses her expertise in ethnography to craft rich descriptions of communities in which she's involved. Karen L. McComas's "Cow Tale: A Story of Transformation in Two MOO Communities" describes how electronic media offer opportunities for professional renewal; weaving critical consideration of collaboration with her own experiences in an educational MOO, McComas demonstrates how collaborative relationships among online characters can be both professionally and personally meaningful. In the section's last chapter, "Humanities Scholarship, Computing, and the Library: The Collaboration That Created the Kolb/Proust Archive," Jo Kibbee and Caroline Szylowicz present a case study of their collaborative work in the creating of the Kolb/Proust Archive at the University of Illinois, emphasizing ways that collaboration enabled the construction of a unique resource, one with significant implications for both offline and electronic collaboration in the humanities; more, Kibbee's and Szylowicz's chapter makes a connection too often ignored in the humanities between scholars of information science and those of other humanities disciplines. T. Lloyd Benson responds to this section's chapters.

The final section of *Electronic Collaboration,* "Electronic Collaboration and the Future," includes a diverse range of perspectives on the future of humanities education as it may be informed and shaped by opportunities for electronic collaboration. In "Imagining Future(s): Towards a Critical Pedagogy for Emerging Technologies," Timothy Allen Jackson argues that the formulation of critical teaching and learning practices can lead to an empowered student population, one well prepared to meet the challenges of the 21st century. Jackson articulates and explores elements of his model of critical pedagogy. Paul J. Morris, in "Critical and Dynamic Literacy in the Computer Classroom: Bridging the Gap Between School Literacy and Workplace Literacy," suggests that writing students in the contemporary academy can benefit from collaborative relationships with industry, especially because these relationships are constructed in electronic media, and surveys current scholarship about such collaborative ventures. Morris describes specific classroom applications that he has employed, and he suggests alternatives for readers who are interested in pursuing similar directions. Demonstrating a likewise industry-informed approach to electronic collaboration in the humanities, Tari Lin Fanderclai explores the open-source software movement in the computing industry in her "Collaborative Research, Collaborative Thinking: Lessons from the Linux Community," arguing that the movement's progress offers important knowledge about the impact of information technologies on any organization, whether in or outside of the academy. Fanderclai adopts Eric Raymond's cathedral/bazaar development model to suggest specific options that humanities teacher-scholars should consider in their thinking about future collaboration. In "Current and Future Research in the Production and Analysis of Electronic Text in the Humanities: Bridging Our Own 'Two Cultures' With Integrated, Empirical Studies," Peter Sands argues for a new approach to contemporary research about electronic teaching and learning in the humanities, one grounded in empirical inquiry, especially because it enables new and important means of textual analysis and interpretation. Sands demonstrates how such a research agenda enables the humanities' two cultures—the quantitative and qualitative camps—may be bridged productively. John Craig Freeman's "Imaging Florida: A Model Interdisciplinary Collaboration by the Florida Research Ensemble" outlines early outcomes of the Imaging Florida Project, an effort born at the University of Florida that attempts to invent and promote new educational practices suggested by electronic media. Freeman specifically suggests that the contemporary academy reflects an evolutionary genre of meaning making best known as postliteracy, and as a graphic artist he performs his argument, crafting his text as a postliterate montage of critical analysis and electronic artifacts. Randall Bass authors this section's response.

We invite readers to join with us in examining together the future of electronic collaboration in the humanities: its promise, its perils, and its character. Only with such a thorough and multivocal conversation can the most careful and responsible futures be forged.

ACKNOWLEDGMENTS

In preparing this volume, we've relied on many colleagues and friends, and we want to publicly thank them here. We offer specific thanks but also mean this acknowledgment to include all of the many generous individuals who contributed in some way to this project. We appreciate all that you have done.

James A. Inman: I'd like to thank members of my family, especially Ralph and Sandra Inman, who have made a difference for me for much longer than this project's lifespan. And I'd like to thank colleagues and friends at the University of Michigan, Furman University, and the University of South Florida, who supported, encouraged, and challenged me. I'm especially grateful to Anne Ruggles Gere, mentor extraordinaire and author of this book's afterword, for her guidance and support throughout this project and others.

Cheryl Reed: I would like to acknowledge students who have been told over and over again that they simply aren't good enough to succeed. I thank teachers who ignore labeling and encourage students to explore their potential anyway. Many of them contributed to this book.

Peter Sands: I want to thank my wife and daughters, Karen, Fiona, and Anya, who always support me even when I get that work-crazed look in the eye. And I'd like to acknowledge the financial and temporal support of the UWM Preparing Future Faculty (PFF) Program, the UWM Graduate School, and the Center for Twentieth-Century Studies (now Twenty-First), who provided financial and other support during the time we worked on this book. The ideas that led me to the PFF program came out of both my graduate seminar in computers and pedagogy, where I suspect I learned more from the students than they from me, and a very rewarding year I spent as a Wisconsin Teaching Fellow. Finally, I'd like to thank the Department of English at UW-Milwaukee, which is an amenable home for someone like myself, interested in too many things.

Together, we would like to thank the authors—smart colleagues who worked diligently in preparing their chapters and who taught us a great deal. We also thank the outstanding publishing team at Lawrence Erlbaum Associates. They have been terrific at all stages of this project.

I

THEORIES OF ELECTRONIC COLLABORATION

1

Collaborative Selves, Collaborative Worlds: Identity in the Information Age[1]

Sherry Turkle
Massachusetts Institute of Technology

> *There was a child went forth every day,*
> *And the first object he look'd upon, that object he became.*

These words by the poet Walt Whitman (1881) capture a relationship that is central to our understanding of technology and identity: We make our objects, and in turn, our objects make and shape us. In the case of computational technologies, we come to see ourselves differently as we catch sight of our images in the mirror of the machine.

In the 1980s, when I first called the computer a second self, these identity transforming relationships were most usually one on one, a person alone with a machine (1984). This is no longer the case. A rapidly expanding system of networks now links millions of people together in new spaces that are changing the way we think, the nature of our sexuality, the form of our communities, thus our very identities. In cyberspace, we are learning to live in virtual worlds. We may find ourselves alone as we navigate virtual oceans, unravel virtual mysteries, and engineer virtual skyscrapers. But increasingly, when we step through the looking glass, other people are there as well.

In the spirit of Whitman's lines, what are we becoming when the first objects we look upon are virtual objects—objects on a computer screen? What are we becoming if the first objects we look upon are representations of ourselves in virtual

---

[1] This text is adapted from my *Life on the Screen: Identity in the Age of the Internet* (New York: Simon & Schuster, 1995).

space, interacting with other virtual representations? These questions lie at the intersection of technology, art, and psychology because, today, the objects on our screens are both aesthetic and deeply evocative.

TECHNOLOGY AND IDENTITY

For more than a decade, I have studied people's online experiences, focusing on their impact on personal identity. Online identity play is perhaps most explicit in role playing virtual communities, such as Multi-User Domains, or MUDs, where participation literally begins with the creation of a persona (or several), but it is by no means confined to these somewhat exotic locations. In bulletin boards, newsgroups, and chat rooms, the creation of personae may be less explicit than on MUDs, but it is no less psychologically real. One IRC (Internet Relay Chat) participant describes her experience of online talk: "I go from channel to channel depending on my mood . . . I actually feel a part of several of the channels, several conversations . . . I'm different in the different chats. They bring out different things in me." Identity play can happen by changing names as well as by changing places. Online services offer their users the opportunity to be known by several different names. It is not unusual for someone to be BroncoBill in one online community, AmaniBoy in another, and MrSensitive in a third. In changing names, people can explore different aspects of self, some of which may be insufficiently explored in the physical real.

People invest their virtual representations with power and presence. In cyberspace, it is well known that one's body can be represented by one's own textual description, so the obese can be slender; the beautiful, plain. The fact that self-presentation is written in text means that there is time to reflect on and edit one's composition, which makes it easier for the shy to be outgoing; the nerdy, sophisticated. Virtual self-fashioning has become a new, populist form of performance art, having things in common with street theater, improvisational theater, commedia dell'arte, and script writing. And as players participate in online world building, they become authors not only of text but also of themselves, constructing selves through virtual social interaction. As one participant in a MUD tells me,

> You can be the opposite sex. You can be more talkative. You can be less talkative. Whatever. You don't have to worry about the slots other people put you in as much. It's easier to change the way people perceive you, because all they've got is what you show them. They don't look at your body and make assumptions. They don't hear your accent and make assumptions. All they see is your words.

In traditional theater and in role-playing games that take place in physical space, one steps in and out of character; MUDs, in contrast, offer a parallel life. One's character or characters can become parallel identities because in a certain sense the virtual games don't have to end. Their boundaries are fuzzy; the routine of playing them becomes part of participants' daily lives. For example, people who experiment with virtual worlds often work with computers at their regular jobs. As they participate in a virtual community, such as a MUD, they periodically put their characters to sleep—remaining logged on to the game but pursuing other activities: they work on their spreadsheets, labor over papers in progress. From time to time, they return to the game space. In this way, they break up their workdays and experience their lives as a "cycling through" between the real world or real life, sometimes termed RL, and a series of virtual spaces.

CYCLING THROUGH: IDENTITY AND MULTIPLICITY

This kind of cycling-through interaction with virtual life is made possible by the existence of what are called windows in modern computing environments. Windows are a way of working with a computer that makes it possible for the machine to place you in several contexts at the same time. As a user, you are attentive to only one of the windows on your screen at any given moment, but in a certain sense you are a presence in all of them at all times. You might be writing a paper on bacteriology and using your computer in several ways to help you: you are present to a word-processing program on which you are taking notes and collecting thoughts; you are present to communications software, which is in touch with a distant computer for collecting reference materials; and you are present to a simulation program, which is charting the growth of bacterial colonies when a new organism enters their ecology. Each of these activities takes place in a window, and your identity on the computer is the sum of your distributed presence.

The development of the windows metaphor for computer interfaces was a technical innovation motivated by the desire to get people working more efficiently by cycling through different applications much as time-sharing computers cycle through the computing needs of different people. But in practice windows have become a potent metaphor for thinking about the self as a multiple, distributed, time-sharing system. The self is no longer simply playing different roles in different settings, something that people experience when, for example, one wakes up as a lover, makes breakfast as a mother, and drives to work as a lawyer. The life practice of windows is of a distributed self that exists in many worlds and plays many roles at the same time.

This notion of the self as distributed and constituted by a process of cycling through undermines many of our traditional notions of identity. Identity, after all, from the Latin *idem,* literally refers to the sameness between two qualities. On the Internet, however, one can be many and usually is. If, traditionally, identity implied oneness, life on today's computer screen implies multiplicity, heterogeneity, and fragmentation.

In the late 1960s and early 1970s, I was first exposed to notions of identity and multiplicity. These ideas—most notably that there is no such thing as the ego, that each of us is a multiplicity of parts, fragments, and desiring connections—took place in the intellectual hothouse of Paris, where the world was presented according to such authors as Jacques Lacan, Gilles Deleuze, and Felix Guattari. But, despite such ideal conditions for absorbing theory, my French lessons remained abstract exercises. When 30 years later, I use my personal computer and modem to join online communities, I experience this theoretical perspective as brought shockingly down to earth. I use language to create several characters. My textual actions are my actions; my words make things happen. I create selves that are made and transformed by language. And different personae explore different aspects of my identity. The notion of a decentered identity is concretized by experiences on a computer screen.

The poet Whitman, who wrote so beautifully of how we are constituted by the object world, also said, "Do I contradict myself? Well, then I contradict myself. I contain multitudes" (1882). Life on the screen, very much in the spirit of Whitman, involves a redefinition of identity as multiplicity. It challenges our notion of psychological health as following from a completely integrated self; rather, it suggests that psychological well-being follows from being able to negotiate flexible transitions among self states or aspects of self. The artistic community is, of course, a crucial actor in creating the experiences, relationships, and environments that support, interpret, and give meaning to these transitions. And a good measure of these environments will be virtual.

A PERFORMATIVE LIFE

I want to illustrate the psychological and aesthetic dimensions of virtual self-representation through a case study of a man I shall call Case, a 34-year-old industrial designer happily married to a coworker. Case spends many hours participating in role-playing virtual communities and explicitly thinks of his online self-fashioning as a form of artistic endeavor. Case's performative life has become a way to experience what might be termed virtual consciousness raising and to work through psychological issues that are difficult for him to address in the physical real.

Case is currently participating in a MUD community (a practice I shall refer to as MUDding) as an elaborately constructed female character whom he calls Mairead. Case has created Mairead in an online community with a medieval theme. She is a commoner who has risen above her class because her lover, a nobleman, has helped her to get a legal education (the flavor of the MUD is medieval—its details are not). In response to my question, "Has MUDding ever caused you any emotional pain?" Case says,

Yes, but also the kind of learning that comes from hard times. I'm having pain in my playing now. The woman I'm playing in MedievalMUSH [Mairead] is having an interesting relationship with a fellow. Mairead is a lawyer. It costs so much to go to law school that it has to be paid for by a corporation or a noble house. A man she met and fell in love with was a nobleman. He paid for her law school. He bought my [Case slips into referring to Mairead in the first person] contract. Now he wants to marry me although I'm a commoner. I finally said yes. I try to talk to him about the fact that I'm essentially his property. I'm a commoner, I'm basically property and to a certain extent that doesn't bother me. I've grown up with it, that's the way life is. He wants to deny the situation. He says, "Oh no, no, no . . . We'll pick you up, set you on your feet, the whole world is open to you." But everytime I behave like I'm now going to be a countess some day, you know, assert myself—as in, "And I never liked this wallpaper anyway"—I get pushed down. The relationship is pull up, push down. It's an incredibly psychologically damaging thing to do to a person. And the very thing that he liked about her—that she was independent, strong, said what was on her mind—it is all being bled out of her.

Case looks at me with a wry smile and sighs. "A woman's life," he states. He continues:

I see her [Mairead] heading for a major psychological problem. What we have is a dysfunctional relationship. But even though it's very painful and stressful, it's very interesting to watch myself cope with this problem. How am I going to dig my persona's self out of this mess? Because I don't want to go on like this. I want to get out of it . . . You can see that playing this woman lets me see what I have in my psychological repertoire, what is hard and what is easy for me. And I can also see how some of the things that work when you're a man just backfire when you're a woman.

Case has played Mairead for nearly a year, but even a brief experience playing a character of another gender can be evocative. William James said, "Philosophy is the art of imagining alternatives." MUDs are proving grounds for an action-based philosophical practice that can serve as a form of consciousness-raising about gender issues. For example, on many MUDs, offering technical assistance has become a common way in which male characters "purchase" female attention, analogous to picking up the check at an RL dinner. In real life, our expectations about sex roles (who offers help, who buys dinner, who brews the coffee) can become so ingrained that we no longer notice them. On MUDs, however, expectations are expressed in visible textual actions, widely witnessed and openly discussed. When men playing women are plied with unrequested offers of help on MUDs, they often remark that such chivalries communicate a belief in female incompetence. When women play men on MUDs and realize that they are no longer being offered help, some reflect that the offers of help that they routinely received when they presented themselves as women may well have led them to believe that they needed help. A college sophomore says that as a woman "first you ask for help because you think it will be expedient. Then you realize that you aren't developing the skills to figure things out for yourself."

In this sense, the evocative nature of online gender swapping has resonance with the work of Shakespeare, who used gender swapping as a plot device for facilitating the reframing and reconsideration of personal and political choices. *As You Like It* is a classic example, a comedy that uses gender swapping to reveal new aspects of identity and to permit greater complexity of relationships. In the play, Rosalind, the duke's daughter, is exiled from the court of her uncle Frederick, who has usurped her father's throne. Frederick's daughter, Rosalind's cousin Celia, escapes with her. Together they flee to the magical forest of Arden. When the two women first discuss their escape plan, Rosalind remarks that they might be in danger because "beauty provoketh thieves sooner than gold." In response, Celia suggests that they would travel more easily if they rubbed dirt on their faces and wore drab clothing, thus pointing to a tactic that frequently provides women greater social ease in the world—becoming unattractive. Rosalind then comes up with a second idea—becoming a man: "Were it not better,/ Because that I am more than common tall,/ That I did not suit me all points like a man?"

In the end, Rosalind and Celia both disguise themselves as boys, Ganymede and Aliena. In suggesting this ploy, Rosalind proposes a disguise that will be both physical ("A gallant curtle-ax on my thigh,/ A boar spear at my thigh") and emotional ("and—in my heart,/ Lie there what hidden women's fear there will."). She goes on, "We'll have a swashbuckling and martial outside,/ as many other mannish cowards have/ That do outface it with their semblances." In these lines, Rosalind does not endorse an essential difference between men and women; rather, she suggests that men routinely adopt the same kind of pose she is now choosing. Biological men have to construct male gender just as biological women have to construct female gender. If Rosalind and Celia make themselves unattractive, they will end up less feminine. Their female gender will end up deconstructed. Both strategies—posing as men and deconstructing their femininity—are games that females play when they participate in MUDs. One MUD participant, a woman currently in treatment for anorexia, described her virtual body this way:

> In real life, the control is the thing. I know that it is very scary for me to be a woman. I like making my body disappear. In real life that is. On MUDs, too. On the MUD, I'm sort of a woman, but I'm not someone you would want to see sexually. My MUD description is a combination of smoke and angles. I like that phrase "sort of a woman." I guess that's what I want to be in real life too.

As You Like It makes another point that is relevant to the experience of identity play in virtual environments. When Rosalind and Orlando meet "man to man" as Ganymede and Orlando, they are able to have conversations about love quite different from those that would be possible if they followed the courtly conventions that constrain communications between men and women. In this way, the play suggests that donning a mask, adopting a persona, can be a step toward reaching a deeper truth about one's self and others. This certainly was true for Case, the graphics designer who plays the online Mairead in MedievalMUSH. For Case, gender swapping was a way of working through significant psychological issues that have limited him in the real.

GENDER TROUBLE

Case describes his RL persona as a nice guy, a "Jimmy Stewart-type like my father." He says that in general he likes his father and he likes himself, but he feels that both have paid a price for their low-key ways. In particular, Case feels at a loss when it comes to confrontation, both at home and in business dealings. Although some men find that MUDding as a female makes it easier to experiment with the collaborative side of their nature, Case likes MUDding as a female because it makes it easier for him to be aggressive and confrontational. Case plays several online "Katherine Hepburn-types," strong, dynamic, "out there" women who remind him of his mother, "who says exactly what's on her mind and is a take-no-prisoners sort." Case says,

> For virtual reality to be interesting it has to emulate the real. But you have to be able to do something in the virtual that you couldn't in the real. For me, my female characters are interesting because I can say and do the sorts of things that I mentally want to do, but if I did them as a man, they would be obnoxious. I see a strong woman as admirable. I see a strong man as a problem. Potentially a bully.

In other words, for Case, if you are assertive as a man, it is coded as "being a bastard." If you are assertive as a woman, it is coded as "modern and together." He says:

> My wife and I both design logos for small businesses. But do this thought experiment. If I say "I will design this logo for $3000, take it or leave it," I'm just a typical pushy businessman. If she says it, I think it sounds like she's a "together" woman. There is too much male power-wielding in society, and so if you use power as a man, that turns you into a stereotypical man. Women can do it more easily.

Case's gender swapping has given him permission to be more assertive within the MUD and more assertive outside of it as well:

> There are aspects of my personality—the more assertive, administrative, bureaucratic ones—that I am able to work on in the MUDs. I've never been good at bureaucratic things, but I'm much better from practicing on MUDs and playing a woman in charge. I am able to do things—in the real, that is—that I couldn't have before because I have played Katherine Hepburn characters.

Case says his Katherine Hepburn-like personae are "externalizations of a part of myself." In one interview with him, I use the expression "aspects of the self," and he picks it up eagerly because MUDding reminds him of how Hindu gods could have different aspects or subpersonalities, while having a whole self.

> You may, for example, have an aspect who is a ruthless business person who can ne-gotiate contracts very, very well, and you may call upon that part of yourself while you are in tense negotiation, to do the negotiation, to actually go through and negoti-ate a really good contract. But you would have to trust this aspect to say something like, "Of course, I will need my lawyers to look over this," when in fact among your "lawyers" is the integrated self who is going to do an ethics vet over the contract, be-cause you don't want to violate your own ethical standards and this [ruthless] aspect of yourself might do something that you wouldn't feel comfortable with later.

Case's gender swapping has enabled his inner world of aspects to achieve self-expression without compromising the values he associates with his whole person. In response to my question, "Do you feel that you call upon your personae in real life?", Case is sure that the answer is yes.

> An aspect sort of clears its throat and says, "I can do this. You are being so amaz-ingly conflicted over this and I know exactly what to do. Why don't you just let me do it?" MUDs give me balance. In real life, I tend to be extremely diplomatic, nonconfrontational. I don't like to ram my ideas down anyone's throat. On the MUD, I can be, "Take it or leave it." All of my Hepburn characters are that way. That's probably why I play them. Because they are smart-mouthed, they will not sugarcoat their words.

Case is a man playing a woman; when women play male characters in cyberspace, there are similar stories of using the virtual to work through issues of the real. For example, some women are drawn into playing men because they find it gives them permission to be more outspoken or aggressive. One young woman, Zoe, puts it this way:

> I played a man [in a MUD] for two years. As a man I could be firm and people would think I was a great wizard. As a woman, drawing the line and standing firm has al-ways made me feel like a bitch and, actually, I feel that people saw me as one, too. As a man I was liberated from all that. I learned from my mistakes. I got better at be-ing firm but not rigid. I practiced, safe from criticism.

Zoe's perceptions of her gender trouble are almost the opposite of Case's. Case sees aggressivity as acceptable only for women; Zoe sees aggressivity as accept-able only for men. What these stories have in common is that, in both cases, a virtual gender swap gave their protagonists greater emotional range in the physi-cal real.

A MORATORIUM OF PLACE

For some people, cyberspace is clearly an environment for acting out unresolved conflicts, to play and replay characterological difficulties on a new and exotic stage. On the other hand, for many other people, such as Case and Zoe, life on the screen provides an opportunity to work through significant personal issues, to use the new materials of cybersociality to reach for new resolutions. These more positive identity effects occur because cyberspace can provide what Erik Erikson (1963) would have called a "psychosocial moratorium" (p. 262), a central element in how Erikson thought about identity development in adolescence.[2] Although the term *moratorium* implies time out, what Erikson had in mind was not withdrawal. On the contrary, the adolescent moratorium is a time of intense interaction with people and ideas. It is a time of passionate friendships and experimentation. The adolescent falls in and out of love with people and ideas. Erikson's notion of the moratorium was not a hold on significant experiences but on their consequences. Moratorium is a time during which one's actions are in a certain sense not counted as they will be later in life. They are not given as much weight, not given the force of full judgment. In this context, experimentation can become the norm rather than a brave departure. Relatively consequence-free experimentation facilitates the development of a personal sense of what gives life meaning, which Erikson called identity.

Erikson developed these ideas about the importance of a moratorium during the late 1950s and early 1960s. At that time, the notion corresponded to a common understanding of what the college years were about. Today, more than 40 years later, the idea of the college years as a consequence-free time out seems of another era. College is often relentlessly preprofessional, and AIDS has made consequence-free sexual experimentation an impossibility. The years associated with adolescence no longer seem a time out. But if our culture no longer offers an adolescent moratorium, virtual communities often do. It is part of what makes them seem so attractive.

Erikson's ideas about stages did not suggest rigid sequences. His stages describe what people need to achieve before they can easily move ahead to another developmental task. For example, Erikson pointed out that successful intimacy in young adulthood is difficult if one does not come to it with a sense of who one is—the challenge of adolescent identity building. In real life, however, people frequently move on with serious deficits. With incompletely resolved stages, they simply do the best they can. They use whatever materials they have at hand to get as much as they can of what they have missed. Now virtual social life can play a role in these

---

[2]For the working out of the notion of the moratorium in individual lives, see Erikson's psychobiographical works, *Young Man Luther: A Study in Psychoanalysis and History* (New York: Norton, 1958) and *Gandhi's Truth: On the Origins of Militant Nonviolence* (New York: Norton, 1969).

dramas of self-reparation. Time in cyberspace reworks the notion of the moratorium because it may now exist on an always-available window.

THE PSYCHOANALYTIC
AND THE COMPUTER CULTURE

Having literally written our online personae into existence, they can be a kind of Rorschach. We can use them to become more aware of what we project into everyday life. Identity play in cyberspace can be serious business because it can become a form of self-knowledge. And the people who make the most of their lives on the screen are those who are capable of approaching it in a spirit of self-reflection. They ask: "What does my behavior in cyberspace tell me about what I want, who I am, what I may not be getting in the rest of my life?"

As a culture, we came to the end of the Freudian century. Freud marked the 20th century with his sensibility and explanatory discourse, and there is much about his approach that seems foreign to our current interests in mind and matter, in psychopharmacology and genetics. It is fashionable to think that we have passed from a psychoanalytic culture to a computer culture—that we no longer need to think in terms of Freudian slips but rather of information-processing errors. But faced with the challenges of cyberspace, our need for a practical philosophy of self-knowledge, one that does not shy away from issues of multiplicity, complexity, and ambivalence, that does not shy away from the power of symbolism, from the power of the word, from the power of identity play, has never been greater as we struggle to make meaning from our lives on the screen. In my view, our relationship to the computer culture and psychoanalytic culture needs to be a proudly held joint citizenship.

2

What's So Democratic About CMC?: The Rhetoric of Techno-Literacy in the New Millennium

Jami Carlacio
Cornell University

> *Technological advance not only subordinates workers to capital, but disenfranchises them. Society has no incentive to teach and they have none to learn the knowledge that would qualify them to participate in the social decisions that concern them. [. . .] [B]y its very nature capitalism requires [. . .] an ignorant and docile labor force tied to highly specialized tasks.*
> —Andrew Feenberg (1991, p. 29)

Despite the relatively short history of computing in the humanities, the ideas and concepts grounding it have undergone significant changes.[1] A new pedagogy that once seemed to promise a more democratic space for students to discuss ideas and that hoped to prepare them to take their place in a more complex and technologically literate workforce is now one that questions the extent to which emerging classroom technologies can deliver on earlier promises. Indeed, the proliferation of computers in composition and other humanities classrooms has meant, for some, a revolution in the way writing is taught; the way information is disseminated and accessed; and the way such technologies have introduced and made possible a new dimension of equality, access, and democratic participation. The scholarly literature of the late 1980s and early 1990s on networked instruction, for example, has promulgated various claims, such as this one: "Once people have electronic access, their status, power, and prestige are communicated neither con-

---

[1]I want to thank Pete Sands for pointing me toward some of the resources I consulted while writing this chapter.

textually . . . nor dynamically. . . . Thus, charismatic and high status people may have less influence, and group members may participate more equally in computer communication" (Kiesler, Siegel, & McGuire, 1984, in Hawisher & Selfe, 1991b, p. 57). Along similar lines, others posited that computer-mediated composition makes possible an "egalitarian discourse" (Day & Batson, 1993, p. 34) and a "strong sense of community" not characteristic of a traditional classroom (Hawisher, 1992, p. 87). The implicit and explicit arguments made here are symptomatic of a rhetoric that has influenced not only our thinking about the way we deploy technology in the classroom but also the way that we fail to think critically about it. That is, these claims have invited us to consider computer-mediated instruction as a democratic space where students may temporarily forget their (unequal) race, gender, or socioeconomic status as they participate in an electronic environment. This democratic rhetoric has informed what Gail Hawisher and Cynthia Selfe (1991b) call "the rhetoric of technology" (p. 56)—an uncritical discourse that sees computers as positive ends in themselves, where "hope, vision, and persuasion" (p. 57) characterize the scholarship in the field.

Classroom experience and our own grappling with the challenges of teaching and working with emerging technologies have since taught many of us that making claims for democracy in the name of technology is not only untenable but is also quite often specious. The more scholar-teachers began working with technology in the classroom—in the form of class listservs, e-mail, instructional software such as Daedalus or packaged Web course programs, and hypertext—the more we discovered that the social and political dynamics of the traditional classroom did not necessarily morph into a networked classroom that was more democratic, either in theory or in practice. In fact, many of us have learned to reassess our original assumptions that electronic classrooms are virtual communities whose egalitarian discourse nullifies markers of gender, race, and class; such views have only served to perpetuate the corollary belief that these environments are also democratic. Whereas a number of computers and composition professionals contended in the early 1990s that the networked composition classroom creates a space for community and potentially equalizes students by erasing these cultural and social distinctions, by the middle of that decade, they were calling for a more stringent critique of such claims, labeling them as democratic myths (Lockyard, 1996; Selfe & Selfe, 1994). Initial scholarly assertions contained the implicit assumption that community, equality, and democracy inform each other, that community necessarily entails equality and democracy. These claims were later circumscribed to focus on the (positive) communal aspect of computer-mediated instruction.

Computer-mediated instruction has demonstrated to many of us that it can contribute in positive ways toward community building in the classroom. As Nancy Knowles and Wendy Hennequin suggest in their chapter in this volume, collaboration in a computer-mediated classroom has the potential to produce an egalitarian environment particularly when the technology changes so rapidly that it requires students and instructors to work collaboratively to find solutions to problems that inevitably crop up. Networked instruction, most agree, does engender a sense of community different from the traditional classroom, but we cannot assume this alone makes the electronic classroom more democratic. This democratic rhetoric is at the heart of the discourse surrounding emerging technologies. Claims such as these have remained unsubstantiated or at best inadequately supported. Because computer technology is becoming an increasingly substantial part of our classroom pedagogy, it is imperative that we gauge its effects carefully. Although we might wish to embrace emerging technologies, we must consider not only the implications of what I call the rhetoric of techno-literacy and how it circulates discursively in both our literature and in our culture generally but also what its rhetoric effaces.

Despite the general move away from claims that networked classrooms engender a democratic space, humanities scholar-teachers have neglected to question what democracy means and how, if at all, we may employ this concept to describe the pedagogical and cultural work it does. Until we foreground the way we are defining and using democracy, we cannot make ethical claims about the liberatory and democratic nature of a technology-rich pedagogy. In this chapter, therefore, I argue for the need to examine the blanket claims made in the name of democracy and technological progress about egalitarian classrooms where race, gender, sexuality, and class continue to differentiate students from one another, despite our best efforts to build community and ensure equality through a well-conceptualized and well-planned curriculum. I first contextualize and define the term democracy; I then examine how it has circulated uncritically, particularly as it effaces gender inequalities; finally, I critique the rhetoric of techno-literacy and uncover the hidden assumptions of a "discourse of crisis" that wants us to believe technological literacy can bridge the socioeconomic gap that divides our students.

WHAT'S SO DEMOCRATIC ABOUT
COMPUTER-MEDIATED COMMUNICATION (CMC)?

Many compositionists have claimed that equality becomes a possibility in the networked classroom when it allows students to explore the multiple ways that identity is constructed. They base their claims on the assumption that postmodern and poststructural theories adequately explain the construction of the subject. Such theories do seem liberatory when we consider their articulation of multiple subject identities as a response to the Enlightenment concept of a unified self. Students might find the electronic environment liberatory as they pursue various discursive identities along axes of race, class, gender, and sexuality. Such a classroom might provide the conditions for the possibility of a democratic space, though democracy itself is not absolutely guaranteed. That is, we cannot assume that democracy presupposes equality because several competing interpretations (read: discourses) of democracy may circulate at any given moment. Moreover, students might occupy multiple subject positions in a democratic community, but these positions do not necessarily enjoy equal status. The term democracy demands a more thorough understanding.

When the term democracy is used uncritically—that is, without considering how it is manifested in social, historical, and material ways—it possesses the power of an empty signifier. Most of us have accepted as common *doxa* that we live in a democratic society, albeit an imperfect one. With the advent of networked communications technology, some have claimed that the potential for citizens to participate in the political governing of America has increased. For example, the former president of the Corporation for Public Broadcasting (PBS), Lawrence Grossman, boldly claimed in 1995 that telecommunications networks "make it possible for our political system to return to the roots of Western democracy as it was first practiced in the city-states of Ancient Greece" (quoted in Barney, 2000, p. 20). Grossman goes on to say that "keypad democracy" is made possible by Americans' increased access to networked communications, which will enable them "to overcome the obstacles of scale that have traditionally thwarted vigorous democratic participation." By making this broad and unsubstantiated claim, Grossman implies that democratic participation is simply at Americans' fingertips; moreover, where other social and political programs might have failed in their efforts to promote democracy, network technologies are capable of surmounting former obstacles and of making possible across-the-board equality. What Grossman and other technology enthusiasts seem to have overlooked are the historical and political contexts that have shaped our understanding of democracy. In ancient Greece, and particularly in 5th and 4th centuries BCE Athens, democracy referred to the "active participation of adult male citizens in the deliberative assembly and the lawcourts" (Kennedy, 1999, p. 20). Male citizens were free to plead their cases in front of a jury of their peers, employing the most rhetorically persuasive arguments available to them. Women were not included in this democratic system; in fact, participation in Athenian political culture was restricted to Athenian men. When proponents of democracy claim for its ideological antecedents that which was developed in the Athenian polis, they may not realize they are promoting an exclusive system based on property ownership (economic stability), education, and gender. As such, claims for the democratic nature of networked technologies become suspect.

If referring to the democratic polis of ancient Greece places us too far in the past, others have found it more convenient to look to its American manifestations, as our Constitutional framers envisioned it. In tracing America's democratic ideals from colonial times to the present, rhetoricians Celeste Condit and John Lucaites (1993) explain that the dissenting colonists who settled here conceived of liberty as the right to hold property. Further, these colonists, in declaring their independence, "were not broadly egalitarian men"; they had indeed "limited their assertion that 'all men' had certain natural rights by emphasizing that such rights could not all be maintained in social compact" (p. 5). This social compact, Condit and Lucaites claim, referred to those men who were politically free because only those who could defend their liberty had earned this right. This framing of equality, under an ostensibly democratic constitution, excluded people of color, most of whom were enslaved and therefore unable to earn liberty; it also excluded women, who were allowed neither to own property nor to participate in the political governing of the new nation. Democracy, in this historical moment, was more closely linked to that of ancient Greece: Men had the power to vote, to govern, to own property, and to exert control over the rights of citizenship. What kind of democracy then, are humanities scholars and teachers, as well as public policy officials, referring to when they make democratic claims for networked technologies? We must consider such claims carefully, acknowledging that democracy is contextual; it is applicable to the cultural and historical conditions that define a particular community. Toward this end, I offer a definition of democracy that fits these criteria and that may be used to measure our contemporary claims about technology as we employ it in our classrooms and in our communities.

In offering my extended definition of democracy, I am drawing from two sources—the first is a radical democracy as articulated in postmarxist theory by political philosopher Chantal Mouffe (1993); the second is a definition posited by Canadian political historian Darin Barney (2000). In a radical democracy, argues Mouffe, there will be no universally agreed-upon definition of liberty or equality because "Justice as Fairness is only one among the possible interpretations of the political principles of equality and liberty" (p. 53). Notwithstanding this lack of universality of ideals, citizens can come together to fight for common principles in a given field. These common principles may be understood as democratic equivalences, where, for example, the continued marginalization of minority groups provides the grounding for the shared principle of fair treatment. More specifically, Mouffe would argue that in supporting antisexist behavior one would also support antiracist behavior because the common denominator here is the prejudicial behavior exhibited toward these groups, although the nature of the prejudice is different. To fully comprehend the potential a radical democracy offers, it is important not to assume that equality is a value shared by everyone in equal ways in all circumstances. At the heart of a radical democracy is identification: Citizens possess (constructed) political identities that alter as each historical, social, and political context shifts. Moreover, as citizens, we may share a political identity while we occupy a range of subject positions. For example, I might identify simultaneously as a woman in a patriarchal society, as a scholar who possesses the cultural capital associated with my academic position, as a democrat, as the daughter of working-class parents, and so on. Put another way, as social agents, we identify with the principles of equality and liberty along axes of race, ethnicity, class, gender, and sexuality (p. 70). Significantly, because identities are fluid and ideas and beliefs shift over time, we must continually reevaluate our stances and redefine them accordingly. A radical democracy neither assumes a universal notion of equality or citizenship or liberty, nor promises one version to be better than another. What is liberatory about Mouffe's definition is that competing interpretations of equality and liberty will always exist and so too will understandings of democratic citizenship. It is a matter of discursive privilege—rhetorical prowess—that determines which interpretations are deemed more useful or more agreeable. In terms of the argument I am making in this chapter, the cultural cache invested in the rhetoric of techno-literacy must be investigated and challenged: Does the application of emerging technologies in our classrooms or the availability of public telecommunications networks serve to democratize our society?

To answer this question, I turn to Barney (2000), who examines claims that a networked society potentially creates a democratic revolution. Drawing on ideas of Canadian theorist George Grant, Barney addresses the presumption that technological progress in modern society is both a universal good and a contributor to a democratic state. This view, he explains, means that "not only are we all equal under the gaze of technological mastery, but we are all the same" (p. 53). This universal, democratic appeal masks the localized and situated differences that render our society unequal. That is, by choosing to see technology and its political and pedagogical applications as all good, we are stripping it of its political nature and failing to examine its impact on us materially. This is precisely what is at issue for Mouffe (1993) in her proposed turn toward a radical democracy: We must interrogate our relationship to equality and liberty (democracy) within social, economic, and political contexts. Whether in classroom, business, or political contexts, our interactions are shaped by our use of e-mail, our participation on listservs, and our ability to access the Internet for information, commerce, or entertainment. Many of us participate in most or all of these activities daily or weekly, and our ability to do so signifies a fairly high level of technological literacy and economic power. Some Americans, however, still lag behind where both techno-literacy and economic ability are concerned. This continued disparity deserves our attention because it reminds us that a universal notion of democracy inadequately describes the material inequities that separate the haves from the have-nots.

A more specifically worded definition of democracy will enable us to measure the democratic claims against the reality. I draw again from Barney (2000), who defines democracy as "a form of government in which citizens enjoy an equal ability to participate meaningfully in the discussions that closely affect their common lives as individuals in communities" (p. 22). This definition contains very specific language that we may use to gauge whether or not technology—access to it, literate ability to use it—is indeed democratic. First, for citizens to enjoy an equal ability to participate means that the legal infrastructure must support their goal and that they are materially capable of gaining access to hardware and software, Internet service providers (ISPs), or other mechanisms for getting online. Second, to participate meaningfully means that citizens must possess the ability to participate in civic and educational matters, whether in local community decision making, e-mailing a senator or U.S. representative to support or oppose an initiative, or MOOing and using synchronous discussion forums in the classroom. Using Mouffe's conception of a radical democracy and Barney's concise definition, I spend the rest of this chapter measuring the rhetoric against the reality.

Several computer- and composition-theorists have indeed realized the complexity of democracy when they have taken it up both within the broader context of the Internet and within the very specific one of the networked composition classroom. Joseph Lockyard (1996), in an essay written for the anthology *Internet Culture,* cautions against a democracy myth, where issues of access prevent "would-be virtual citizens" from "participatory citizenship" (p. 220). In the same collection, Joseph Tabbi (1996) argues, along with Charles Ess, that there is nothing inherently democratic about hypertext or online classrooms because a "democratic impact" depends more on the larger social contexts that frame networked classrooms (pp. 240–241). In other words, the classroom environment does not constitute a separate social sphere, disconnected from the larger social structure. Students—all of us, in fact—carry our gender, racial, and economic histories into the classroom. Tabbi does admit that many professionals in the field of composition studies are aware of asymmetries of power and unequal access, and he applauds efforts of computer-mediated composition theorists, such as Cynthia and Richard Selfe (1994), who argue for a stringent critique of the power relations operative in any classroom. Selfe and Selfe especially caution readers to be wary of the myth that suggests computer classrooms are "linguistic utopias" (p. 483).

Selfe and Selfe's cautionary advice comes only 1 year after computer and composition theorists Michael Day and Trent Batson (1993) made such egalitarian claims. They contend in their essay "The Network-Based Writing Classroom" that students and teachers are equal participants in the electronic classroom and that students share a "strong sense of community" and "participate more" because they feel more "companionship and warmth" than they would in a traditional classroom (p. 334). The communal nature of networked instruction was in fact discussed frequently in the late 1980s and early 1990s. Hawisher (1992), for example, cites researchers in computer and composition studies who claimed that students see themselves as part of a community and as having a "sense of belonging and comradeship" (p. 87). As students find themselves engaged in more frequent participation, she notes, the result is an "egalitarian discourse not characteristic of traditional classrooms" (pp. 88–89). Citing Cooper and Selfe's study (1990), she explains that students in online conferences "open themselves to the divergent views of their classmates" (p. 89). Computer-aided instruction may indeed engender a community, though it might not necessarily foster an egalitarian atmosphere. Although our students benefit from collaborative learning programs that emphasize the social construction of knowledge, such as Daedalus Integrated Writing Environment's (DIWE) Interchange, we cannot reasonably claim that such a program provides a democratic environment where students will automatically have an equal ability to make their voices heard.

As I have tried to make clear, the theme of democracy—and its corollaries, egalitarianism, and community—pervades the discourse that has shaped computers and composition over the last 10 or so years. Much, though not all, of the discussion about electronic communication has not been carefully interrogated, nor has its role been adequately evaluated in our pedagogies. However, a growing body of feminist research questions the egalitarian claims made in the name of networked instruction and, especially, their impact on women. Several studies from the early 1990s through today point to the inherently undemocratic nature of electronic communication. Although Lester Faigley claimed in 1992 that his students using pseudonyms during Interchange sessions in 1988 felt "liberated," other research suggests that markers of race, class, and gender are not erased, even when participants use pseudonyms. Selfe and Meyer (1991), in "Testing Claims for On-Line Conferences," confirm this point. They originally hypothesized that computer-mediated composition is more egalitarian than traditional instruction and tested whether or not socioeconomic, ethnic, and gendered distinctions disappear online, as studies such as Faigley's suggested. They found, in fact, that computer-mediated composition "wasn't as egalitarian as [they] might wish" and that gender cues (markers) not only remain but are more significant than those of race or class (pp. 185–186). Similarly, the students in Knowles and Hennequin's collaborative electronic classroom (see chapter 6 of this volume) chose to forego race and gender identity markers because they believed this would allow them to participate more freely in online class discussion. Dagmar Corrigan's and Simone Gers's contribution to this volume also supports this claim (see chapter 10). In designing their cyberfriends assignment, they paired same-sex students to avoid "any potential problems with sexual harassment."

These studies, conducted separately over a period of nearly a decade, all suggest that online communication is often gender distinguishable and characterized as unequal, where women experience silencing and harassment more often than men. Linguist Susan Herring (1996b) argues persuasively that online styles are in fact gendered: On one hand, men exhibit more aggressive behavior, are more prone to flame (i.e., compose rude and offensive prose), and dominate list discussions; on the other hand, women tend to be more supportive and to use an indirect discourse style. Such differences cast doubt on the claim that CMC is more democratic if we understand that its function is to provide the opportunity for various voices to be heard. Because gender is identifiable and continues to be situated in gendered contexts, we cannot reasonably argue that computer interaction erases gender identity. Composition theorist Lisa Gerrard (1997) explains in a *Kairos* Web text that women have always been identified with their bodies, and the supposed bodyless, soundless woman is more of a myth than a reality. We cannot assume that women or men no longer possess a body because even virtual bodies are real. Hawisher and Patricia Sullivan's (1998) recent research supports this— women in their study perceived themselves fully as women, both online and offline (p. 181). In her study on women's experience in online communities,

Pamela Takayoshi (1994) also found this to be the case. In larger Internet contexts as well as in the classroom, she argues, we cannot assume CMC will engender a more collaborative environment: Not only do such environments produce a depersonalizing effect, but they also quite often undercut the success of group work when members fail to achieve consensus in their decision making (pp. 23, 30–31).

Just as networked communication presents challenges for women on the Internet, communication also remains problematic in the classroom—in the form of student–student and student–teacher interaction. For example, female instructors who attempt to decenter authority in the classroom may simply undermine any authority they have. Instead of sharing authority with their students, these teachers might unwittingly be reproducing unequal power relations. This unequal power relationship becomes clearer when viewed through a Foucauldian lens. In her feminist analysis of power in the classroom, Christy Desmet (1998) explains that when the female teacher attempts to abdicate her authority in the classroom, her power is simply dispersed—not negated—which reinforces, if not intensifies, its presence:

> In Foucault's analysis, the disappearance of the judge—the diffusion of the judge's power and responsibility throughout an impersonal structure—doubly disempowers those caught in the "pedagogical machine." Feminist teachers who unilaterally refuse to play judge in their classrooms therefore may occupy the panopticon unwittingly. (p. 157)

This panoptic effect, then, ostensibly reinscribes the power relations in the classroom rather than ameliorates them. Instead of eradicating the violence of pedagogic power, the move to distribute it results in a different kind of violence that, as Lynn Worsham puts it, "conceals and mystifies relations of domination" (in Desmet, 1998, p. 158). Instead of clarifying the power structure in the electronic classroom, the dispersion of authority actually masks it.

The effects of power in the classroom manifest themselves in other ways as well. That some students exhibit disrespectful behavior online suggests that computer-mediated composition in some ways is inherently alienating because there exists a differently constructed physical distance between writer and audience, where the machine acts as mediator. The alienation might be useful insofar as it provides a necessary critical distance from technology. This critical distance allows us to question the power structures that inform and support technology that would otherwise remain transparent. Yet it is precisely this reified—or transparent—quality of CMC that compels us to place under close scrutiny the nature of social relationships and our ability to glide back and forth between computer-mediated composition and face-to-face interaction. Despite the persistent belief that social oppressions—in terms of race, class, and gender, for instance—are muted if not erased when we move from the traditional to the computer classroom, and that communication styles change from one social context to the other, research of online communication environments tells us this simply isn't true.

My experiences in two electronic classrooms support claims that gender is marked (or distinguished) and received differently online. During the spring of 1998, I was a member of a graduate class listserv, where out of a total of 13 participants including the male professor, only 4 were men. They contributed, however, to about one half of the listserv discussions and tended to display more aggression in their posts than did the women. When the men disagreed with a point made by anyone in the class, they tended to deny a claim's validity outright, whereas the women tended to acknowledge and affirm a point before (or if) they disagreed with what was posted.[2] Likewise, in our classroom MOO discussions, gender identity was evident, even when we used pseudonyms. When I reviewed the transcripts, I found that the women in the class asked more questions, and the men made more assertions (roughly 43% of women's posts were formulated as questions, whereas only 33% of men's were questions).[3] Gendered communication styles were also evident in the composition class I taught that same semester, where my students and I used Interchange. As I reread those class transcripts, I found, as in my graduate seminar, that men participated more and exhibited more aggressive online behavior. In fact, one male student sent me a condescending message that specifically questioned my knowledge on a particular subject in a way that had never occurred in our face-to-face conversations. The alienating nature of the machine apparently liberated him, creating a space for him to challenge my authority on the one hand and to feel free to openly disrespect me on the other.

Though students may exhibit negative online behavior at times, positive behaviors also emerge. Many students do in fact participate more online and even find it liberating. Some of my female students have told me they feel like they can say more using Interchange than they can when the class is engaged in face-to-face discussion. The increase in class participation also means more students are talking to each other online, students who may otherwise not interact at all face-to-face in the classroom. Yet such a change in classroom dynamics does not necessarily imply that this pedagogical space has become more democratic because some voices remain silenced. That is, several female students in my class posted only a few sentences during an entire session. Although my experience dates back to my 1998 classes, recent research on gendered communication styles on a listserv indicates that such patterns remain the same. In her recent *Kairos* Web text, "Online Commu-

---

[2]The most vocal and aggressive student online was, ironically, the least vocal student in face-to-face discussions. In listserv conversations, this particular participant often called others' ideas into question, though he refrained from flaming or addressing anyone inappropriately. On one occasion he writes, for instance, "I can't help admitting that I found it spectacularly egregious to advance a skepticism toward theory *in an e-mail which advances a theory of reading*!!!" I gather from conversations with others in the course that although they find this person in real life to be friendly and accommodating, they found his posts to be somewhat condescending.

[3]These numbers account for posts that engaged directly in the discussion; that is, I did not count whispers, hellos, or other material unrelated to the discussion at hand. Though it is beyond the scope of this study, a study of the MOO transcript would reveal a gendered split in types of questions asked—ones that seek knowledge versus ones that exhibit authority.

nities, Self-Silencing, and Lost Rhetorical Spaces," Kelly Kinney (2001) presents a draft of her study of a listserv exchange between students at two universities. Surprisingly, she found that although women outnumbered men by a ratio of 51 to 13, men on average posted more messages than women. She also discovered that by the end of the quarter in which she conducted the study, the overall communication style on the listserv was predominantly male, using what she describes as "argumentative, agonistic, adversarial language." Correspondingly, a minority of males dominated the list with "adversarial discourse," so several participants withdrew from the discussion, some of whom were women.

Kinney's research, as well as those previously published, point to the need to continue to study the effects of gendered online discourse before we rush to proclaim the medium itself to be democratic. Although it is important to acknowledge that some voices are silenced and to ask why, at the same time I do not wish to imply that students who do not participate in online discussions necessarily lack agency; some have argued, including Kinney, that self-imposed silence might be empowering. Nevertheless, if women feel intimidated or harassed, they will feel less willing or able to participate freely. It is therefore essential that we question both the democratic and egalitarian claims for online communities and the way we invoke these terms. When we integrate technology in our classrooms, we need to expose the real inequalities that its use tends to hide.

Democracy, and the possibility of equality, implies both consensus and competition for discursive privilege. Because our society has historically been shaped by patriarchy and racism, where women's and minorities' voices have been silenced and our discursive privileges limited, none of us can ethically claim that either the traditional classroom or the networked classroom creates a democratic space where women, people of color, and those of lower socioeconomic status have equal consideration with (White, middle-class) men. The struggle for the right to speak should not entail oppressing some while privileging others if we accept that the networked classroom is that space where both students and teachers share the same ethico-political ideal for liberty and equality. I argue, with Takayoshi (1994), that "patterns of interaction deeply entrenched within a patriarchal system cannot be undermined simply by offering access to a new medium" (p. 32) because this would imply that the technology itself, not its users, has the power to effect a change in social relations. We need instead to change the existing economic and sociopolitical structure and acknowledge the way it has operated historically before we can employ computer pedagogy as an egalitarian or democratic site of resistance.

A VIEW FROM BELOW: THE RHETORIC OF TECHNO-LITERACY

Although we ought to consider the value that networked technologies have added to humanities instruction and to our society as a whole, we must continue to foreground our use of the medium itself. Some anecdotal evidence might serve to illustrate my point. In their study involving 10 different computer-mediated classes taught by 10 different instructors, Hawisher and Selfe (1991b) found the instructors insisted that nonverbal as well as socioeconomic cues were absent from their classrooms; moreover, when these teachers talked about technology, they often commented that students' collaborative interaction was distinctly different from what transpired in the traditional classroom. These instructors failed to mention, however, the potentially negative effects that can occur (pp. 58–59). The authors also discovered that most of the instruction was still teacher-centered and that students were not always engaged in collaborative learning and writing activities (pp. 60–61). Hawisher and Selfe admit that anecdotal evidence alone cannot justify the conclusion that the discourse related to networked instruction reflects idealized visions of student-centered classrooms, where computer technology enhances the social construction of knowledge. It does, however, suggest that we should look more closely at the rhetoric that envelops computer-aided instruction. We might find, for instance, that power differentials continue to exist between students and teachers and that networked classrooms reproduce unequal relations. These power structures shape student participation in at least two ways. First, during synchronous conferences, for example, where instructors oversee students' discursive activities, students might feel policed and may thus limit their discussion to topics about which they feel safe.[4] Second, instructors often award points for class participation, causing students to feel they must say something to receive credit. Such disciplinary structures, to borrow Foucauldian language, vitiate the democratizing potential of a networked classroom.

When we refuse to acknowledge our own power as teachers, when we fail to consider the racial, gender, and socioeconomic differences among our students, and when we naively believe that the external social contexts that inform our differences alter considerably once inside the classroom, we participate in a rhetoric of techno-literacy. We must therefore examine the conditions that allow this rhetoric to exist and then work to demystify it. This discourse is imbricated in the ideological structures of late capitalism. The proliferation of networked classrooms has greatly increased since the early 1990s. In the space of a few short years, universities unearthed money and redrew budget lines so that students could "get connected." In a 1997 article, Faigley describes the revolution that has made computers a ubiquitous presence in college classrooms. He explains that administrators at the University of Texas (and elsewhere, to be sure) "believe that college students should be able to use the media of literacy that they will likely use in their later lives" (Faigley, 1997, p. 35). This remark is an instance of the way in which corporations, whose presence on the Internet has

---

[4]This is certainly not always the case. Some instructors allow students to use pseudonyms, which at times invites flaming and other inappropriate (sexualized) comments. Still, in many instances, instructors are actively involved in DIWE Interchange conferences. For a more thorough discussion of these, see Faigley, 1990.

become pervasive in the form of Web sites and advertisements, have begun to influence our classrooms and our curricula. Faigley's remark and indeed the proliferation of online classrooms have paved the way for the widespread belief in the importance of possessing techno-literacy, the ability to use computer technology to improve one's learning and to become more productive in an increasingly information-oriented economy. In 1998, the National Endowment for the Humanities (NEH) launched a grant opportunity hoping that by the year 2000 "all U.S. schools will be connected to the Internet and that most eighth graders will know how to log on." The NEH grant promises, moreover, to "jump start the process by which U.S. schools and their teachers become competent, comfortable, and creative with these new humanities materials and technologies. This, of course, will directly affect student learning and achievement" (p. 1). The promises are bold, but the reality has fallen short. Not all schools and students are connected, and, consequently, a gap persists—defined largely by race and socioeconomic status—in computer usages and literacies. In the race for techno-literacy, some are consistently left behind.

I want to make this connection clear. As use of the Internet spreads, with college students comprising a rapidly growing user base, both the demand for the new technology and the need for technologically literate consumers grows. At the same time, public universities' budgets, fueled in part by corporate partnerships (read: interests), are spent on equipping classrooms with the latest technology. Selfe (1998b) notes that a number of schools across the country have cut aid in areas such as the arts to redistribute money for technological improvements.[5] We are robbing Peter to pay Paul. Already we see many incoming college students with technological savvy: They enter our colleges and universities fluent, to varying degrees, in e-mail and Internet Web site use. Providing them with the opportunity to increase and refine their literacy so that they may contribute to the production–reproduction cycle is what feeds capitalism. Andrew Feenberg would argue, as the epigraph at the beginning of this chapter suggests, that students are not being educated to do critical work; rather, they are being trained to participate in a production process whose conditions make possible capitalism's success.

Today's students, those fortunate enough to have access and the ability to participate in a variety of technologies—from using e-mail to participating in Internet chat room discussions to downloading their favorite music and burning tracks on CDs to play on their CD-ROM drive—already realize the value of techno-literacy. To determine students' abilities to develop what she calls critical computer literacy, Barbara Blakely Duffelmeyer (2001) studied 140 first-year composition students to discover the extent to which students understand the way their attitudes toward technology are culturally constructed. She found, not too

---

[5]See Selfe's "Peril" (1999b) for more detailed information on this. I wish to point out that although her numbers refer to elementary and secondary education, this trend continues into the university level.

surprisingly, that many of them simply accept technology as a neutral or transparent force that makes things easier and faster. This group made up 37% of the total and were not in the majority. More students—in fact, 52% of them—expressed a more skeptical attitude toward technology, evincing an ability to negotiate and to recognize multiple worldviews about technology. Although many of these students seemed to understand that a gap exists between those who do and those who do not possess the ability or the means to use technology, some wanted to "feel more positive" about it, according to Duffelmeyer. The minority of students in her study expressed an oppositional attitude, but this too was laced with "hopefulness that someday they will feel better (more 'hegemonic') about it." This research suggests that students consider it important to be able to use technology and to have access to it, despite any personal misgivings they may have about being left behind in the digital revolution.

The more we rely on computers to teach and to learn, the more they seem natural to us, like many of the students in Duffelmeyer's study. Selfe (1999b) explains that we "no longer feel the need to articulate [our use of technology] . . . we may also be allowing ourselves to ignore the serious social struggles that continue to characterize technology as a cultural formation in our country" (p. 415). For many of us, technology is a part of our everyday lives; for those who remain skeptical, just wait—it will be. It is precisely because of technology's slippage into the daily, taken-for-granted aspect of our lives that we must question not only the ways in which we use it and participate in its discursive production but also the impact it has on our material lives. The rhetoric of technology makes transparent the ideology that informs our culture: It feigns an innocent interpretation of the world. Its discourse, that is, shapes how we understand and address the problem of social inequality. In the classroom, this becomes a particularly urgent concern when we pretend that our race, gender, and class histories need no longer be an object of analysis. Cultural *doxa* wants us to believe network technology is a democratizing force, and such a belief sustains the rhetoric of techno-literacy.

As English studies and humanities professionals, we must consider both how we participate in this rhetoric and how we can critique it. Selfe and Selfe's "Politics of the Interface" (1994) examines one instance of technology's hegemonic discourse—the computer interface. The authors explain that computer interfaces act as maps that "order the virtual world according to a certain set of historical and social values that make up our culture" (p. 485). These maps exemplify the social power relations that characterize late capitalism in their virtual representation of White, middle-class values: ownership and opportunity (pp. 485–486). Moreover, they work to reproduce "asymmetrical power relations [that have] in part shaped the educational system" within which we work (p. 485). Selfe and Selfe further claim that the desktop icons function as ideological signs not only of White, middle-class professionalism but also of the "commodification of information" because desktops sport icons such as folders that hold information for future retrieval (p. 486). Whether we speak of a global or a national market, we must acknowledge that computer technology has facilitated the movement of enormous amounts of information from financial reports to health care to personal information. Herbert Schiller (1996) explains that information sharing can contribute to the social good, but more often it is merely traded as a commodity. On one hand, the exchange of information can help organizations trade knowledge about public health issues, it can facilitate decision making, and it can "promote the development of science and invention that are socially beneficial [as well as] organize historical experience for meaningful contemporary reflection and use" (p. 35). On the other hand, he describes how public companies, even public libraries, buy and sell information. Schiller explains that more often libraries are losing their social character as they mutate into "information-processing instrument[s]" where administrative systems specialists are replacing traditional librarians (pp. 35–56). Information is bought and sold for a price: We must ask, then, who sets the price, and who is able to pay it?

One might very reasonably argue that desktops representing the storage and retrieval of information, displayed in English, do not in themselves represent the hegemonic discourse of late capitalism. Indeed, most college students speak English, notwithstanding their native or nonnative status, and are familiar enough with icons that most of us determine to be everyday items. Yet these critical objections might just underscore the argument that desktops are a sign of class privilege—both materially and discursively. The method of information storage and retrieval is indicative, Selfe and Selfe (1994) argue, of a hierarchically based set of principles that govern rationality. It is interesting to consider, too, how desktop icons label our digital file refuse—"trash" for Mac operating systems or "recycle bins" on PC desktops; either way, they resemble trash cans, representing Western culture's throwaway mentality. Carmen Luke (1996) points out that "only in Western societies affluent enough to generate garbage can the concept of trash have any meaningful association with choice to discard excess." She remarks further that from the perspective of poorer nations that "fee[d] off first world refuse . . . the garbage can carries very different cultural meanings." These desktops, in other words, promote certain ways of understanding and interpreting the world, ways that actually reinforce the dominant social structure. Some computer interfaces preclude alternative ways of knowing, such as intuitive knowledge, and these elisions might prevent access for others whose cultural backgrounds differ from the norm (Selfe & Selfe, 1994, pp. 488–490).

Like any discourse that becomes naturalized, the rhetoric of techno-literacy has succeeded in promoting a tacit acceptance of the democratic potential of CMC. It has been able to function as a social good that many teachers have accepted as a pedagogical boon. No doubt technology has contributed immensely to the teaching of writing—I can attest to the success of DIWE's Interchange and Invention, for example, in my classrooms. At the same time, however, it is dangerous to ignore the material and nonmaterial differences that shape our student(s') bodies. In fact, one of the most significant "successes" of the rhetoric of techno-literacy is manifested in its almost total eclipse of the issue of access. The widespread belief, then, that computer technology will (or has) become the key to building a democratic classroom community feeds a virtual metaphor that masquerades as a truism: Students who are able to use computer technology to improve their learning and their productivity (both in school and beyond) stand a better chance at succeeding in the global information economy. This is no doubt true. Techno-literacy promises to redress social inequality, making it more difficult to address the existing inequalities with which many students still live. Possessing the requisite literacy to compete fairly in a technologically complex society is important, but the promise associated with this ability falls short when it is expected to somehow narrow the gap between the more and the less economically fortunate. It therefore continues to be an insufficient solution to a pervasive problem: Not all of our students are technologically literate, nor are they all online.

EVERYBODY'S GOTTA HAVE IT

Americans' access to the Internet grows substantially with each passing year. Reports indicate that more than half of Internet users are relatively young and somewhat financially secure—most are between the ages of 30 and 50 and draw incomes ranging from $25,000 to $50,000 annually. A 1998 *Business Week* (I/ PRO) poll finds that 73% of Internet users have some college education, 44% of them having completed their degrees.[6] These statistics seem encouraging, yet they fail to account for differential levels of access, participation, and ability. To prepare students for participation in the U.S. and global economy, primary and secondary schools are increasingly emphasizing computer technology instruction, as the NEH Millennial Schools project I described indicates. Yet the democratization of the classroom can only be achieved when all students receive the opportunity to be trained on and have access to computers. We know, however, that this will not be the case for those incoming freshmen who attended middle and high schools lacking in these and other resources. In the past 2 years alone, colleges and universities providing Internet access and some form of instruction using emerging technologies has increased substantially. A 2000 Campus Computing Project survey reveals that more than 59% of all college courses use e-mail, compared with only about 20% in 1995. Nearly one third of college courses also have a Web page, a more than threefold increase since 1996. More colleges are also reporting the use of course management software, indicating that with each passing year, most students will not only be technologically literate by the time they graduate but also that more of them will need to enter college possessing some degree of techno-literacy to participate fully.

---

[6]Internet statistics vary. I reviewed several different sources and found the CyberAtlas and *Business Week* survey numbers to be in the middle range.

Possessing the necessary skills will be easier for some, however, because wealthier elementary, middle, and high school districts offer students more instruction than those in districts where poverty levels are higher. For example, the U.S. Education Department's National Center for Educational Statistics (NCES) reported that school classrooms (as opposed to schools as a whole) with Internet access increased from 3% to 77% between 1994 and 2000 (p. 2). Although this number suggests the possibility that more precollege students are gaining technological literacy, it only applies to smaller and suburban schools. The survey also found that by 2000, only 64% of classrooms in high poverty areas were connected; moreover, the student-computer ratio in these classrooms reached 17 to 1 in 1999; by 2000, this ratio decreased even further, to 9 to 1. In more affluent districts, as few as 6 students share one computer (p. 3). And although it is important that levels of access are increasing, we must remain concerned with the differing abilities of educators to teach with networked technology. Although more than 80% of teachers claimed in a survey conducted in February of 2001 that computers and Internet access improve the quality of education, two thirds agreed that technology is not well integrated in their classrooms. Further, although they employ the Internet primarily for research, most teachers use their classroom computers less than 30 min per day. In fact, nearly 80% of teachers indicated lack of time as the reason for not logging on, and 44% admitted they did not know how to use the Internet (Pastore, 2001b).

If the inequities facing our students are not addressed now, similar patterns will continue to present themselves in our adult population. A 1988 study from Vanderbilt University helps illustrate my point. Researchers Thomas Novak and Donna Hoffman (1998) revealed that though African Americans represent the largest growing demographic of those planning to buy computers in the near future, only 29% of them, versus 44% of Anglo Americans, have home computers. Although this discrepancy may not seem large, overall more Whites (59%) use the Internet than do African Americans (29%). Hoffman predicts that fewer of the latter will be prepared for the 21st century, one that clearly demands a techno-literacy that will include Internet access and the ability to negotiate the rapidly changing circulation of information.[7] More recent data suggests the continued presence of a digital divide. As of August 2000, the total share of households with Internet access reached 41.5%, with the biggest gains among those whose income and education levels are average. Black and Hispanic households with Internet access reached 23.5% and 23.6% of the national average (Pastore, 2000a). Race and income continue to be a factor, even within different ethnic groups. A University of Massachusetts survey found that among residents in Boston, New York's Harlem and Brooklyn neighborhoods, and in Newark, New Jersey, and Hartford,

---

[7]Paul Morris in this volume indicates that 30% of Hispanic households now own computers, up from 14% in the year before; although this is encouraging, the Hispanic population still comprises a minority of user/owners (see his essay in this volume).

Connecticut, 56% of the respondents earning less than $40,000 annually admitted they knew very little to nothing about the Internet. As income levels rise, however, familiarity does too. Of the African Americans surveyed, 44% earning less than $40,000 said they were unfamiliar with the Internet, yet only 15% of those earning more than $40,000 admitted their lack of Internet knowledge. The study also reveals a direct correlation to education level, where those with college degrees were 30% more likely to have access to a computer (Pastore, 2001a).

Within the last few years, more reports such as these have surfaced, indicating a growing awareness among the public and private sectors that we must train not only students but also adults to be technologically literate, and we must work to decrease or to make disappear the margin of unequal access that currently exists among Whites and people of color, between high- and low-income Americans. In his "virtual" call to arms in the late 1990s, President Clinton envisioned a goal that would put a computer within reach of "every single child" (in Selfe, 1999b). He stood poised to pour billions of dollars into a project designed to prepare children (who will later become our students) to become technologically literate, economically productive citizens. This is an important goal: Who among us would not want our children to become productive participants in the global economy? Yet, if we examine the rhetoric of this goal, we find that technological literacy are the buzzwords to make tomorrow's leaders "more productive" where they will have the "opportunity to grow and thrive" in the "new knowledge-and-information-driven economy" (Riley, in Selfe, 1999b).

These excerpts, taken from former Secretary of Education Riley's pronouncements, signify the government's desire to educate our children to replicate the capitalist model of production-reproduction-surplus value. In his address, President Clinton did not explicitly explain how we will train the teachers to teach the children, nor did he call for a critical interrogation of the function of technology and information production today. We must ask, therefore, whether this initiative will bring about a democratization of computer access, widespread literacy, and positive social change, or whether we will be training school children today to replicate the existing inequalities of the social structure tomorrow. Clinton's rhetoric of technology implies the belief that computers—and technological literacy—will act as the panacea for our social ills. We can better understand Clinton's commitment to this project by comparing some figures. He called for an expenditure of only $2 billion over a 5-year period for the America Reads project and only a small fraction of the $11 billion annually he proposed for the technology literacy initiative for other programs[8] (in Selfe, 1999b).

President Clinton's—and many Americans'—wholesale embrace of this rhetoric is reminiscent of a parallel event that occurred more than 25 years ago when the media promoted a literacy crisis. *Newsweek*'s 1975 article "Why Johnny Can't Write" cemented in the American consciousness the fear that America's schools

---

[8]Selfe gives a more thorough account of these expenditures in her 1999 article ("Peril").

were failing in their efforts to prepare children for participation in a competitive economy that demanded literate workers. In "Literacy and the Discourse of Crisis," compositionist John Trimbur (1991) analyzes the effects of this literacy crisis and argues that the worried middle class wanted an explanation for their economic angst. He explains that upwardly mobile middle-class Americans needed a scapegoat to blame for their unease regarding their children's economic future. The supposed literacy crisis was an attempt at deflecting criticism of the "faltering" educational system and at "certify[ing] the success" of the meritocracy while it legitimized the "unequal outcomes of . . . the minorities, the poor, and the working class" (p. 279). The lesson we can learn from this is that like techno-literacy, the new democratic equalizer, the literacy crisis of the mid-1970s functioned to assuage the public's fears by increasing pressure on schools to teach children how to read and write. Just as emphases on linguistic correctness were heralded as the cure for the economic blues then, the large-scale effort to promote technological skills and to provide access to computers in schools and colleges is the new response to the same need to dominate the production and reproduction of information in the new millennium.

We must remember that computer access is not an across-the-board equalizer of race, class, or gender. Selfe (1999b) explains, for instance, that "computers continue to be distributed differentially along the related axes of race and socioeconomic status" and that poor students and students of color have less access to computers and training (p. 420). What is significant is the correlation between race and socioeconomic status. Less affluence means less access, and people of color comprise a larger percentage of those who fall into low-income brackets. Further, this correlation is played out in the workplace where, Selfe notes, Hispanic and Black workers are less likely than Whites to use a range of computer applications. This suggests that computers are not an easy answer to the structural inequality that characterizes and in fact enables capitalism. The promise of universal access nevertheless sustains the rhetoric of techno-literacy and keeps hidden the material inequalities that such a discourse promulgates.

CONCLUSION: AN INVITATION
TO RHETORICAL CRITIQUE

Computer technology may change the level of literacy among our population, but it will not change the ratio of literate to nonliterate people. As I have discussed, we must first understand the conditions that have made possible existing socioeconomic inequalities before we can adequately remedy them. The revolution in computer technology and the literacies it bears will have an enormous impact at every level of our society, but computers alone cannot address the social and economic issues that result from global multinational capitalism. English studies and humanities professionals, I am sure, believe in the value of computer-aided instruction. We all hold out the hope that our classrooms will be democratic sites of learning, where knowledge is socially constructed; where students collaborate on an equal basis, and race, gender, and socioeconomic status are acknowledged in less stigmatized ways; and where both we and our students can learn to use emerging technologies in socially responsible, critical ways. At the same time, we find ourselves up against a powerful rhetoric that makes promises it cannot hope to keep and is not, in fact, in a position to do so. The rhetoric of techno-literacy masks two crucial points. First, the larger social contexts that shape our lives are reproduced in the classroom (networked or not); second, mere machines are unable to solve the deep structural inequalities that capitalism has exacerbated and will continue to reinforce as we move forward in the 21st century.

Humanities and composition studies teachers have taken notice of these issues and have worked toward demythologizing the democratic promise of emerging technologies. The essays in this volume represent a discursive and potentially material effort toward this end. Christina Prell's chapter on service learning in this volume, for example, offers us an important cautionary tale of the use of Web-based learning as it connects with a composition and service learning course. At the same time, her experience and insights offer us hope that we can use our technological knowledge and resources to effect some change in our communities. Although she admits to discovering several problems related to user ability, sustainability, and the availability of appropriate hardware and software, Prell does offer specific and concrete ways in which we might share our technological expertise with interested communities and thus facilitate a positive relationship that answers an important need.[9]

Concrete actions must be paired with critical engagement. That is, I believe we must continue to rhetorically critique assumptions that technology as a medium or technology as a literacy is inherently democratic. We can continue to work to demystify the discourse that has framed our interpretations of reality, and by doing so we will perhaps change our pedagogical approach in the networked classroom. That is, when we begin to examine our assumptions about teaching with technology, we might discover how our rhetoric has shaped and is shaped by our practice. When we critically examine the discourse that has framed emerging technologies, we necessarily invite ourselves to rethink what it means to teach in this environment. Some of the best scholars in the field have begun to do this, and we must extend the work they have initiated. We can follow the leads of Selfe, Hawisher, and others who call for educators in the humanities as well as professionals in both commercial and government sectors to adopt a critical perspective on our use of

---

[9]A project similar to Prell's was conducted by Alison Regan and John Zuern (2001) at the University of Hawai'i Manoa in 1997. Of special note are the insights gained from their project, which they discuss in "Community-Service Learning and Computer-Mediated Advanced Composition: The Going to Class, Getting Online, and Giving Back Project." Both students and teachers came to realize that literacy is more complicated than learning to read or write (p. 10) and that "functional print literacy is no panacea for social inequity" (p. 13).

technology and to examine the way we participate in and perpetuate the circulation of its discourse. Surveys continue to be conducted and reports continue to be written, emphasizing the importance of technology in our society, both for those who have access and for those who do not. And yet reports are worth little if we fail to effect concrete changes in our own classrooms and our own communities. We must therefore continue to investigate the ways in which we may employ emerging technologies in creative and pedagogically ethical ways. As critical teachers, we have much more work to do to expose the discourses that sustain social and economic inequality and that sustain the rhetoric of techno-literacy. We can only be absolutely sure of one thing: Computer technology does not a democracy make, but a critique of its rhetoric is a positive step in that direction.

3

Computer-Mediated Communication as Reflective Rhetoric-in-Action: Dialogic Interaction, Technology, and Cross-Curricular Thinking

Rebecca J. Rickly
Texas Tech University

SITUATING THE WRITING CLASSROOM

Douglas Ehninger (1972) maintains that the "notion that rhetoric is something *added* to discourse is gradually giving way to the quite different assumption that rhetoric not only is inherent in all human communication, but that it also informs and conditions every aspect of thought and behavior: that man himself is inevitably and inescapably a rhetorical animal" (pp. 8–10, emphasis mine). Seeing rhetoric as integral to who we are and what we do in our daily activities—including what we do in the classroom—provides a basis for exploring how we might rethink our communication both in and outside of the classroom, particularly as it is enhanced through technology. The idea that rhetoric is more than style, language, or discourse is certainly not new.[1] George Kennedy (1999) defines rhetoric even more broadly as "a form of mental and emotional energy" involved in any communicative act (language is not a requirement for these acts; rhetoric is the driving impulse behind communication, which may be written, oral, visual, or physical; pp. 3–4). Yet in our Western culture (and, I'm sad to say, in humanities departments), we still tend to locate rhetoric in language, specifically in language meant to persuade. If we think about writing as part of a larger rhetorical process, how-

---

[1]See, for instance, Britton, Burgess, Martin, McLeod, and Rosen, 1975; Christensen and Christensen, 1978; Corbett, 1971; D'Angelo, 1975; Kinneavy, 1971; Berlin, 1984, 1987; Berthoff, 1981, 1984; North, 1987; and Kennedy, 1998, for some prominent authors who have championed a broader definition of rhetoric.

ever, which involves mental and emotional energy for communication to take place, we start to see beyond our own formalistic education into a broader, more rhetorically based mode of learning that involves dialogic interactions between disciplinary knowledges—lifelong learning. In the last 10 years, the notion of rhetoric informing and shaping writing outside the English classroom has allowed renewed emphasis on writing across the curriculum (WAC) and writing in the disciplines (WID), expanding the ways we view and teach communication in and outside our discipline. Even so, the logical counterparts of WAC and WID—thinking across (and in) the curriculum through conversing across (and in) the curriculum—have received far less attention. Can we situate this kind of cross-curricular writing rhetorically, in terms of understanding how effective communication can occur in specific contexts by looking at the rhetoric—the energy behind the utterances—and reflecting on how this particular rhetorical situation is like/impacts/is different from others? I believe that thinking and conversing across (and in) the disciplines are rhetorical acts that should play a large part of what we do in the composition classroom before (and as) we write. In this chapter, I explore how computer-mediated communication, or CMC, might prove to be both a vehicle and a hindrance to this kind of rhetorical repositioning of thinking—writing and conversing (and subsequent reflecting and learning)—in our writing classrooms, both within and outside the humanities.

My writing courses have always centered around class discussion and peer response as forms of thinking/conversing across the curriculum as much—if not more than—writing. It is in the environment of class discussion and the critical reflection of texts-in-process, I believe, that some students begin to examine what (and why) they think, modeling the kind of reflection-in-action that Donald Schon (1987) describes. After engaging in conversations about ideas, focusing in on and trying out specific topics, students are able to write about a topic, considering the other voices they have heard or read. If the process continues to include peer response, in which students read and respond to student-generated texts as intelligent readers, another layer of sophistication is added, which the author can choose to include, alter, or disregard, depending on his or her own take on the rhetorical situation. After such a process, the final product is likely to contain more depth of analysis, more considered perspectives, and a greater understanding of both the specific rhetorical situation, as well as how it might impact others than if the writer were to write without this type of interaction. When students can turn to one another in solidarity and community, reinforcing one another's ideas and getting feedback from a less intimidating source than the instructor, they gain strength—and so does their writing.

However, in most 1st-year composition classes, this dialogic interaction occurs as students are just beginning their academic journeys. Most of the students in these classes don't know much about their prospective discipline, and those who do have some experience in their chosen field aren't yet immersed in the writing of that discipline. Yet I believe that thinking and conversing across the curriculum does—or at least can—occur in these contexts. Students' differing abilities, their views, their experiences in the workplace, in school, and in personal life, all inform their thinking, their conversational patterns, their ability to read and respond critically to texts and situations, and, subsequently, their writing. Some would argue that these students need focused, field-dependent content to write for a potential academic audience as well as a sophisticated understanding of a very specific disciplinary audience, yet I would maintain that what these students need even more than content is exposure to multiple perspectives, modeling for the students the act of reflection as it pertains to rhetorical situations.

Because students come from very different backgrounds, they bring with them very different perspectives. The conversations and thinking that occur when students make public these perspectives in a discussion parallels the cross-curricular thinking and conversing that can happen in a WAC scenario. Let me give an example: When I taught junior high school students in a self-contained curriculum (one where I taught all of the eighth-grade courses, from language arts to math to social studies), I was able to help students relate ideas from one course directly to another in class discussion and in subsequent writing assignments. When we studied the 1920s in social studies, I asked students to research and present oral reports on important figures from that time in art, music, literature, and science that our book did not mention. From these reports, we identified at least one or two texts we then read in language arts, an artistic conception we looked at in art class, a scientific principle or breakthrough we examined more closely in science, and so forth. The conversations in each of these isolated subjects was enriched by the background information we all shared, but it was also made complex by the different perspectives some of the students brought with them due to extended research on a particular subject (e.g., those who researched and gave an oral presentation on a topic), interest in a specific subject area, previous experience with cross-curricular thinking, and so forth. One young man, a quiet, bright, introvert who didn't always fit in, confessed in a journal entry that he wished to be a great rock guitarist one day. He was assigned to research and report on Paul Robeson for his history project. When he found out that Robeson was, among other things, a great musician, he became fascinated with the topic, and his report culminated with a scratchy recording of a few of his songs, and his public declamation that he now wanted to play Robeson's kind of music. His enthusiasm infected the others, who brought different agendas and experiences to the table (the young man giving the report was White, but the African American students in the class were interested in Robeson's activism, for instance; others were impressed by his sports career; still others admired him for integrating traditional spirituals and ethnic music into the mainstream). Although different people carried different information away from this report, they all had expanded their rhetorical awareness of the culture of the early 1900s, and they all had revised somewhat their existing rhetorical situations.

In college composition courses, this kind of cross-curricular dialogue is a bit less overt, yet in this setting, too, teachers can encourage students to bring to class discussion what they know, what they have experience with, and what they are currently learning in other courses. The rich environment of the composition classroom results not just in writing, thinking, and conversing across the curriculum but, we hope, learning across the curriculum. Students will begin to examine, critically, how rhetorical situations are forged, assessed, broadened, integrated, and revised by examining the communication—and rhetorical energy—behind them.

But oral class discussions can be problematic. We've all known times when students come to class underprepared, when only one or two class members are willing to participate, or when the students seem to be content to let the instructor tell them what to think. Too, an especially good class discussion is literally gone once the class is over; students can usually remember only bits and pieces, or they have a vague (and often incorrect) notion of what was said to reflect on or use in subsequent discussions and writing assignments. Recent advances in computer technology, however, have allowed for radical changes in the composition classroom, particularly in the realm of class discussion. Local area networks, or LANs, link computers together and, in effect, link people together. Wide area networks, or WANs, do the same kind of thing on a broader scale, linking students and computers together on the Internet. Students using LAN software or WAN technology are able to conduct conversations, share ideas, view each other's work, and even work collaboratively on a project using CMC. In essence, this technology has changed the computer from a tool used for individualization to one used for socialization, where students in a composition classroom form a community of writers as they share perspectives and experiences, learning from instructors, each other, and themselves as they enter and participate in the larger academic community.

Too often, however, writing programs are simply either given limited access to a computer lab or must make do with the software and hardware available to them. When possible, English departments (and particularly writing programs) should take an active part in studying their needs and making recommendations, both short term and long term, after first examining their underlying assumptions, current practices, and projected desires—their rhetorical situation (or, as is more often the case, situations). For instance, if a program purports to value communication, they should first be able to identify *why*, articulating desired goals and outcomes. Then the members must ascertain ways that they would like students to be able to communicate both online (in a LAN- or WAN-based setting) and traditionally. Finally, participants must distinguish which of these communication methods they will likely still value in 5 and 10 years, recognizing their individual, departmental, and institutional goals and constraints, as well as the theoretical underpinnings, that they operate within. Similarly, if a program values collaboration, then this goal must be overlaid onto the previous one, and short- and long-term projections made. Only after these overlaid objectives are realized can a program begin to discern which software programs will best meet their needs (and I suggest looking at software first because it changes more rapidly), both now and in the future, and only then should hardware needs be addressed.

The goals of communication and collaboration mentioned previously were not chosen randomly. As in the previous examples, these concepts are highly valued in writing programs around the country. During the mid- to late 1980s, theorists such as Kenneth Bruffee, Andrea Lunsford, and Lisa Ede began to challenge the efficacy of stand-alone composing, arguing instead for a theory of collaborative thinking, learning, and writing in which knowledge is socially constructed by members of a group. In literary theory, Stanley Fish (1983) advocated a theory of the interpretive community, by which groups of readers are "persuaded to a set of assumptions" that "all of its members see" (p. 189). Fish posits that communities of readers not only agree on what constitutes a text, but their definitions are continually reshaped by new members, and "when and if that community is persuaded" to see things another way, previously agreed upon definitions "will disappear and be replaced by others that will seem equally obvious and inescapable" (p. 189). Similarly, members of what Joseph Harris (1989) has called discourse communities share ideas and create communal knowledge, becoming "real groupings of writers and readers, that we can help 'initiate' our students into" (p. 15). Harris notes that although "interpretive community" is a label usually describing more of a "world-view, discipline, or profession," a "speech community" generally refers to more specific groupings, such as "neighborhoods, settlements, or classrooms" (p. 14). Yet even "speech community" doesn't completely define what students do in the context of a classroom. Instead, Harris offers the term discourse community to describe what happens in academia, where communities consist "not of speakers, but of writers and readers" (p. 15). A discourse community, then, involves speech acts, literate acts, and acts of communal interpretation. And a discourse community is exactly what we hope occurs in the context of our writing classrooms, through technological or other means.

SYNCHRONOUS CMC

Synchronous CMC, communication in which many individuals are able to "talk" at once (and be "heard" at once), is gaining repute in the composition community as a means for students to develop electronic discourse communities. The model for synchronous CMC, or real-time, conferencing, according to Diane Langston and Trent Batson (1990), is conversation—several people speaking at once, each participant able to hear and respond to anything at anytime (p. 143). Yet participation in synchronous conferencing goes beyond the metaphor of the conversation because a conversation is traditionally dominated by one or two individuals speaking at once, especially in the context of the classroom. The potential advantages of synchronous conferencing over traditional oral classroom discussions are important: The participants cannot be physically singled out, no one can be interrupted, everyone is ensured a voice on the network, and a transcript of the discussion can be made available for future reference and reflection.

Marilyn Cooper and Cynthia L. Selfe (1990) attribute the success of CMC to three influences: "the synergistic effect of written conversation, dialogue, and exchange; the shift in power and control from a teacher-centered forum to a student-centered one; and the liberating influence of the electronic medium within which the conferences occur" (pp. 857–858). One particular vehicle for CMC available in many LAN-based classrooms around the country is InterChange, the synchronous or "real-time" conferencing program which is part of the Daedalus Integrated Writing Environment (DIWE).[2] DIWE is truly an intelligent writing environment, a collection of programs that enables students to engage in virtually every aspect of their writing and learning process online. It comes complete with a word processing program, a revision program, a heuristic program and an asynchronous e-mail program, a citation management program, a class assignment program, and a program manager. I found, however, when my writing classes met in the computer lab only once a week, the program I took advantage of most often was InterChange. This CMC program allowed all class members to participate in what might be seen as a huge, text-based conference call, except that in InterChange, unlike a phone call or a traditional oral discussion, no one is interrupted or cut off; as soon as a student types and sends a comment or response it is sent to a bulletin-board-like queue, which every class member can read. The resulting text-based class discussion can be somewhat difficult to follow, however, because the discussion often unravels into subtopics, or threads. The nonlinear nature of the discussion, though at times confusing, can help subvert the traditional hierarchy of the traditional classroom, allowing for more student-to-student interaction. Too, the anonymity of InterChange encourages participation by those students not comfortable with being physically singled out in the sometimes new, sometimes uncomfortable academic environment. This collection of sometimes disparate thoughts that often weave together as a cohesive tapestry provides a metaphor for what we hope students bring to the composition classroom: a variety of shared experiences and insights in writing. With the proliferation of Internet connectivity, such LAN-based programs are now WAN-based: Multi-User Domains (MUDs), Internet relay chat (IRC), Instant Messaging, and other chat-based programs, now offer new ways of breaking down classroom walls and connecting students—students who would have never had the chance to interact before—together dialogically.

---

[2]The Daedalus LAN-based software now has an online component—Daedalus Online—and it competes with other excellent Web-based educational programs, such as Blackboard, WebCT, Web Course in a Box, and so forth. Although I am now "married to the mob" in that my husband, Locke Carter, was one of the original programmers of Daedalus, the examples I use in this chapter occurred before I had more than a pedagogical relationship with Daedalus.

CMC AS A TOOL IN THE CLASSROOM

A flurry of articles and chapters outlining specific applications of CMC in writing classrooms has surfaced in the past few years.[3] After nearly 2 decades, CMC as a classroom tool is becoming common enough that instructors want more than theoretical models. Although theorizing is still important, teachers (and students) are examining their own contexts and developing specific applications for electronic communication. As might be expected, however, most of these applications center around the writing classroom, or perhaps around connecting students in different writing classrooms. I explore some of these models, looking at how CMC is being used as a reflective, rhetorical tool, particularly one that fosters thinking that crosses boundaries: physical, cultural, and curricular.

In "Advanced Composition Online: Pedagogical Intersections of Composition and Literature," Linda K. Hanson (2000) describes her use of CMC in conjunction with oral discussions in an advanced writing class for English majors in what she calls an integrated pedagogy. Her goal was to facilitate and model the kind of community-based discussion and writing that she feels is common to those in the humanities. Although pedagogy is bounded by decisions that have to do with relationships among power, authority, and knowledge, teachers must nonetheless make these decisions in an informed manner. Hanson notes the following:

> In composition classrooms, students find themselves at the interstice between academic study and communal participation. Social, aesthetic, and ethical values may be safely distanced from the personal when a text is being studied; but the same issues of value challenge students—and faculty members—when they must commit themselves to words in print. Teaching grammatical and rhetorical forms, teaching to models, teaching processes, teaching plot, character, theme, figurative language—each can successfully permit avoidance of engagement with either writing or reading tasks. (p. 215)

By diversifying both writing and reading experiences in the classroom, not only are the students' learning opportunities expanded, but reader–writer relationships with the literary text are allowed to move naturally to the forefront as student readers themselves participate as writers in a limited discourse community, modeling activities their professors engage in. Hanson calls what she does in the classroom using CMC communal conversation, a mode where "those who are knowledgeable shar[e] their expertise with those who are not, shaping and refining their own knowledge in the sharing" (p. 216). Her classroom sounds vaguely

---

[3]See, for instance, three recent collections of essays: *The Dialogic Classroom* (1998a), edited by Jeffrey R. Galin and Joan Latchaw; *Electronic Communication Across the Curriculum* (1998), edited by Donna Reiss, Dickie Selfe, and Art Young; and *The Online Writing Classroom* (2000), edited by Susanmarie Harrington, Rebecca Rickly, and Michael Day.

like the apprenticeship model in the sciences, or the internship model springing up in higher education across the nation, acknowledging the "impact of inclusion on one's motivation to learn" (216).

One of Hanson's (2000) pedagogical goals for this class is for the classroom community "to model the larger scholarly community in building a sense of common purpose and in developing an awareness of writing as interaction with a particular audience" (p. 218). Her use of CMC, then, parallels the model for writing and communicating valued in the humanities: Students are given the opportunity to engage in solitary writing and reading, as well as in the social dimensions of writing and reading (most of which are enhanced by technology). Writing assignments include personal responses and expression as well as an opportunity for a more public statement. Her syllabus includes a requirement for creative writing—in imitation of the literature studied—peer critiques and synchronous conversation through InterChange, oral discussions of literature and InterChange transcripts, a reading response log, and more formal academic-type papers.

But what evidence do we have that students are learning to be professionals in this class, or that CMC is helping students to take responsibility for their own learning in a classroom based on communal conversation? Hanson (2000) cites sample texts that students have constructed in this class where "students become conscious of themselves at the points of intersection between reader and text, writer and text, reader and writer, they test the boundaries of the forms and genres they receive" (p. 224). Her examples include a final academic paper written in the form of an InterChange discussion to include many viewpoints, yet still recreated artificially the spontaneity of a synchronous discussion, and a woman who wrote a short story from a different perspective imitating the short story "The Yellow Wallpaper." Her use of *imitatio* in this classroom—both on a small scale, as in the response to a short story, and on a larger scale, as in the recreation of the dialogic interaction between multiple authors by one student—harkens back to the Greek rhetorical tradition, balancing imitation and modeling with thinking and construction that are allowed for (and encouraged by) technology.

But increased connectivity allows CMC to go beyond the individual classroom, allowing for communal conversation between students in disparate disciplines. In the following example, we see how CMC has been used in the collaboration between a computer science class in software engineering and an English class in developing instructional materials. Because the creation of software is an activity that involves a great number of people with wide areas of expertise, three professors at Texas Tech University thought it odd that students did not get experience working in this kind of diverse grouping until after graduation. As a result, they combined three required courses in very different majors—marketing, computer science, and English—and allowed the students to work together to produce and analyze a product. The computer science students focused on developing a project—a piece of software that a medical company had requested. The English students focused on creating the users' manual and help files, and the marketing students performed a marketing analysis for the project. Susan Mengel and Locke Carter (1999) describe their linked courses:

> The students were given guidelines in collaboration, as well as an opportunity to meet and discuss the final product, how it might look, what the requirements would be, and so forth. They were placed in small groups so they could work with different aspects of the software engineering project, and they were encouraged to divide their tasks according to titles (project manager, web development, etc.) or according to self-identified skills. Because the classes met at different times, it was necessary to communicate electronically (though a few students augmented this communication with face-to-face meetings). (p. 1394)

The English instructor set up a listserv for the class; as might be expected, the computer science students were used to e-mail, and they used the list frequently. However, the English students were overwhelmed with the amount of e-mail generated on the listserv and preferred other means of interaction (such as face-to-face meetings and telephone calls). Once students became comfortable with dealing with the onslaught of information and with asynchronous e-mail as the standard for communication, however, a new problem arose: Information was being "lost" due to the use of private e-mail among the students. The instructors once again mirrored real-world collaborations and set up two separate e-mail lists: one for the client, project managers, and team leaders and one for team members in all classes. This hierarchy, although unnerving at first, put the responsibility onto the leaders of the class projects—not the instructors—to disseminate information, to answer questions, and to make sure communication occurred effectively. Asking students to communicate hierarchically rather than along disciplinary lines helped the students became more active participants in constructing (and reflecting on) a real-world rhetorical situation and thinking across the various curricula they had been immersed in.

CMC AS A MEANS TOWARD REFLECTIVE RHETORIC

My own experiences with CMC programs such as InterChange and Internet-based chat programs in the classroom have been anecdotally impressive: Several students, who confided to me one-on-one that they felt marginalized in the classroom due to race, gender, age, or handicap, found a voice here and interacted with their peers, generating more conversation than they had in the traditional oral setting. Did these environments, as Cynthia L. Selfe posited, invite "multidirectional, multivoiced exchanges and products" (1990, p. 129)? And did this seemingly more egalitarian interaction carry over into the traditional classroom, changing the dynamic of the traditional class discussion?

Before we begin to make claims, however, about how CMC changes the essence of group dynamics, we must first consider which kinds of group dynamics are integral to successful classroom discussion, then determine whether the electronic medium encourages such dynamics. Russian theorist Mikhail Bakhtin (1981) sees dissonance as vital in successful interaction. To be successful, a conversation must have both centripetal, or unifying, forces as well as centrifugal, or dissipating, forces ("Discourse in the Novel" pp. 270–273). The unity of the group will not be challenged by these two seemingly conflicting forces, however; rather, it will go through processes of reclarification. This notion is furthered by examining Bakhtin's theory of heteroglossia, or "many voices." The many voices of group members need all be present for a dialogue to occur. These voices are first created by the speakers themselves, then perceived by other members of the community. The voices may not be true to the creators, and the perceivers may construct a voice completely different from what the author intended; yet, again, it is this tension that keeps the group dynamic successful.

Although Bakhtin was referring to literary voices in his discussion of heteroglossia, I believe that the concept can be applied to CMC because it necessitates both writing and reading—two extremely literary practices. Students take on a voice, or a persona, in synchronous conferencing programs like InterChange (and more so in MUD environments) more easily than they can in traditional oral classroom discussions. There they are limited by many factors: physical placement of people (teacher standing, students sitting) or the classroom itself (chairs in rows, bolted down), gender, race, age, and so on. The seeming anonymity of the computer screen alleviates some of these physical factors (while perhaps creating others), which prevent students from freely taking on a voice in oral class discussion. Computer-mediated communication allows students not only the anonymity of the computer screen but also, for some, the opportunity to take on a completely new persona, one they would be unable to express in a physical setting. For example, during an online discussion where students were to discuss something that had a profound influence on their lives, one student, on his own, logged out of InterChange, then logged in again under a different name: Drop Dead. This persona made the following comment during a lull in the electronic textual conversation:

Drop Dead:

death has come to take your thoughts and soul. It will not leave till it has everyones [sic]

Interestingly, the student who made this comment (and two similar subsequent ones) was also an active, productive member of the discussion about paper topics.

Yet he wanted to try on another voice, a conversational persona, to shake up the status quo of the electronic discussion, and he felt safe enough to do so in the InterChange environment in the context of a classroom conversation. I don't believe he would have done anything like this had the conversation been oral.

Of course, the kind of playful risk taking described here isn't always positive; this behavior can disintegrate into flaming, which takes many forms. It can be merely off the topic, as in the post made by Drop Dead, or it can be hostile and aggressive. Yet I believe flaming has, for the most part, gotten a bad rap; I can't help but wonder if the occasional off-topic variety isn't healthy—perhaps even necessary—for students to synthesize what has been said. In oral discourse, lulls and non sequiturs in a conversation are common, perhaps even necessary for the participants to gather steam for a new round of discussion. During these times in oral conversations, the topical conversation either disappears altogether, or a tangential subject (often humorous or inconsequential) is introduced. When the conversation once again returns to the original topic, participants are refreshed, and often the topic is attacked with renewed vigor. Similarly, I believe, this dissipating occasional flame or off-topic comment on InterChange then allows for a subsequent unifying conversation with renewed depth and energy, allowing for a more reflective engagement with the rhetorical situation at hand, as well as those yet to come.

ISSUES TEACHERS MUST CONSIDER

Because the nonlinear interaction inherent to CMC is different from the traditional oral discussion patterns that teachers of composition are familiar with, those who teach with technology must learn to manage these discussions differently. The role of the instructor here is one of facilitator, not director. The authority automatically attributed to an instructor by his or her mere physical presence disappears online, so setting up very clear, recognizable tasks in the medium itself encourages unity; then, once students feel comfortable online, a form of dissonance can be introduced, in forms such as difficult questions, topic changes, or off-topic remarks.

Teachers also model desired student behavior, first in the prompts they use and then in the questions they ask and how they ask them. By asking a question directly of a particular student, teachers model genuine interest as well as the student-to-student dynamic we hope to inspire in our class discussions. By being specific in their question or comment, instructors model the kind of rich, descriptive oral text that they would like students to be engaging in. Finally, by addressing specific students, especially those who might tend to wander off topic, teachers remind students that they are part of a larger discussion, gently urging them to remain on task, not by nagging or admonishing them but again by taking an active interest in the text they've created and what it means for the collaborative text being created by the class.

Although such discussion strategies are not altogether foreign to teachers, traditionally there are familiar, institutionally based ways of interacting in a classroom. According to Glynda Hull, Mike Rose, Kay Losey Fraser, and Marisa Castellano (1991), "the structure consists of a tripartite series of turns in which a teacher *initiates*, a student *replies*, and the teacher *evaluates* the student's response—the IRE sequence" (p. 301). This sequence presents a pattern of discourse that too often pervades the writing classroom. Hence, the following predictable scenario plays out in the classroom even today:

> In the initiation, or opening turn, the teacher can inform, direct, or ask students for information. The student's reply to this initiation can be non-verbal, such as raising a hand or carrying out an action, or it can be a verbal response. In the evaluation turn, the teacher comments on the student's reply. (p. 302)

This particular rhetorical situation is one familiar to most of us: formalistic, hierarchical. The power differentiation between student and teacher is great; the teacher has power, students attempt to gain it through correct response, and they are assessed accordingly. The physical aspects of this hierarchy are, to some extent, diminished online. Because the teacher seems to be no different than other class members in CMC, with only (perhaps) a name on a screen to distinguish himself or herself from anyone else in the class, students are less likely to see the teacher as authority figure and follow the socially prescribed patterns of interaction described previously and more likely to listen to/read what others have to say, forging new boundaries for their evolving rhetorical situation(s).

THE POWER (AND PERIL) OF CMC

Do students conduct themselves differently in an electronic environment as opposed to a traditional oral class discussion? More important, do students engage in more dynamic, student-centered discourse—allowing for what Bakhtin calls heteroglossia—in online discussions? Researchers Marilyn Cooper and Cynthia L. Selfe, in "Computer Conferences and Learning: Authority, Resistance, and Internally Persuasive Discourse" (1990), describe network forums as spaces that "allow interaction patterns disruptive of a teacher-centered hegemony" (p. 847). They describe traditional classroom discussions as centering on the teacher, who "maintains control of the topic, the direction, and the pace of the discussion." Students who are willing to accommodate the teacher's goals and agenda are rewarded, and "those willing to demonstrate their knowledge publicly . . . are rewarded with smiles and nods from other students and of the teacher" (p. 852). The authors describe students' participation as competitive and one-sided. On the other hand, they argue the following about networked computers:

> [Networked computers are] powerful, non-traditional learning forums for students not simply because they allow another opportunity for collaboration and dialogue—although this is certainly one of their functions—but also because they encourage students to resist, dissent, and explore the role that controversy and intellectual divergence play in learning and thinking. (p. 849)

Lester Faigley (1990) concurs that the many voices present on a networked computer discussion "act out Bakhtin's principles of dialogism," the movement recalling the opposition between the "monologic centripetal forces of unity, authority, and truth" as well as the centrifugal forces of "multiplicity, equality, and uncertainty" (p. 293).

Jerome Bump (1990) found an increase in the number of students who participated in class discussions when using synchronous conferencing software, thus creating a higher level of individual learning (p. 55). He also suggested that because students have more time to formulate their arguments, they will be more deliberative in their comments. And because students have constant access to the instructor (as well as one another) in the context of a synchronous conference, students technically have constant individualized instruction (p. 56).

In my own more recent study of four basic writing classrooms, two that used InterChange and two that did not, there was, indeed, a noticeable difference in dynamic. In the oral discussions that were transcribed and analyzed, there was no student-to-student interaction. The conversation was directed at the teacher, reminiscent of the IRE approach described earlier. However, more than 75% of the transcribed InterChange discussions were made up of students interacting with other students individually or the class as a whole (Rickly, 1995, pp. 108–110). InterChange appears to decrease the sense of hierarchy students feel in the traditional classroom, encouraging them to see themselves—and others—as authorities, rather than investing the authority with only one member of the class—the instructor. This type of recursive, interactive group instruction resembles George Hillocks's (1986) description of the environmental mode of instruction, which, according to his meta-analytical study, has proven to be far more effective than traditional lecture classes for teaching writing.

Although these findings are exciting, we need to look beyond the increased student-to-student dynamic in synchronous CMC and examine critically what function it serves. Is it more than just idle chatter? Again, the instructor plays a large role in how fruitful a discussion can be. Task-oriented discussions, which are nonetheless flexible, seem to be the most productive, the most likely to be democratic in nature, and the most likely to lead to fertile, critical thinking/conversing across the curriculum. And integrating InterChange into a multivocal pedagogy—one that values a variety of ways of interacting—makes the fabric of the writing course that much richer. InterChange supplements, rather than supplants, oral class discussion.

Carolyn Handa, editor of the pivotal book *Computers and Community: Teaching Composition in the Twenty-First Century* (1990a), warns that "innovation cannot come from simply adopting a new technology; rather, a new pedagogy must be developed, one which takes into consideration the potential differences in context and content created by the new technology" (Handa, 1990b, pp. 169–170). "Writing teachers," Handa (1990b) states, "need to *build collaborative techniques consciously into their pedagogies*" (p. 160, italics added). But if we build it, will they come?

If the charge of the university—and, more specifically, the writing classroom—is to help create responsible citizens who can speak, write, and think well in and outside of their chosen discipline, then it is vital that the instructor understand the fabric of the writing course: its theoretical aims, institutional constraints (e.g., equipment, layout, and policies), and how this all relates to the instructor's own self-defined goals. The energy that drives the communicative impulse must be acknowledged, examined, and critiqued if students are to go beyond using technology to mirror traditional, formalistic educational scenarios where learning is rote and systematic. If we want students (and teachers) to be aware of the significance of rhetoric—and rhetorical situations—we must help them (and us) learn to listen to each other, to value diverse opinions, to rethink according to not just their own insights but also to integrate these with the insights of others. By encouraging a variety of perspectives, then providing a nonthreatening technological forum for these voices to be heard, the instructor is more likely to encourage thought, reflection, and learning. The result can be a rhetorically based communal conversation that doesn't have to conform to institutional boundaries but allows both students and teachers to think across the curriculum via reflective dialogic interaction.

4

Electracy for the Ages: Collaboration With the Past and Future

James A. Inman
University of South Florida

In the 1997 movie *Amistad*, kidnapped and imprisoned Africans find themselves in American courtrooms, being the subjects of what sadly is termed a property dispute. The legal dueling brings both sides to the Supreme Court of the United States, where the Africans are represented by former President John Quincy Adams, played by Anthony Hopkins. Before the hearing begins, Cinque, played by Djimon Hounsou, explains to Adams that he will draw strength from his ancestors, who will be with him in spirit for support and strength, and Adams, in fact, picks up on this idea of invoking ancestry by making his case not only on the facts of the matter, which are sufficiently blurred by this time, but also on the way America's forefathers themselves would have spoken to the issues at hand. That is, Adams convinces the justices that the case is precisely not about property and instead is about the human values on which the United States was founded. Had Adams not pursued the approach Cinque's faith in ancestry showed, then perhaps the Africans would not have been freed once again and given passage back to their homeland.

I begin with *Amistad* because it makes evident the central claim of this chapter: that collaboration occurs across generations, not just in the same real or virtual place at the same time. In both Cinque's and Adams's minds, the ancestors really are present in the courtroom, and they make a significant difference, whether finally it is the strength their presence gives or instead the lessons they taught in the past. When we think about collaboration in the humanities, we envision most often a group of people sharing ideas and responsibilities, and we usually are interested in examining their group dynamics or the products of their work. At first glance, such a focus makes sense; clearly collaboration, as a concept, involves multiple voices, if not multiple presences, and I do not dispute that model here. But it's my view that too rarely do we understand the other sorts of collaborative possibilities available—especially those that span generations. The group meeting scene for collaboration cannot alone represent the sum total of collaborative opportunities available for contemporary humanities scholars. As I will show, we must broaden our understanding of collaboration to include careful study of the past and the future. Although this new view of collaboration does not overtly require specific people to be engaged, the implication is certainly there. After all, people shaped the past, just as they will shape the future. In this way, I am at least in some measure always talking about collaboration among individuals, whether among ancestors or those we imagine will inhabit the future.

The definition of collaboration developed in this chapter extends and even challenges previous definitions. Most prominent of such previous definitions has been the work of Kenneth Bruffee, especially in his book *Collaborative Learning: Higher Education, Interdependence, and the Authority of Knowledge* (1993). For Bruffee, collaboration means participating in an extended and ever-expanding dialogue with peers, with the aim being a degree of democratic consensus. Not surprisingly, scholars have since challenged Bruffee's definition. Lisa Ede and Andrea Lunsford (1983) argued convincingly that genuine democracy in collaboration is a myth, not a realistic expectation, by showing how collaboration can reflect social stratification. Pushing Bruffee's sense of collaboration further still, John Trimbur (1991) suggests in several writings that *dissensus* also offers a reasonable basis for collaborative work. Despite the differences in opinion, though, all of the scholars cited in this paragraph share a view of collaboration as interactional, dependent on the ability of people to work together directly in some measure. My argument in this chapter, however, is that people do not have to interact in the same era for their work to be bidirectional. I can learn a great deal from people I've never known, just as they can benefit from my attention to and interactions with their ideas or prospective ideas. In this way, collaboration is transactional, as well as interactional, and thus able to span generations.

It's important for readers to understand that I am talking about careful and responsible collaboration across generations, not a cursory look to the past, the future, or both. If I was to become interested in women's roles in the development of computer technology, for instance, it's not enough for me to name Grace Murray Hopper or Alice Burks and list a contribution or two from them; instead, I should imagine my learning about them as collaboration and take on the responsibility of understanding their identities and the broad contexts around their innovations as well. The act of collaboration, then, comes from thinking about more than individual ideas without context; to collaborate with the past and future, individuals must develop detailed context-based understandings of any ideas they consider, and they must further influence and be influenced by those ideas. Reporting that Hopper pioneered COBOL and computer compilers becomes much more useful and interesting if I also know that she rose to the rank of rear admiral in the United States Navy and had a reputation for being tough but fair and enthusiastic. The same is true of any study of Burks. I can't reasonably report that she was a primary writer of documentation for ENIAC without also knowing her husband's prominent role and the culture of innovation in the University of Pennsylvania's Moore School of Engineering at that time. And, of course, much more still is important to learn about Hopper and Burks, if we are to collaborate with them and their contexts in the spirit I describe in this chapter.

In considering the potential influence of cross-temporal collaboration, which here is the term I give any collaboration across time, I want to stipulate that what we think about the past and future tells us more about the present than anything. In her *Intimate Practices: Literacy and Cultural Work in U.S. Women's Clubs* (1997), Anne Ruggles Gere tell us specifically that "history or what we say about the past has to do with the present more than with what happened at another time" (p. 269), and I don't think we have to stretch much to understand that the same is true about the way we describe the future.[1] So the sort of collaboration I am constructing may be reflexive to some degree, tapping into personal visions of various eras but also figuring in the way our sense of other times is determined, at least in some respects, by the way that our communities and social spaces also construct those times. Put simply, understanding the past and imagining the future require us to think about our own values and those of the world around us. As Gere reminds us, any historical contributions that are not included in mainstream American history are threatened with erasure from public memory, an observation that powerfully demonstrates the way politics can influence what we know about ourselves, our current era, and other people in other eras. Indeed, *Intimate Practices* reveals an impressive catalog of club women's activity around the turn of the 20th century; these women shaped the American society of that time and thus of our time now. That their contributions have been forgotten for so long suggests that even consequential and important aspects of any era are subject to erasure, if not identified and supported actively.

This chapter illustrates cross-temporal collaboration by examining Gregory Ulmer's concept of electracy (1997). Electracy, for Ulmer, is an apparatus for meaning making that is "to computing what literacy is to print." The concept includes all electronic technologies, meaning that it includes not just computers but also film, video, television, and more. Its particular focus is on what media produced by these technologies bring to our lives, not on what the technologies themselves mean or represent. More, as an apparatus, electracy affects all levels

---

[1]I choose Gere's project because I believe it has particular resonance for this chapter's focus on social and cultural issues writ large, but readers who want resources more specifically focused on technology might investigate the following titles: *Civilizing the Machine: Technology and Republican Values in America, 1776–1900*, by John F. Kasson (1976); *America by Design: Science, Technology and the Rise of Corporate Capitalism*, by David F. Noble (1977); and *The Machine in the Garden: Technology and the Pastoral Ideal in America*, by Leo Marx (1967).

of meaning making, from reading and writing to teaching and learning, and even to thinking itself. It is important to distinguish electracy from other terms, such as computer-based literacy, Internet literacy, technology literacy, digital literacy, electronic literacies, metamedia literacy, and even cyberpunk literacy.[2] None of these other terms have the breadth electracy does as a concept, and none of them draw their ontology from electronic media exclusively. As Anne Wysocki and Johndan Johnson-Eilola (1999) write, "Too much is hidden by 'literacy,' . . . too much packed into those letters—too much that we are wrong to bring with us, implicitly or no" (p. 349). Electracy also implies an evolutionary trajectory between it and literacy. Ulmer (1997) explains, "In the history of human culture there are but three apparatuses: orality, literacy, and now electracy. We live in the moment of the emergence of electracy, comparable to the two principal moments of literacy (The Greece of Plato, and the Europe of Galileo)." Clearly, Ulmer envisions the shift to electracy as being a monumental moment in human civilization. This sort of theorizing affords us an opportunity to draw connections across time periods.

This chapter outlines electracy for the ages, which I mean to represent anyone's collaboration with the past and future to inform work with electronic media in the present. Coming to know electracy for the ages means not just learning about electracy or about other times in our lives, whether the past or the future, but also the way all of these characterizations converge meaningfully to offer us

---

[2]For a discussion of computer-based literacy, see *Literacy Theory in the Age of the Internet*, edited by Todd Taylor and Irene Ward (1998); *Handbook of Literacy and Technology: Transformations in a Post-Typographic World*, edited by D. Reinking, M. C. Mckenna, L. D. Labbo, & R. D. Kieffer (1998); *Page to Screen: Taking Literacy into the Electronic Era*, edited by Ilana Snyder (1998); *Literacy and Computers: The Complications of Teaching and Learning with Technology*, edited by Cynthia L. Selfe and Susan Hilligoss (1994); *Literacy Online: The Promise (and Peril) of Reading and Writing with Computers*, edited by Myron Tuman (1992a); and *Word Perfect: Literacy in the Computer Age*, by Myron Tuman (1992b).

For more information about Internet literacy, see *Internet Literacy*, by Fred T. Hofstetter (1998) and *The Challenge of Internet Literacy: The Instruction-Web Convergence*, edited by Lyn Elizabeth M. Martin (1997).

Technology literacy resources include *Technology and Literacy in the Twenty-First Century: The Importance of Paying Attention*, by Cynthia L. Selfe (1999a); *Teaching Literacy Using Information Technology: A Collection of Articles from the Australian Literacy Educators' Association*, edited by Joelie Hancock (1999); and *Literacy, Technology, and Society: Confronting the Issues*, edited by Gail E. Hawisher and Cynthia L. Selfe (1996).

For more on digital literacy, see *Digital Literacy*, by Paul Gilster (1997).

Information about electronic literacies may be obtained from *Electronic Literacies: Language, Culture, and Power in Online Education*, by Mark Warschauer (1999) and *Electronic Literacies in the Workplace: Technologies of Writing*, edited by Patricia Sullivan and Jennie Dautermann (1996).

For a discussion of metamedia literacy, see "Metamedia Literacy: Transforming Meanings and Media," by J. L. Lemke (1998), in *Handbook of Literacy and Technology: Transformations in a Post-Typographic World*.

Cyberpunk literacy comes from "Cyberpunk Literacy; or Piety in the Sky," by William A. Covino (1998), in *Literacy Theory in the Age of the Internet*.

a richer and more productive understanding of electracy. Too often, I believe, we want only to know a term's definition, and we try immediately to assimilate it and classify it, instead of engaging it critically and provocatively. Electracy, as I see it, cannot be dismissed so easily. Because it requires us to acquire and think about contextual knowledge, such as how electronic media were crafted in history or how the social implications of widespread technology use influenced human activity, it stipulates to us that we must practice cross-temporal collaboration. In this sense, we might envision electracy as a holistic concept, one with the stability to operate as a subject for intense conversations, whether or not conflicts occur. Practicing electracy for the ages, finally, means placing oneself among binaries that exist in our past, present, and future; we must take a middle-ground position between the sorts of oppositional stances that have taken away some of our in-person collaborative possibilities in recent years. Readers must pursue an informed and savvy sense of the large-scale issues that shape what we know about the past, present, and future, and they must influence and be influenced by those issues.

ELECTRACY AND THE PAST

Collaborating with the past contributes meaningfully to our sense of electracy by sharing valuable lessons about the social impact of any technologies we might employ. Some scholars have argued that the rise of computers differs fundamentally from the emergence of previous technologies. In some respects, they are right; however, the potential differences do not mean that we cannot learn from the past. Instead, as I demonstrate in this section, it seems that the ways we think and talk about the past have everything to do with contemporary computing and thus one aspect of the development of electracy.

We can begin with 19th century industrialization, especially as it defined and reified particular social structures, because the economic realities of that era show that a move toward electracy is not without considerable cost, both literally and metaphorically. Put simply, the emergence of industrial technologies made possible an increased schism between the working class and the ruling class; here use of the terms suggests *working class* as those citizens (and children even) who were forced by their social standing to endure oppressive working conditions and schedules to support themselves and their families and *ruling class* as those people who were not forced into factory working conditions, either because they controlled those conditions as administrators or because they were born wealthy (royalty, for instance). Without detailed interrogation, we can see that this sort of social structure perpetuated the proletariat-bourgeoisie dichotomy identified by Karl Marx. Although the social roles suggested by Marx may not be themselves directly applicable to electrate practices, the spirit with which Marx's theories were crafted and received remains relevant. Using electronic media requires access, and access requires either affluence or privilege, at least with regard to the newest technologies, unless federal or other monies can be leveraged for additional public points of access. The matter becomes even more complex, however. Many jobs require experience with electronic media and further require such media to be used effectively on a day-to-day basis. Although the access issues are notable themselves, then, they become substantially more compelling when they define at least in some measure the lives individuals may lead.

To assess the public consciousness in the late 19th century industrial era, we can turn to literature, as many leading publications reflected the techno-centric view of that day. We might begin with Mark Twain's *A Connecticut Yankee in King Arthur's Court* (1917), in which the Boss, a character from the future, effectively destroyed a vision of King Arthur's golden age by introducing yet-to-be-invented technologies; had the Boss not brought contemporary technologies into Camelot, so the implication goes, it would have been spared. Twain's plot line clearly suggests that new technologies may corrupt or even destroy the most golden eras of life. Stephen Crane's *The Red Badge of Courage* (1895), authored at about the same time as Twain's *King Arthur,* includes language that often portrays armies as machines, reflecting both the tenor of the American Civil War and the ethos of the American 1890s, in which Crane crafted his novel. Just as the Union and Rebel armies advanced and threatened assault in the novel, so new and potentially problematic technologies advanced in Crane's world at the time of his writing. We might also point to Henry Adams's *The Education of Henry Adams* (1918), in which he wonders aloud, after a visit to the Paris Exposition in 1900, if we have begun to place more faith in technologies than religion, which has been much more often an aspect of human culture informing faith and constructions of values. Adams's point can be illustrated by a simple scenario and question: If a group of people climb aboard a frightening roller coaster, would they hope to themselves that the track and ride vehicles would work, or would they turn to their religious faith? Collaboration with literature and other popular texts in the 19th century brings to light a voice of resistance, or at least one of critical engagement. In the various texts of that era, we see authors grappling with many tough questions still evident today: What should we make of the incredibly rapid advance of technologies? Might they destroy aspects of life we hold dear? And, of course, authors in the late 19th century had no easy answers, as we do not today.

Building on what we saw in turn-of-the-century technologization, let's move now to the rise of the personal computer in the mid to late 20th century, again with an interest in examining the sociopolitical issues at stake. Collaborating with the past in this way offers much of value. As a test case, let's explore the emergence of networked computer technologies, such as the Internet and the World Wide Web. As many scholars have noted, the United States government developed a system called ARPANet between World War II and the Vietnam War; the design theory at work was that the government needed a way to move information from computer to computer to prevent data loss in the event that a computing site was threatened by attack. In essence, then, the first sorts of information transferred between computers via telecommunications technology were military.[3] After World War II, when ARPANet was conceived and implemented, the world was still recovering from the shock of two global conflicts. The diversity and range of warfare technologies had demonstrated that no country or location was safe from attack, and events like the Cuban Missile Crisis showed Americans in particular that even their mainland was not safe from attack. Such realities inspired both resistance to and fear of technologies in general, so if we think about the way networked computers were first born, it makes sense that the potential of electracy was somewhat obscured then. In many respects, it would not be surprising for many people to still distrust technologies as a result of their conception in conflict, whether minorities who fear additional majority oppression or those in the majority who believe their power would be threatened by a technologically sophisticated minority population. These fears intersect with contemporary concerns about the military, such as the degree to which women still have not been fully integrated into military units. And, of course, on a basic level, military technologies kill. What we can take finally for contemporary electracy is a conundrum: The military has the scientific expertise and funding level to achieve great technological successes, but these must incorporate attention to access and diversity to have a foundational impact on our world.

Even when computer technologies did not develop in the military-industrial complex, they demonstrated other social structures with implications for electracy. If we collaborate with the past and examine the emergence of computer technologies as presented in current histories, what we learn is that most of the developments were crafted by a group of White male innovators who met sometimes in California garages to share ideas but who also moved from company to company, aligning themselves with whatever corporate interests would offer the most consistent and lucrative support of their ideas. Included in this group are Bill Gates and Steve Jobs, founders of the Microsoft Corporation and Apple Computers. Instead of challenging these histories here, which I think can be done in the same manner as Gere's recovery of turn-of-the-century club womens' literacy and cultural work, what I want again to think about is contemporary electracy. If it is widely believed that computers were developed by White males, is it any wonder that women and minorities sometimes fear their potential impact? Does it surprise us that women and minorities were not generally among the first to use computers in either workplace or social settings? What we need to understand, then, is how the sorts of White-male-centric histories that currently are in place affect electracy if people do not all feel comfortable with the supporting technolo-

---

[3]See *Transforming Computer Technology: Information Processing for the Pentagon, 1962–1986* by A. L. Norberg and J. E. O'Neill (1996).

gies. More generally, we need to think about the way electracy can be perceived as an elitist or majority practice, where minorities or other oppressed social groups are left to fend for themselves in what can only be seen as a frighteningly technological world, one where opportunities for hands-on experimentation and learning have become increasingly scarce, thus perpetuating social disjunction.

Collaborating with the past—influencing and being influenced by it—also enables a consequential voice of resistance to be heard in the 20th century, as well as diversity in critical approaches to technology. The subject shared across many 20th century philosophical concerns proves to be technological determinism, which refers to the view that technologies write the way we write our worlds. That is, our lives are determined causally by the technologies we select and employ. However, scholars often too rapidly essentialize the idea of technological determinism without exploring in more detail the different projects in which it is invested and centered, the sort of broader perspective for which this chapter argues. An illustrative example is the relationship between Martin Heidegger and Herbert Marcuse, who studied together for a time in the Frankfurt School but separated with the rise of Nazi power: Heidegger stayed in Germany, while Marcuse moved to New York. The splitting was more than geographic, and it is these factors that demonstrate how determinism is a complex response to the rise of various technologies. In Heidegger's "The Question Concerning Technology" (1977), for example, he argues that technology means much more than simply a machine or electronic device: One of Heidegger's chief concerns, in fact, is that humans themselves take an instrumental view amidst the revealing of technology, meaning that we are more interested in human perception and understanding than on the influence of technology. Marcuse (1964), on the other hand, writing generally with a neo-Marxist perspective, voices a concern for the welfare of industrial nation-states in a technological world; he fears that the means of capitalist production could be taken away from humans by the sorts of power invested in the industrial push toward technology. Although it would not be prudent to suggest that only Heidegger's and Marcuse's different worlds—Nazi Germany and New York—resulted in their different views, the influence is clear, and comparing their works demonstrates the rich and diverse possible fabric of resistance to technology. Contemporary scholars can take heart from such examples. Not only does well-articulated resistance prove distinct from ill-conceived, reactionary views, but it also assumes an important place in the path of any technology's advance. The best model of electracy, then, is inclusive of advocates and resistors, not exclusionary of one or the other.

Finally, I want to point to some scholarly pursuits that have developed as a result of technologies and other innovations in the 20th century and still quite relevant for any understanding of contemporary electracy via this broad model of collaboration articulated in this chapter. First, we can point to cybernetics, a line of thinking perhaps most famously detailed by Norbert Wiener is his *Cybernetics; or Control and Communication in the Animal and the Machine*, published in

1948. In essence, the principal concern of cybernetics is constructing a systems-level examination of life processes as information cycles, especially as life processes benefit from causal cycles with feedback loops. In other words, Wiener and other systems theorists want to understand what we do in the world as information exchange, not as anything more particular. Their sense is that taking a large-scale perspective gives us the most promising chance of understanding life processes. A simple example would be the way new computers might influence the lives of children in Tampa, Florida, where I live. Small donations of computers would not appear as significant in the large-scale cybernetic model unless their reach far exceeded their numbers, but sizeable donations may well prompt sea changes in computer use patterns in Tampa. Recently, especially in N. Katherine Hayles's *How We Became Posthuman: Virtual Bodies in Cybernetics, Literature, and Informatics* (1999), the loss of individual knowledge about system subjects has been called into question, but, nonetheless, what's important is that cybernetics emerged as a new way of understanding life. Cybernetics, as a field of inquiry, proves particularly interesting because it frames humans and other living beings in the same way that we often frame electronic technologies—by their capacity to process and distribute information. Problematic or not, it is an interesting means of thinking about the scale of electracy in life systems.

Another important line of thinking to emerge was diffusion theory, here defined as the means by which innovations are considered and adopted by social groups. The finest text of this school of thinking is Everett Rogers's *Diffusion of Innovations*, first published in 1962 and now in its fourth edition (1995). *Diffusion*, in essence, traces the way an innovation develops through communication channels, and one of its principal strengths is its interdisciplinary foundation; Rogers himself argues that we can see diffusion theory's roots in disciplines from agriculture to education, from medicine to law. For Rogers, any diffusion process necessarily includes four key elements—the innovation itself, communication channels, time, and a social system—and these together offer the sort of broad perspective for which this chapter argues. If we take, as an example, the videocassette recorder, then we can see the way diffusion works. The innovation may be understood not just as the recorder technology, but also as the way it is constructed socially and culturally in various contexts. In defining communication channels, the key is to think broadly; these include interpersonal interaction in many forms and across a range of media. Channels associated with the videocassette recorder might include popular media, such as newspapers and television, because they each reach a wide audience. Rogers's attention to time is also key. The rise of the videocassette recorder took a number of years and still continues in some respects, and the competition being created now by DVD technology complicates the mix. Last, social system indicates the importance of social networks to diffusion. When the videocassette recorder was diffused into the University of South Florida's campus community, then it required communities of faculty, staff, and students to adopt it for it to have visibility on campus, be brought into various family networks, and be shared with the community. Such complexity demonstrates the importance of the diffusion model; its interdisciplinary foundation is robust enough to persist. Cybernetics and diffusion theory help us understand how mainstream academic and social communities began thinking constructively about emergence and proliferation of advanced technologies, just the sort of approaches to demonstrate the way contemporary people might think broadly about electracy in their lives.

From what we learn when we collaborate with the past, we can acquire and contextualize important information about electracy, from a historical sense of the development of electronic technologies to an understanding of the ways scholars began to think about representing a technologized society, as well as to what literature and popular culture can share with us about perceptions and portrayals of electronic technologies. Learning about the past through this chapter's broad definition of collaboration can push us to know more about what we do with electronic technologies in our lives now. Because these events are situated in the context of electracy, then it's all the more important that we understand the fluid interactions between different technologies' developments, including the complex worlds in which they operated and continue now to operate.

ELECTRACY AND THE FUTURE

As we collaborate with the future and think about electracy, we cannot fear utopian visions. Simultaneously, we cannot be afraid to engage such visions critically and completely to see what they tell us about our world. The prize of such efforts is a collaborative enrichment of contemporary electrate activities through an understanding of projected directions into the future. In this sense, collaboration is between those who will inhabit the future and us, no matter that those of the future will likely be much different than we imagine them today. Again it's critical to imagine this collaboration as bidirectional; we influence and are influenced by future possibilities.

A useful starting place is to examine public policy, especially in terms of any values ascribed to electronic technologies. Chief among the documents we should examine is the Goals 2000: Educate America Act (United States Department of Education, 1996b), which was introduced and adopted in 1994 and which now has been amended and revised several times, as well as integrated into state-level education reforms projects. Of particular interest is Section C of Title II of the act. This section is titled "Leadership in Educational Technology," and it sets out six subsections of recommendations, including providing a national vision for the use of technologies in education and remedying inequalities by offering all American schoolchildren access to the technologies they will need to become successful 21st century citizens. The national vision, first, forwards the sort of centered approach that was missing for entirely too long. Although I do not want to overemphasize its role, I do believe that having an expressed interest and value of educational technologies by the American federal government gives their use a much stronger status, which can push more teachers and students to take advantage of their possibilities. At the same time, however, scholars are already beginning to locate critical issues in the federal plan, especially in terms of its attention to social and cultural inequities, which seem still to be persisting. One of the most recent examinations is Cynthia Selfe's *Technology and Literacy in the Twenty-First Century: The Importance of Paying Attention* (1999a), in which she articulates challenging questions about government initiatives for technology literacy and outlines roles that actors in education (i.e., teachers, parents, students, administrators) can play. As Selfe suggests, any vision of a technological future must also have built-in social responsibility. This sort of realization is a crucial aspect of what we do with technologies and where we imagine them taking us. Although these public programs and commentaries do not take on electracy as their approach, they do forge an increased awareness by citizens of the ways we need to think about contemporary technologies, some of which are the electronic media that shape electracy.

We can also collaborate with the future via organizational policy, particularly in the professional statements of committees charged with recommending ways to understand and value electronic scholarship and activity. The Modern Language Association's Committee on Computers and Emerging Technologies in Teaching and Research, for example, in 1996 published a document offering suggestions both for hiring committees and job seekers and for tenure and promotion committees with regard to evaluating electronic scholarship and activity. Although the document puts forward important information linking the sorts of offline professional assessments that have been conducted for years with the new online and offline assessments now required by faculty making productive use of technologies, it remains relatively conservative, suggesting that any activity should be included in a dossier, without stipulations about the possibility of an electronic form. It further suggests that candidates must ensure that their activity meets well with departmental standards, a challenge perhaps more difficult for scholars pursuing substantial electronic activity, where community and interinstitutional projects are more the norm. In contrast is the Conference of College Composition and Communication's *Promotion and Tenure Guidelines for Work with Technology* (1998), a document that places overtly more responsibility on senior scholars and evaluators themselves to learn about and succeed with electronic technologies. A brief excerpt demonstrates this difference: "It is important that the candidate's work be evaluated in the medium in which it was produced. Printing off web pages . . . is a poor substitute for evaluating those pages online." Here, we learn not just that candidates' activity should be evaluated in its natural medium, but we also learn the value judgment that not examining such activity in its natural medium means an innately irresponsible review procedure. These and other organizational guidelines clearly begin to etch a future for technology-based professional activity in the humanities by making visible the way that each person values (or doesn't value) technology. More, these documents validate a suite of electrate practices that should become increasingly valuable to us over time, such as requiring senior scholars to examine computer-based activity of tenure and promotion candidates online. Such activities certainly promise to expand what we know about widespread electracy in terms of who uses electronic media productively.

Where literature was an appropriate means of gauging public consciousness in the late 19th century, many different media dot the ideospheric horizon in the contemporary era, and among these movies seem to have a particularly influential role. Collaborating with the future visions in these popular media promises to reveal more important information for a broad understanding of electracy. In *The Matrix*, released in 1999, the character Morpheus, played by Laurence Fishburne, leads a team of computer geniuses, including Neo, played by Keanu Reeves, to learn that the world they believe exists is instead only a mental projection. The reality of human existence is that we are being bred as sources of energy for artificially intelligent techno-beings, who won a war with us and now have enslaved us. At one point, Morpheus demonstrates to Neo the truth of his statement by removing him from the machine's control and demonstrating how the matrix requires a plug-in to the computer software that is its composition. *The Matrix* offers an engaging and progressive theory, one far more developed than some of the other movies that look at the future, such as *The Fifth Element*, wherein Bruce Willis plays a cab driver of the future, or any of the *Batman* movies, which show an advanced Gotham City saved by Batman again and again. At the same time, we do have other movies that offer a darker view of the future, including one which demonstrates the potential for a negative impact of technologies. One of the simplest to follow is *Christine,* based on Stephen King's novel of the same name, in which a car becomes possessed and kills people. Also based on a book by King, *The Lawnmower Man* tells the story of an inventor who tests a virtual reality system he is designing on his simple groundskeeper; the battle that ultimately rages is a struggle for the mind of the groundskeeper, between the inventor and a dark force bent on world domination. What proves interesting about many of these movies is the way they champion subversive behavior. That is, the future seems always to be about control, and the heroes reject the possibility of being controlled, instead configuring themselves as agents against any and all forces that seek such control. Given the strong effect of movies on our contemporary world, their portrayal of electronic technologies seems to have considerable potential to inform both what we know about electracy and the way we think and talk about it as a human practice.

Another way of thinking about the influence of electronic technologies on our future is to explore how those technologies may be changing our very existence. Indeed, the human body has been a subject of much interest in recent humanities scholarship, and one of the most active pursuits is in imagining and defining and posthuman condition. In their *Posthuman Bodies*, published in

1995, Judith Halberstam and Ira Livingston explore how posthumanity and postmodernity are intertwined:

> Posthuman bodies are the causes and effects of postmodern relations of power and pleasure, virtuality and reality, sex and its consequences. The posthuman body is a technology, a screen, a projected image; it is a body under the sign of AIDS, a contaminated body, a deadly body, a techno-body; it is, as we shall see, a queer body. (p. 3)

Picking up on the idea of a techno-body, we can turn to the mid-1980s as a time when cyborg identity first was made prominent in the academy. Donna Haraway's "Manifesto for Cyborgs," first appearing then (1985) and later published as a chapter in her *Simians, Cyborgs, and Women: The Reinvention of Nature* (1991), serves perhaps as the touchstone for this identity claim, though NASA was pursuing bioengineering with laboratory rats as early as 1958. NASA's project was to imagine how human bodies could be engineered for extended space travel; the rat studies were the first steps. In this sense, Haraway's project is even more important because it articulates a nonmilitary-industrial-complex view. Readers do not have to imagine themselves as posthuman or cyborg to understand the impact these categories' existence makes; instead, what we need to know is that productive thinking can come from thinking about technologies and humans working together, whether in hybrid form or not.

By collaborating with the future, as I have discussed, we can learn a number of important lessons that impact the way we think about electracy and electrate practices now. Although we might imagine our own future as cyborg or more generally as posthuman, we can clearly see that thinking about technologies as partners offers a different rendering of their influence on our being and our activities. More, we can see how the influence of electronic technologies has begun to emerge in popular culture, especially cinematic portrayals like *The Matrix,* wherein technology is the text upon which human life is written, more or less. We also have examined policy implications, both for national and international domains and for organizational guidelines and statements; from these, we can learn about how our values may or may not intersect with the professional challenges of the future. Electracy, as a concept, clearly is enhanced by these lessons from collaboration with the future—both the way we ourselves influence them and the way we are influenced by them.

ELECTRACY FOR THE AGES

Across the past and future, I have surveyed here some of the ways I see collaboration as temporally constructed, instead of spatially and demographically defined. That is, effective collaboration does not always involve group activity; instead, as I hope I have demonstrated, anyone who takes into consideration the past and the future in assessing the present indeed practices collaboration by referencing what others have done or will do. What matters, again, is that such collaboration is fluid and bidirectional.

Because Ulmer's electracy has been a centerpiece of analysis in this chapter, it only makes sense that we configure our own understanding of cross-temporal collaboration within what he envisions for the future of electrate practices. Coming to this sort of understanding involves reading across Ulmer's projects: the Florida Research Ensemble, the Electronic Learning Forum, and Imaging Florida, each designed to take advantage of communicative media and technologies to increase the agency of scholars in education and, indeed, of people in the world. First, emerAgency is the organization that reaches across these projects; configured as a consulting or outreach agency, it seeks to forward the causes of education, not by following the rules and all the proper and official channels but instead by making its own rules in acting as a collaborative agent. Electracy, then, is not the single emphasis of Ulmer's activity, though I believe it's a highly important element; instead, as he reminds us, electracy is only the apparatus that allows the sort of educational and intellectual reform he envisions to happen. Our interest here is in taking full advantage of the possibilities available for electronic collaboration in the humanities, and in this project we also understand that our aim requires us, even compels us, to take on significant responsibility. Electracy, finally, is the means by which we can begin revolutionizing what we know about how electronic media afford us a venue for collaboration in the humanities; we need to continue to push forward and to analyze both the results and implications of any move we make.

One of the most well-articulated discussions of this sort of professional responsibility is in Patricia Sullivan and James Porter's *Opening Spaces: Writing Technologies and Critical Research Practices* (1997). In this book, the authors' larger project is to forge conversational spaces between theoretical research practices most often associated with postmodern critique and empirical research activities; *Opening Spaces* indeed makes this case well, offering readers specific ways, such as mapping, for critical research practices (their theoretical/empirical model) to emerge more prominently into discussions of research, especially in how they can share with us important information about methodologies. Beyond this focus, however, Sullivan and Porter draw on feminist research activity to craft an ethical dimension to their argument. In particular, *Opening Spaces* reminds us that our obligation to examine cultures and population should be with an eye toward improving the life or standing of those subjects involved, not just for furthering academic knowledge. For Sullivan and Porter, this sort of ethical dimension emerges from the conversation between theoretical and empirical stances, where critique serves as an ethical assessment technology, more or less, by not allowing empirical activity to forget its obligations. Whether we ultimately want to configure our cross-temporal collaboration as a critical research practice, we can certainly take from Sullivan and Porter's book a more detailed sense of how we can be responsible in our professional practice.

To close, I want to return to the movie *Amistad*. While I cited it initially simply to demonstrate the nature of cross-temporal collaboration, we can learn more from it. As Gere cautions us to examine critically what remains and flourishes in the public memory, so we can look to the plot of *Amistad* as an example of the potentially problematic realities of any sort of approach like this chapter has advocated. In the Supreme Court hearing, it was Adams's oration that won the day for the Africans, not Cinque's bravery or spirit, no matter that it was his view of collaboration that brought Adams to the winning argument. And in the movie's final scenes, it is White military officers who free other African slaves, not the slaves' own ingenuity and effort. Although the movie's focus requires a portrayal of America's social system before the Civil War, its depiction is no less horrific, as we realize just what violences can be brought on innocent individuals by factors they cannot control. Finally, then, *Amistad* not only shows us that collaborating across generations can yield significant results but also that the choices we make in characterizing the times and places with which we collaborate have a host of critical issues around them. We cannot just neutrally look at the past, present, and future; instead, we must examine their convergence critically and responsibly.

5

Collaborating Across Contexts: Rethinking the Local and the Global, Theory and Practice[1]

Radhika Gajjala
Bowling Green State University

Annapurna Mamidipudi
Dastkar Andhra, India

Date: Mon, 25 Oct 1999 10:18:16 -0400 (EDT)[2]

From: "t.k. smalec" <tks201@is9.nyu.edu>

To: sa-cyborgs@lists.village.virginia.edu

thissuddenrushedswirloftangledintentsstruckmeassalientthefirsttim
earoundbutcaughtinthelensoftranslationinthetransitsofinterpretati
onthemessagechangessometimes.sometimesforthebetterandsometimesfor

---

[1]The authors would like to thank members of our families both in the United States and in India, for their active participation in this collaborative across-contexts effort. In addition, we thank the transnational research cluster and ICS writing group (both funded by a grant from the Institute of Culture and Society at Bowling Green State University) members for helpful insights and bibliographic suggestions that enriched our writing. We also thank members of various discussion lists for helping us understand some issues related to communicating across contexts online. We especially thank Peter Sands for his encouraging, helpful feedback and careful reading on several drafts of this chapter and James Inman for his patient editing and encouragement. Radhika Gajjala takes full responsibility for any mistakes or misrepresentations. Versions of this chapter and parts of the conversations around which this chapter is centered were presented at the National Communication Association Conference 1999 (held in Chicago, IL, USA), the International Cultural Studies Conference 2000 (held in Birmingham, UK), the American Studies Association Conference 2000 (held in Detroit, MI, USA) and the International Communication Association Conference 2001 (held in Washington, DC, USA). The section on the veiling of the subaltern is adapted from Radhika Gajjala's dissertation *The SAWnet Refusal: An Interrupted Cyberethnography* (1998).

[2]This chapter includes several email messages, which are distinguished from chapter text by the use of an alternate font.

```
worseandsometimesjustchangesthatonethingissure.doesanyonehearder
ridatalkinthedistanceofthisconversation?derridatalksofmorpholog
yofgrammatologysoftaccentsandglyphs.derridasaysnotmuchnewnothing
thatthoseonthislisthavenotalreadysaid.hetalksofthewaysthat"mean
ings"travelthroughtimeandtraverse.transitthroughtimeandthroughs
pacemeansanticipatethemultiplewaysinwhichothersuntieones"own"sens
eandmakenew.alwaysanticipatesandwiches:thesinewylayersofnostalgia
amnesiadistortionrepressionandpersonalmemorythatreaderswillbring
to"ones"texts.alwaysanticipateghosts.intentionisnotsomethingstabl
   eorfixedbutratherasnakeinthegrassaspringajackintheboxan"
   abracadabra"anewsongnewsoundsfullofdistortion,yetthere.
      anoldsonganoldsoundandoldhold,stillthere.


       On Sun, 24 Oct 1999, cyberdiva wrote:

   > agreedisagreeagreedisagreedisagreeagreemymindisnowin
    > aswirlofbinaryframingIdidnotintendbutwhathasagree
    > disagreegottodowiththeissuesathandIwonderwhatmy
       > intentreallyisdoyouknowperhaps?
```

How can we form discursive and action-based networks between the local and the global and between place-based practices and virtual practices? In collaborating to write and work together to co-compose (with several others online) and design electronic spaces/networks for dialogue and action, we encounter numerous issues relating to the (im)possibilities for building discursive and material networks that "would be most conducive to [a redefinition and reconstruction] from the perspective of the multiple cultural, ecological and social practices embedded in local models and places" (Harcourt, 1999, p. 44). What is the relationship between the building of online networks for communication and communities of production within everyday contexts? The relationship between the global and the local, the analogue and the digital is of central importance to our ongoing dialogue.

One of us, Annapurna Mamidipudi, approaches this collaboration from her experience as a worker for a nongovernmental organization dealing with handloom weavers in a South Indian village. She works with a group of volunteers who are trying revive the old technology of vegetable dying and cotton handloom weaving in a few villages of South India. The other, Radhika Gajjala, approaches the collaboration from her academic work regarding cross-cultural dialogue and the expression of women's identity among virtual communities of diasporic postcolonials. In her efforts to examine diasporic and transnational digital subjectivities, she engages in the production and maintenance of Web-based and e-mail-list-based interactive e-spaces.

Both of us negotiate histories of colonialism, nationalisms, definitions of culture, and various binaries produced within narratives of modernity. We are engaged in processes of production that appear to be at the two extremes of the premodern to digital continuum framed by a modernist, linear telos of progress, a capitalist technological imaginary, and a colonial past (Robbins, 1995). Annapurna's concerns center around the creation of sustainable systems of production in the context of her work with handloom weavers, whereas Radhika's center around the creation of dialogic, action- and place-based online networks in the context of her work in the digital and academic spheres. We both negotiate tensions between theory and practice within complex tangled webs of historical and ongoing material/discursive formations. Together, we are situated within an increasingly digital and transnational economy.

This chapter is organized around some of the dilemmas and concerns that arise in the context of negotiating local and global modes of production in a climate of transnational corporate hegemony. In our efforts to articulate our concerns from different contexts and in coming together in dialogue via e-spaces, we are confronted with various issues related to the (im)possibility for indigenous coalitions via Internet-enabled electronic spaces. Our previous investigation into these issues and dialogue via e-mail discussion lists, personal e-mail and telephone conversations resulted in our published article entitled "Cyberfeminism, Technology and International 'Development' " (1999). That print collaboration was part of our attempt to interrogate binaries such as traditional–modern and theory–practice produced within modernist framings of technology, culture, and development by juxtaposing apparently contradictory contexts of production ("old" and "new" technologies). We were attempting this to open up analytical categories and social spaces (both online and off) that do not merely pose differences and binaries but help aid in the articulation of processes for the relational, interconnected, and contextual use of technologies. Our engagement led us to examine contexts related to apparently different modes of production—one related to the production of handloom fabrics and the other related to the production of digital spaces against the grain of the "continuist, progressivist myth of Man" (Bhabha, 1994, p. 237).

Form is a major concern for us both in trying to collaborate within this present chapter and in collaborations within cyberspace and in our local contexts. Both in cyberspacial environments and within everyday practices, our collaborations extend beyond ourselves and draw on community resources while they are embedded in hierarchies. Our individual and community-based subject positions emerge within various contexts in multiple voices in relation to these social spaces both online and off. The portrayal of the heteroglossia of each of our individual and collaborative subject positions in print format is therefore a major dilemma. In response, we play with form in a variety of ways. For instance, we share some text (clips from e-mail exchanges and some quotes) in different fonts, while part of the writing is in the form of parallel narratives or dialogue exchanged in relation to a subtopic. Thus, we have interspersed our discussion with blocks from e-mail exchanges between us as well as some relevant quotes from related scholarship. The main unifying point for both our narratives is a critique of scientific progress and its top-down approach to technology transfer/diffusion within current unequal power relations between first-world and third-world contexts.

It is a well-known fact that first-world ideologies and decisions often pass for global and third-world contexts. Struggles for autonomy and insistence on real choices are labeled as local and indigenous. There are various connotations to the terms global and local, depending on the context and discipline used. Within the context of this chapter, the following definition by Tomlinson (1997) is apt:

> Globalization in its most general and uncontroversial sense . . . refers to the rapidly developing process of complex interconnections between societies, cultures, institutions and individuals world-wide. It is a process which involves a compression of time and space . . . shrinking distances through a dramatic reduction in the time taken—either physically or representationally—to cross them, so making the world seem smaller and in a certain sense bringing human beings "closer" to one another. But it is also a process which "stretches" social relations, removing relations which govern our everyday lives from local contexts to global ones. Thus, at its highest level of generality, globalization can be understood . . . as simply "action at a distance." (pp. 170–171)

Action at a distance and stretching of social relations have increasingly come to characterize processes of production within the current (post?)modern frameworks of economic globalization, wherein information industries and digital technologies play a crucial role. However, the dispersal of social and economic activities represent only a partial account of the socioeconomic restructuring currently in progress. As Sassia Sasken (1994) points out, "even the most advanced information industries have a production process" (p. 1). There is a "spatial dispersion of economic activities" leading to the centralization of top-level management and control within power nodes situated in some cosmopolitan (global) urban centers (cities) of the world (p. 1). These urban centers are networked and connected by vast place-based infrastructures. Thus, an "economic configuration very different from that suggested by the concept of information economy emerges, whereby we recover the material conditions, production sites, and placeboundedness that are also part of globalization and the information economy" (p. 1). Despite the virtualization of the discourse surrounding globalization and ideas related to the global village and so on, however, digital existence, digital finance, and digital globalization are indeed very place based. Dominant imagery concerning global economic processes and the use of digital technologies emphasize instantaneous digital transmission of money and information, thereby fostering ideas about the "neutralization of distance through telematics" (p. 1). Material processes, activities and place-based infrastructures central to the process of economic globalization are generally overlooked. Abstract, often ahistorical models that do not engage issues regarding hierarchies embedded within these spatial configurations are presented by theorists who have not sufficiently acknowledged the importance of the everyday contexts of production processes within advanced information economies.

Arturo Escobar (1999) and Arif Dirlik (1998) argue that contemporary theories of globalization tend to assume a binary and an almost irreconcilable opposition between the local and the global while marginalizing place-based practices in debates concerning the local and the global. They suggest that the term glocal, as defined by Dirlik, might better help us conceptualize the relationship between the local and the global.[3] As Dirlik points out:

> Instead of assigning some phenomena to the realm of the global and the others to the realm of the local, it may be necessary to recognize that in other than the most exceptional cases these phenomena are all both local and global, but that they are not local and global in the same way. (p. 7)

The categories of local and global are interdependent (i.e, one category cannot exist without the other), and they exist within glocal configurations and relations that occur within unequal relations of power. In the context of the production of theory, the power-scale—ideologically, materially, in terms of access to the structures of power, access to global capital and technologies—is heavily framed within a developmentalism that is materially and ideologically tipped to the advantage of the global, the digital, and the virtual haves (Saldana-Portillo, 1997). Thus, although knowledge and theory are produced about contexts that are on the have-not end of the glocal scale of power, the production and circulation of theory remains situated within a very global heavy framing. Therefore, a related question for us is the following: Although the Internet remains situated within a global matrix, how are the voices from different locales to be heard as distinct and politically as well as materially situated within this virtual sphere?

What does negotiating the glocal imply for populations situated within the not-so materially and culturally privileged locales of the south and north hemispheres? Any attempts to answer these questions must take into consideration the unequal power relations between the global and the local, as well as the role that the perceived location of theory plays in the articulation and implementation of solutions. We need to explore if there is indeed a "widening gap between our analytical constructs" (p. 61) and actual experiences and negotiations that lead to ac-

---

[3]According to Dirlik (1998), the term " '[g]local' expresses cogently . . . the hybridity of the global and the local." He writes:

> the question of the local cannot be marginalized without an equal elimination or marginalization of the global, which restores to the problematic of local/global a symmetry that is missing from most discussions. If the local is not to be conceived without reference to the global, it is possible to suggest that the global cannot exist without the local, which is the location for its producers and consumers of commodities, not to speak of the transnational institutions themselves.

> The question then is not the confrontation of the global and the local, but of different configurations configurations of "glocality." Instead of assigning some phenomena to the realm of the global and others to the realm of the local, it may be necessary to recognize that in other than the most exceptional cases these phenomena are all both local and global, but that they are not all local and global in the same way. (p. 7)

tion and resistance within specific material contexts. If so, how do these contexts "challenge theory to catch up with lived realities?" (Ong, 1997, p. 61).

Scholars like Escobar and Dirlik speak of place-based practices, but taking their critique further, we might ask the following: what role do time, culture, and "old" modes of production, such as rural farming practices in southern locations, play in glocality? In our past exchanges regarding old technologies and new, we have come to realize that analytic categories and boundaries are confused and blurred as various nongovernment organizations and activists attempt to revive indigenous knowledges and ecologically friendly rural modes of production (Gajjala & Mamidipudi, 1999). Within an analytical framing (global theories) that privileges modernity, these attempts at revival and rediscovery of old modes of production are constructed as attempts to go backward in time. They are conveniently placed (museumized) in alternative or tokenized slots, often commodified and appropriated within a romanticized notion of multiculturalism, diversity, and difference. However, the actual experiences of field-workers (local practices) is that these ancient traditions and modes of production are fluid, negotiated, and renegotiated within modern contexts of unequal power, global capital and cultural flow, everyday struggle, pain, humiliation, and poverty.

Consideration of time in relation to modernity, postmodernity, and premodernity, then, adds another dimension to discussions concerning glocality and digital technologies. Sohail Inayatullah and Ivana Milajevic (1999) argue:

> Cybertechnologies . . . create not just the rich and poor in terms of information, but a world of inattentive time and slow attentive time. One is committed to quick money and quick time, a world where data and information are far more important than knowledge and wisdom. (p. 77)

When the world is divided into quick time and slow time, the local and global are further complicated, leading to questions regarding theoretical analysis and wisdom in relation to speedy practice, in turn leading to the question of binary framings of theory versus practice. Is the notion of a stable, enduring wisdom that is contextually situated within practices and theories of production merely a nostalgia for a mummified past? In any case, it cannot be denied that digital technologies have speeded up time for the Westernized, urban elites while the majority of the world's population does not even have access to the digital sphere. "Critical commentary," therefore, "is not a matter of merely being pessimistic or optimistic but a matter of survival" (Inayatullah & Milajevic, 1999, p. 78).

RADHIKA GAJJALA

Digital production is situated within processes of production, despite much of the dominant imagery associated with it that often disconnects digital existence from place-based and embodied processes of production. It is this rhetorical and discursive disconnection and ideological displacement perhaps that erases power structures and exploitative practices embedded in the production of the digital economy. Within such a discourse, notions of complicity, accountability, and responsibility are distanced from digital contexts producing unaccountable hierarchies and irresponsible extreme individualisms.

Thus, what is often lost under the flurry of excitement regarding the so-called revolutionary possibilities of cyberspatial existence and Internet networks is the following:

> Cyberspace not only exemplifies but today actually shapes greater political economy of which it has become a critical part, [while] the Internet is catalyzing an epochal political-economic transition toward . . . digital capitalism—and toward changes that, for much of the population, are unpropitious. (Schiller, 1999, pp. xvii–xviii)

It is within such a context of digital production that I strive to produce online networks for communication that might potentially lead to collaboration and dialogue on issues related to third-world and black feminisms[4]. My quest to understand communities of resistance and communities of diasporic postcolonials and transnationals, as well as communities of women (whether feminist or not) in relation to place-based contexts, leads me to revisit questions related to the formation of community.

When does community "happen"?[5] How are social relations shaped? In addition to communication and cultural identity, what are the factors that lead to the formation of community? Indeed how do prevailing definitions of communication, culture, and identity influence how we approach notions of community? In my quest to understand the possibilities for the creation of dialogic online networks, I am faced with questions related to the definition of community. Most discussions of community in relation to online networks (virtual community) tend to focus on the discursive, cultural, and communicative aspects of being a community. Most often these are placed in a "cultural" sphere that is implicitly separated from the material practices of economic production and power negotiations that arise therein.

---

[4]I use the term in the sense that feminists such as Chandra Mohanty use it. Mohanty (1991) defines the term *third-world women* as "a term which designates political constituency, not a biological or even a sociological one . . . What seems to constitute 'women of colore' or 'third-world-women' as a viable oppositional alliance is a common context of struggle rather than color or racial identifications. Similarly it is third world women's oppositional political relation to sexist, racist, and imperialist structures that constitutes our potential commonality" (p. 7).

[5]The present chapter was presented at a panel at the National Communication Association Conference in Chicago on November 6, 1999. Discussion between the panelists and members of the audience contributed to some of the questions raised in this section. I thank Alberto Gonzalez, Steve Jones, Michelle Rodino, Joseph Schmitz, Jonathan Stern, and Greg Elmer for the insights the discussion helped provide.

Through Annapurna's and my continued discussion regarding the nature of community and our engagement within real-life and virtual communities, we have, however, come to agree that community is very centrally shaped (although not completely determined) by the modes of production in which the members of the community engage. Even as community occurs within a set of material practices, the discursive realm that simultaneously performs sets of communication practices necessary for the maintenance of community cannot be separated from material communal practices. Communication between members of the community and the defining of relationships (family, friendships, and so on) occur in relation to the prevalent modes of production and are in turn embedded in hierarchies. This realization led us to explore the connection between the modes of production associated with each of our contexts to try and understand influences that might shape the formation of dialogic action-based networks online.

ANNAPURNA MAMIDIPUDI

Most modern technology comes as a result of a problem-solving approach to lived Western situations. It is thus clear that unless the problem is articulated correctly, the technology/solution is not applicable. What are the variables that shape the articulation of a problem? Even to attempt this question, we need people who are confident and willing to question existing situations or solutions that are imposed. Although, on the one hand, concepts that are taken to be universal need to be questioned, it is imperative that alternative viewpoints have legitimacy. Notions of work, luxury, and education seem to have a kind of universal meaning that excludes some of the traditional understanding of older societies.

Until the 19th century, cotton textiles had been India's premier industry. The reputation was built on the following three issues:

- the skill of the farmers who produced different varieties of cotton,
- the ability of men and women from different communities to work together and process the cotton into fabric, and
- the wealth of knowledge regarding dyes and techniques that added aesthetic value to utility.

However, the synthesis of chemical dyes almost 100 years ago in Europe has had a calamitous effect on traditional Indian dyeing practices. Except in very small isolated pockets, natural dyeing practices, which were the pride of the textile industry in this country, have been totally replaced by the chemical ones. The further effect of the spreading wave of modern science has been the creation of the perception of traditional technology as outmoded, resulting in almost total erasure of knowledge of the traditional dyeing processes within these communities. But the European documentation of these local practices had been initiated only in the

18th century. This process was further continued indigenously in the early 20th century in a bid to preserve information about practices that seemed to be going into extinction. In essence, this pattern meant that knowledge that had been firmly in the domain of practice of the artisans now was converted into textual information, shifting the ownership of the knowledge squarely into the lap of those able to study, rather than do.

Until recently the web that linked various communities in the activity of producing fabric was intact in Chinnur, a village in South India. Chinnur is a cotton growing area, which is why, until recently, cotton weaving was a major source of livelihood for people in the area. While attempting to help the surviving weavers in the area, we found that yarn availability was one of the critical factors involved. Most of the cotton grown was being bought by the spinning mills and spun into yarn that was available only in the market in Hyderabad, more than 400 km away. In trying to resurrect the older connections to hand spinning, local Reddy women were approached by the weavers. A large quantity of raw cotton was bought and distributed to these women, who, until 10 years ago, spun in their leisure in exchange for sarees, which were woven to specification. But the scheme failed because the understanding of leisure activities did not include spinning, which had been replaced by the television. Also the women of higher caste did not see any reason why they had to interact with the weavers who were lower down in the caste structure because cultural interactions centuries old had disappeared under the more rigid caste system taboos.

The idea was revived almost 10 years later in Chirala, this time with politically aware weavers who understood the need to revive connections with spinners if yarn monopolies of the spinning mills were to be broken and rates made affordable to handloom weavers. Today engineers in premier institutions are working with weaver organizations toward a small scale mode of production of yarn where small motors are attached to the traditional hand-spinning machines to make the activity economically viable for women who are housebound and can only weave in the leisure time available to them (between taking care of their children and homes). The process of producing yarn from cotton in large spinning mills in India has undergone very little change and has been totally imported. Most of the technology and machines used by spinning mills in Adilabad and Chinnur are from the Manchester spinning mills from the late 19th and early 20th centuries, when the industry moved to America.

AXES OF COMPLICITY/RESISTANCE

As our words and thoughts merge, where does "she" begin, and where do "I" end?

```
i am propogating the use of technology in a certain way..which
is not relevant in the west..doesnt even exist..because of the
vastly different contexts of how worlds like poverty, education
```

```
literacy, state..etc are percieved here and there . . . is there
tho a universal concept to the theory i am developing on tech-
nology here for the artisans that is applicable there . . . the
internet is proving to be an equalising tool of some kind.. and
it is maybe drawing lines down the segments in a less geographic
manner . . . is there a chance of learning from the internet
about how issues and directions create negative power struc-
tures...tho the manifestations there and here may be differ-
ent...
```

I've been thinking about —your own "complicity" in the global that you criti-
cise and what I am calling global capital- for example, your function in training the artisan may be to "discipline" (in a foucauldian sense) the artisans
into being more efficient servers of the global market (there are contradictions, even as you do this . . . and these contradictions are what I am interested in)- are you really learning to speak to the artisans ways or are you
teaching them to make themselves more palatable (not that this is undesirable on a personal need for access level)—if all that is happening is that
vegetable dying becomes "hip" within the existing consumer culture, nothing changes about the exploitative nature of global market structures . . .
(take for example the rhetoric of "multiculturalism" and "difference" here in
the US - etc.)[6]

```
when people have access to the outside what seems to happen is
that they dont ground their local properly and so a lot of peo-
ple doing very interesting things are doing them now physically
here but virtually outside . . . which means that it has no roots
or forum or accountability here . . . .
```

```
> dear r, i dont know that i can theorise on this . . . but i
> will try . . . it is do with my objections to the internet.
> as a system that is elitist and does not finally feed into my
> own work place. i will myself become an object in your system
> . . . if what i say does not have legitimacy beyond the
> novelty value of my location . . . lines are being drawn down
> new 'classes' and now one way to obtain access to western
> resources is through the internet. and being visible on it. it
> can be a spring board in a very personal agenda sense . . .
> but what is worse is if the location itself is 'objectified'
> in a way that takes away legitimacy . . . i will continue to
> write on the internet as long as i am rooted in my work.
```

as long as i am aware that i am creating a space that is legitimate and a dialogue that is 'real'. I chose to speak for my community . . . this is valid as
long as my community allows me to be a spokesperson or i hold account-

---

[6]Though this text is also an e-mail message, it has been formatted in a different font than previously indicated to separate it from the e-mail text immediately above it. The two fonts readers have encountered to this point interchange as needed throughout the remainder of the chapter text to enable
readers to continue discerning different message texts.

ability for my words in a real sense here. because you on the other side of a phone line have no access that i don't give you . . . .

Are we ventriloquising the voice of the "subaltern"?[7] Do we speak as "representative" others?

Are we indeed "veiling the subaltern"?

> Today, with globalization in full swing, telecommunicative informatics taps the Native Informant directly in the name of indigenous knowledge and advances biopiracy.
>
> —Gayatri Spivak (1999, p. ix)

> New technologies have readapted and reinforced systems for capturing the voices at the margins so smoothly that the systems have escaped notice along with the voices.
>
> —Belausteguigoitia Ruis (1999, p. 23)

Statutory warning: The present work is not about "the subaltern." It is about the privilege of being able to speak, to write. Yet it is also about the silences—the unsaid and the cannot-be-said. It is not only about what "position[s] of authority we have been given,"[8] and at whose expense we speak, but implicitly it is also a questioning into how we might be able to negotiate from within our speech and our silences to transform or disrupt hegemony. It is about negotiating from within the hegemonic and about attempting to disrupt hegemonic narratives. It is about resistance and complicity. Once again, we emphasize that this work is not about the subaltern. But it is about not remaining silent in the face of the imagined subaltern at the same time as it is about trying not to silence voices of dissent. It is important for us to examine the speaking roles we are assigned as well as the location from which we speak. Although the nonsubaltern third-world woman, as the "other" of the Western woman, finds a point of entry into the hegemonic sphere, in itself enabled by a history of relative cultural and material privilege, she must remember that her speech could be used as representative of a subaltern who is not located within the same sphere of material and cultural privilege. As Deepika Bahri (1996) points out:

> [This] 'other' (who has not spoken so far, only been spoken about) [who] begins to gaze at herself in the hope of reopening examination . . . must acknowledge the power of this gaze, the context of its production, the privilege implied in the right to speak at all, as well as the limitations of that can be known or said. (p. 51)

---

[7]The term subaltern is a term used by the Subaltern Studies Group, an interdisciplinary organization of South Asian scholars led by Ranajit Guha (1998) and is "a name for the general attribute of subordination in South Asian society whether this is expressed in terms of class, caste, age, gender and office or in any other way."

[8]This phrase is taken from a response to Radhika's dissertation project (Gajjala, 1998). See http://www.cyberdiva.org/erniestuff/sanov.html

Working class diasporic populations and a majority of the men and women within the real third-world (locations not necessarily defined by the geographic separations between the first and third worlds) locations pay the price for the discourses produced by bourgeoisie diasporic postcolonials, who are viewed by the Westernized world as ideal informants because of their/our ability to translate ourselves and our "other" so that we fit appropriately within hegemonic structures of power and thought. Discourses thus produced perform a veiling function in relation to the materially and culturally underprivileged subaltern populations.

We must assert, however, that the solution is not for the relatively privileged third-world subject to silence herself in an effort to avoid speaking for the subaltern. Although suggesting a questioning of the roles assumed by and assigned to the third-world woman, the feminist, or both, within the Western hegemonic sphere, our intention is not to dismiss her speech as untrue and invalid. We suggest that we should clarify the difference in the speaking locations of third-world subjects (whether feminist or not) situated within a setting of material and cultural privilege that is not available to subaltern women situated elsewhere geographically, materially, or culturally. Our intention is not to deny the struggles of diasporic third-world women. A member of one such postcolonial e-space points out:

> The question about "what is *our* right to speak?" while it *looks* like a question that places the Subaltern/Other on the map, doesn't after all produce the spaces *from* which the Subaltern could speak. . . . Instead of ending with "From what position of authority would *we* speak?" I'd phrase question more explicitly as "who is paying the *price* for this authority and *whom* am I taken to be speaking *for*? *What* did we *say* when we were given/took up the authority to speak? And *how* did we say it?" ("C" as cited in Gajjala, 1998)

And we write:

> The Internet will be a more colourful, exotic place · for us with women like Venkatavva flashing their gold nosepins, but what good will it do them? As it is at present, the Internet reflects the perceptions of Northern society that Southern women are brown, backward, and ignorant. An alternative, kinder, depiction of them which is also widespread is that they are victims of their cultural heritage. Is being exposed to such images of themselves going to help Southern women by encouraging them to fight in dignity and self-respect, or will it further erode their confidence in their fast-changing environment? (Gajjala & Mamidipudi, 1999, p. 14)

CONCLUSION

How do we resolve the contradictory sentiments of seeing the Internet as a panacea to the problems of south; of thinking that on the contrary, it may even be bad for us; and of asserting that this doesn't mean we don't want it? We need to study processes of empowerment and work out how it is to be done in the context of the Internet.

While case studies abound for the failure of this process, development workers in particular would not regard it as fair (or politically correct) to down-play the potential of the Internet to empower women like Venkatavva in South and North. We cannot say, "I won't give you the Internet, for your own good." (Gajjala & Mamidipudi, 1994, p. 15)

Collaborating across contexts in interactive cyberspaces is not restricted to one or two people but is open to whole groups of people who travel through these spaces. Some heteroglossic and even chaotic encounters are a part of being online. The designing and composing of cyberspaces is an ongoing, interactive, and sometimes unpredictable activity. For us, the most key issues arise from an engagement with both new technologies and "old" technologies (their histories, the processes involved), the modes of production and community within which they are shaped, and those that they in turn shape.

Response

Stephen Tchudi
University of Nevada, Reno

I was a bit surprised to receive James Inman's invitation to contribute to this volume "as a humanities scholar with significant collaboration and electronic media experience/s," for I see myself as something of a cynic and possibly even a Luddite when it comes to electronic progress in our profession.

My doubts about the electronic revolution go back to the 1960s, when, as a graduate student, I carried decks of IBM cards over to the campus computing facility, placed them in the 50-ft trough, queuing up with stacks of similar white cards (data) and orange cards (programs). I would return the following morning to find my job stalled at the midpoint, my cards returned to me with a printout identifying which of the thousands of cards in my deck had brought my computer research to a halt. At the time, I was reasonably skilled at using a Monroe mechanical desktop number cruncher and once was offered a fellowship with the psychology department for no better reason than that a professor happened to spot me pecking away on the Monroe and thought I might be useful to him in cranking out chi-square calculations. So it was difficult for me to see the great advantage of computers.

At about the same time in my education, my skepticism toward machines with plugs was deepened when a national manufacturer of electronic goods delivered, free, to our College of Education, a laboratory full of "teaching machines." In my undergraduate days, I had experienced a paper-and-pencil pedagogy called programmed learning, based loosely on behavioral theory and involving the completion of innumerable multiple-choice items, each incrementally a bit more complex than the last, each including a bit of new information as well as a question. As undergrads learning about the Greek philosophers (perhaps not the best choice for programmed learning), my fellow students and I discovered that the quickest way out of the labs was simply to read all the statements (or stubs), treating them as text rather than questions, and to take the mastery test. Forty-plus years later, my high-school-age son, studying geometry on the highly touted NovaNet learning system, did me one better: "The tests are self-explanatory and most of the answers are self-evident," he told me, "so I just take the test first and if I pass, that's that."

In the 1960s, as I inspected the new programmed learning machines in the basement of Old College, and particularly when I tried out some of the programs, I discovered these electric gizmos to be pouring old wine in new bottles, and bad wine at that. It was the same old traditional content—English grammar, dates in American history, the multiplication tables—that one found in standard textbooks of the era. Thus when I read in Marshall McLuhan's claim in *Understanding Media* (1964) that we approach new media "in a rear view mirror," the electronic teaching machine seemed to me to fulfill McLuhan's worried prophesies. And as I thought about McLuhan, I considered that not only do we look backward with the content of new media, but also for every new electronic form of communication there may well have been a paper-and-pencil preexisting alternative, which might have been reasonably satisfactory. Naturally, such an argument can be invoked by the Luddites for any new medium, and I can imagine someone saying in past millennia, "Paper and pencil? I prefer the permanence of stone and a good sharp chisel."

A decade or so after the teaching machines (which came to sit, mostly unused, in the Old College basement), my antitechnology predispositions were reinforced by the onset of the home computer era in the early 1980s. My first computer was a Radio Shack Color TRS-80, archrival of the Commodore, and it was then I learned to distrust the word *simply*, as in the instruction, "Simply compare the input jack on your television set to the five drawings below, select the one closest to yours, and as appropriate, insert a G-9 cadibulator or 6-prong lolapazooza." To this day, *simply* is red-flag computerese to me, for I know it means (select the best possible answer below):

A. The hardware or software "simply" won't work as advertised.

B. Your hardware will "simply" not resemble any of the diagrams in the one-size-fits-all instructions.

C. You'll have to call a help number where your request will be denied on the grounds that you "simply" forgot to register your product or because that problem is "simply" not covered by either the help desk or your warranty.

D. You'll "simply" have to wander the halls of academe hoping that somebody has an answer or knows somebody who knows somebody to help.

Even as I was struggling with my "Trash 80," I was hip enough and interested enough to attend a conference on word processing, where I heard a neo-Luddite warn the computerphile audience that it was sacrificing personal freedom for computational convenience. "For every font in your word processor," he declared, "you lose a degree of freedom of expression." With paper and pencil, he pointed out, you can draw any character any size in any font you can imagine; with a computer, you have no such flexibility. Although the audience wanted to lynch this bringer of bad news, his point stayed with me, leading me to weigh gains and losses. In fact, when I swapped in my TRS-80 for a then-new Macintosh, I realized that slick as it was, the Mac had hidden its inner workings from me. On my TRS-80 I could program in Basic from the opening screen. The Mac came preprogrammed with a system that was powerfully user friendly, with icons for files and folders, with a mouse, and with a smiley face that welcomed me every time I cranked up. Yet gone was the access to the system that had allowed me (in a bit of McLuhanesque rearview mirroring) to convert some of my old ditto handouts into a modest program that would quiz students on their personal interests and print out a laundry list of topics for them (Tchudi), complete with phony personalized statements, "Congratulations (Name), you should probably think about writing a paper on (Topic)."

Now, I do not claim that my skeptical observations about the flaws of electronic media are original or even completely fair, but I remain devoted to them, first because I mostly figured them out for myself and, second, because most of them grew out of significant and unpleasant encounters with the growing pile of electronics and peripherals that now clog my workspace.

Nor would my skepticism be novel to the writers whose work I reviewed for this collection, for these writers, too, mix great caution with modest claims about the application of 21st century media in the classroom.

For instance, in "What's So Democratic About CMC?" Jami Carlacio has written a powerful warning about the profession's claims for the power of electronics to democratize the classroom, urging us neither to fall victim to the excesses of our own "rhetoric of techno-literacy" nor to become blind to the failures—the losses of freedom—imposed by those tools. In "Collaborating Across Contexts," Radhika Gajjala and Annapurna Mamidipudi ask, "How can we form discursive and action-based networks between the local and the global?" They wisely conclude that our profession needs deep research and nonsimplistic claims about "the modes of production and community within which [new and old technologies] are shaped, and those that they in turn shape."

James A. Inman draws on Gregory Ulmer's term *electracy* and carefully analyzes media history to challenge simplistic views of technological literacy where competence in running machinery is inappropriately conflated with higher skills of literacy that are independent of that machinery. In discussing CMC "As Reflective Rhetoric-in-Action," Rebecca J. Rickly is enthusiastic in her support for computer-mediated communication education, yet chooses (like Inman) to emphasize the fundamentals of literacy instruction: helping "to create responsible citizens who can speak, write, and think well in and outside of their chosen discipline."

In her essay "Collaborative Selves," Sherry Turkle, like Gajjala and Mamidi-pudi, raises at least as many questions as she answers, framing her key questions not in terms of media or technology but in language of the fundamentals of human identity. Turkle looks to Walt Whitman, not Bill Gates, to discuss how media and identity evolve symbiotically, not necessarily in healthy ways.

I think it is interesting, then, to take a paper-and-pencil look at the content of the essays in this collection. Suppose CMC didn't exist? Suppose we didn't have computers and other electronic gew-gaws available in schools? What would our aims and goals be? What are the pedagogical (and other) themes in this volume? In particular, I've wondered, "What if this book were written without any reference to electronic media at all?"

The answer, to me, seemed very clear. The themes, issues, problems, and topics that emerge in one essay after another are these: collaboration (predictably, given our topic), information access, gender and political equity, student independence, interdisciplinarity, critical thinking, citizen and community participation, and democratic education.

Not surprisingly, each of these themes emerged in education well before Eniac, and each has coevolved with electracy in the past 3 to 4 decades. For example, in graduate school, while I was learning not to fold, spindle, or mutilate my IBM cards, I also learned of emerging group and collaborative learning approaches, such as the open classroom, free school, and brainstorming and other group processes. (It should be said, too, that in the same era I learned of behavioral objectives, the power of standardized testing, and the alleged values programmed into learning and teaching machines, all of which operate against the collaborative classroom and all of which have coevolved with electronic efficiency and technological vengeance.)

The contributors to this volume could thus be accused of doing precisely what McLuhan warned against: putting the content of older or existing pedagogies into the new media. Or, as Carlacio implies, our profession may have been guilty of wishful thinking that whatever we want for education at the moment, whatever we see as genuinely viable and supportable goals, will be enhanced if not powered by electronics.

Again, each of the themes I've listed has old technology, paper-and-pencil analogs. Collaborative learning, as I've suggested, has roots that are much longer than those of group e-mails and listservs. I could argue that such exchanges could be seen in my 1948 second-grade Valentine box, where Miss Hall had each of us write a Valentine for every member of the class. (Charles Schultz has even hinted at problems of access and participation in that old-fashioned paper-and-pencil medium when Charlie Brown annually receives no Valentines from his classmates.) We can see the genesis of cross-cultural communication in pen-pal programs, and I think my elementary school teachers were quite right to urge this sort of rudimentary globalization on us.

With computers, e-mail, discussion groups, online syllabi, the Web, faxes, and phone cards, our dreams and visions of collaboration and multidimensional communication can be enabled so that, with electricity and electracy, we can do these things, to quote the opening of the *Six Million Dollar Man* series: "better, faster, cheaper." But there are also those losses of freedom to be considered.

For example, in grad school I also flirted with the idea of becoming a professional cartoonist and did hand-drawn, individually water-colored comic Valentines for my mother. Recently, I confess, I logged onto Yahoo.greetings. com and sent her an electronic Valentine from Yahoo's limited list of options, penning a lame message, "Happy Valentine's Day, Mom." Faster? Yes. Cheaper? Yes (not taking into account the cost of my computer, hers, and the World Wide Web connection). Better? By no possible stretch of the imagination (which was not, of course, stretched at all by my venture into Yahooism).

Well, enough of my anecdotal and satirical evidence. I would now like to argue that an ancient paper-and-pencil discipline, rhetoric—originally a voice-box and oratory discipline—can not only offer guidance for ways out of some of the problems created by overly enthusiastic adoption of new media but can also offer productive ways of thinking about media issues.

I will begin by characterizing what I think Rhetoric is all about. (I will capitalize Rhetoric from this point on to emphasize that I am talking about the discipline, not the lowercase word *rhetoric*, which is synonymous with idle palaver, political trickery, and the lingo of used-car salespeople.) At its best, Rhetoric is simply the study of discourse, discourse that can include palaver and propaganda but can and should also include inspirational speech, computer manuals, love letters, mathematical formulae, art, music, and dance—any form of symbolic communication, including electronic. Roughly, Rhetoric is to discourse as Linguistics is to the sentence (although linguists engaged in discourse analysis and rhetoricians working on style would disagree with that simplistic division of the world into sentences and beyond sentences). Learning from the linguists, we need to recognize that Rhetoric is principally *descriptive* rather than *prescriptive*, whereas many discussions of rhetoric and Rhetoric founder because they move from observation to perceived imperative, from what happens with language to what language users ought to be doing.

Capital *R* Rhetoric leads to conclusions and observations about the patterns (sometimes inappropriately called rules) of lowercase rhetoric, whether palaver, e-mail, or hate mail. A good Rhetorician can suggest and describe the characteristics of that language but should certainly not attempt to legislate the content or morality of rhetorical matter. Rhetoric should be nonprescriptive in its declarations, just as Linguistics remains noncommittal on, say, the morality of a person who uses *ain't*. Refusing to be prescriptivist removes us from Plato's trap of legislating the morals, ethics, and philosophical character of Rhetoricians and actually banning them from the Republic for their loose minds and lips.

Such distinctions—Rhetoric/rhetoric and description/prescription—may be self-evident to many readers and seem simplistic to others, yet our profession has frequently gotten itself into pedagogical and philosophical difficulty by failing to make them.

Thus, when Carlacio talks of "the rhetoric of technology" that "feigns an innocent interpretation of the world," the claims are based on Rhetorical observations about the speech and writing—the lowercase rhetoric—of humanities teachers. The *R/r* distinction allows us to recognize that although techno-rhetoric currently may be dedicated to promoting our professional issues of democracy, collaboration, and empowerment, a very powerful alternative techno-rhetoric could easily be generated for the cause of using of technology to standardize thought and information content, to regulate the information received by the citizenry, to Big Brother daily life. As a humanist, I personally vote for the former—the use of technology to empower our students—but as a Rhetorician, I should not, as Carlacio shows, impose my personal political value system on technology by claiming extraordinary powers of democratization for it.

A Rhetorician, of course, can never achieve the objectivity of the neutral, objective, and quite mythical scientist. We are biased in what we choose to investigate and the tools we select for our investigation. We use rhetoric of our own in promoting our ideas and views of the truth, in arguing for perceived virtues and values. But by attempting to isolate Rhetorical analysis from either the language of palaver or the underlying motives for that palaver, we can bring clarity to discussions of electronic technology, what it does versus what we want it to do.

In this respect, I think the Rhetorician of electronic collaboration needs to investigate carefully another set of multiple-meaning terms: form and content. While drafting this essay, I was also in the process of attempting to master a fully featured Web design program, Dreamweaver. I was intrigued to observe—Rhetorically—that although the Dreamweaver manual writers frequently refer to *content,* they have no interest whatever in that content, in what the user puts up on the Web. Thus, the manual is concerned strictly with form.

Now, in the Web world, form is quite precise and limited in meaning. Form is, quite literally, what the program will permit you to do: to make windows, to create frames, to pour in textual content, to import image content. The Dreamweaver manual is a complete and total Rhetoric for the program itself.

Would that Rhetorical analysis of everyday discourse could be so complete, for even though complex, a computer program is finite. That is, although Dreamweaver can produce endlessly complex webs, it can only do so within its programmed form. It may be able to do stunningly nested pop-up windows and morphed graphics, but if you want it to shine your shoes, you'll have to wait until a programmer decides to include that in the next numbered version. Let me label Dreamweaver's concept of form, then, as Static Form.

In real-world discourse, including the content one pours into a Web site, there is no neat division between content and form; indeed, our profession has engaged in endless chicken-and-egg discussions of how content and form interrelate—whether form follows or proceeds function, whether formal structures can be identified independently of their ideational content.

At one extreme we could claim that form and content are completely isolable (like the Dreamweaver model). We might create a catalog of forms, genres, modes, and their characteristics—a taxonomy (or bestiary) of the kinds of discourse produced in the world: print forms and nonprint and even, if we chose, artistic and creative forms, such as ballet, painting, and sculpture. Such a Rhetoric might make for interesting study, but it would be very difficult to compile, if only because, in the end, the forms of real-world discourse are mostly generative or dynamic, not static. If, for example, we were successful in describing the formal constraints of *poem* (or even a subgenre, such as *narrative* or *epic* or *haiku*), tomorrow's poet would predictably produce something that breaks the rules or defies the conventions.

In composition and literature classrooms we do have teaching materials that attempt to freeze form in time: These would be the writing handbooks that list patterns of writing development (*comparison, cause and effect, examples*) and literature introductions that are divided by the major genres: *poetry, fiction, drama*, and occasionally *nonfiction*. I would argue that such texts have diminished in importance over the past couple of decades because of their static treatment of form. Writing-as-process and literature-as-response call for more dynamic Rhetorics.

However, to recognize that form and content are linked, if not inseparable, creates its own problems. If we were to take inseparability literally, we would replace our bestiary of rhetorical forms with one of a different sort: a catalog of every bit of discourse ever created in the history of humankind, a list, growing by the minute, by the instant, of human utterances. For if form and content are truly inseparable, then the Rhetorician can make no abstract generalizations about the characteristics of discourse; he or she would largely be limited to cataloging utterances in chronological order. The universe could theoretically contain as many Rhetorics as there are, have been, or will be utterances in the universe. This catalog, too, would be static in the sense that it would simply freeze each piece of discourse as a specific bit of language at its place in time, an infinite library.

A middle ground is not easy to discover, unless we take a dynamic view of discourse where the Rhetorician is rigorous, yet appreciates that discourse (driven by the human mind) is generative. Any isolated bit of discourse—say a graffitum or the Declaration of Independence—has its individual Rhetoric but more important, at least for purposes of pedagogy, is for the Rhetorician to look at groups or clusters, a corpus of utterances, that will allow for some generalization. The Rhetorician could choose to investigate the dynamic of form and content using a variety of methods, which might include, say, Francis Christensen's (Christensen & Christensen, 1978) "generative rhetorics" of the sentence and paragraph, Kenneth Burke's (1950) pentad, with its dramatic grammars and rhetorics of writer, audience, and motive, or Stephen Toulmin's (2001) logical schema of looking at data, warrants, claims, and discourse forum. One might employ a variation of new critical/eclectic criticism seeking recognizable patterns and themes in discourse (Brooks & Warren, 1938). One could employ traditional classical Rhetoric with its interest in ethos, pathos, and logos to discover the relationships between writer (or speaker) and audience (Aristotle).

Or one could borrow further from Aristotle and treat the Rhetoric of a discourse—or of electracy—as "a survey of the available means" of discourse and persuasion. One might also look to the Aristotelian notion of invention, not so much as a series of heuristics or algorithms for the discovery of arguments but as a model for the present-day inquiry into the processes of discourse: How does one go about making an effective speech? A 500-word theme? An essay? A poem? A Web page? A PowerPoint display? A Nobel-prize winning chemistry paper?

Such Rhetorical analysis of the composing process can be especially useful for teachers, linking discourse and its making back to some of our fundamental aims for education, which, as identified earlier, seem presently to stress collaboration, information access, gender and political equity, student independence, interdisciplinarity, critical thinking, citizen and community participation, and democratic education.

And that, in the end, is what we are pragmatically seeking to do as Rhetoricians and teachers of discourse within the humanities: to say something interestingly descriptive and helpful about graffiti and declarations, about palaver and oratory, about e-mail and Web pages, and about collaboration, electronic and other.

The English teaching profession has a growing body of evidence that teaching static content and form doesn't move students very far along the spectrum of literacy. That's why students too often come to English class believing that the essence or definition of a poem is something that rhymes, why they come to us believing that every paragraph has a topic sentence and supporting evidence, but lacking skill in the process of essay-ing, why they can vaguely remember that it's a simile—or is it metaphor?—that uses "like" yet utterly lacking confidence in their own native ability to respond to poetic or metaphoric language.

I think we must also make the same distinction as we look at the forms of electronic literacy. If we take and teach form as static, we fall into the trap exposed by Inman: mistakenly identifying the ability to send and receive messages as constituting literacy, be it computer, Internet, technology, digital, electronic, multimedia, or even cyberpunk literacy.

A few teachers might argue (as did Francis Christensen in his self-named generative rhetorics) that mastery of a discourse form may, in fact, lead to generative content. That is, if one teaches the rigid form of the sonnet, students will often generate original and inventive sonnets. If you teach Dreamweaver to a group of students, pretty soon they will be generating amazing content, turning the program on its ear, tweaking it in ways never dreamt of by the program weavers.

But too often, mere focus on form teaches a limited kind of literacy, such as business writing courses that teach business forms but do not remark on, say, language and business ethics, or tech writing, or document design courses that produce technically brilliant materials of arguably dubious content (e.g., perfume advertisements).

The distinction, then, between static and generative or dynamic form and content brings me back to what I see as the overarching theme of the essays that I reviewed for this collection: What's fundamental? Principally, what are we aiming for in language/humanities education and, second, how can electronics aid us?

In this respect, I must, in the end, somewhat disagree with Inman's argument in favor of electracy for the ages. As intriguing as the concept of electracy may be, it seems to me to accept a false analogy between oral and print culture: that they represent some sort of evolution of which the electric culture is the next phase. But in the end, oral language, the spoken word, what Andrew Wilkenson (1990) called oracy, seems to be very clearly the basis of various spin-off competencies, both literacy and electracy. If anything, we need to become even more fundamental than that, recognizing the rich interplay between oral language and the generative impulse that leads to it. Print literacy and electracy obviously do have generative elements, and I share the perspective of those who argue that print changed the way we think (Ong, 1982) and those who show that electronic communications systems are dramatically altering how we perceive and think and interact (Postman, 1985).

Yet, I believe we need to move in the other direction, not outward to the tools of communication but inward, to the mind and expressions of language users. As we review the historic contributions of such writers as Piaget (1926), Vygotsky (1978), Langer (1957), and Bruner (1990), I think we might have to become advocates of "Brainacy," high skill and competence in generative symbol reception and manipulation—a mosaic of skills about which we know painfully little but which is clearly prior to oracy and certainly prior to electracy.

What would a class in Brainacy look like? At present, it might resemble a good school or college writing class, with an emphasis on democracy, collaboration, empowerment, and expression. Such a class might be focused on some of the overarching formal concepts in language: discourse, comparison, narrative, argument, essaying. Or it might be a course focusing on generative rhetoric and grammar, exploring and practicing how we make meaning. Or it might be a course that doesn't discuss discourse at all, simply involving a group of people in trying to solve an intellectual problem or to create political activity.

I assume it would be a course focusing on dynamic language reception and production, not a series of lectures about grammatical and rhetorical form. It might include a truckload of electronics: computers of all sorts, Webs and Web sites, discussion groups, e-mails. On the other hand, it might be intriguing to ban the electronics from our class. Instead of collaborative e-mails, the central focus in the class might be a variation of the Valentine's box or the graffiti wall, to which every student contributes something every day, with class discussion based on response. Or we might ban paper and pencil for the class and make ours an oral Valentine's box; or we might even move past the concept of "class" and revert to a peripatetic school (or nonschool), meeting in coffee shops or the lobby of city hall for purposes of productive thinking and talk.

I don't intend to seem facetious here; I do, in fact, believe that such a paperless, electronics-less engagement of minds and voices could be powerfully educational. Even better, however, would be to create such an interesting and discourse-centered kind of class yet to include as well all the tools and technology that could make it even more effective: possibly a cyberspace peripatetic school, possibly a Web discussion group, maybe a networked gathering of politicians or citizens.

I believe that the kinds of issues and themes being debated in education, those underlying this book, are not only crucial but also have the potential to make a difference in education over the coming half century. But if the discussions are linked too closely or even misperceived as being causally related to electronic media, they may wind up as a footnote in a technology manual rather than the central chapters of the history of educational reform.

II

STUDENT COLLABORATION AND ELECTRONIC MEDIA

6

New Technology, Newer Teachers: Computer Resources and Collaboration in Literature and Composition

Nancy Knowles
Eastern Oregon University

M. Wendy Hennequin
University of Connecticut

Current scholarship on computing in the fields of English and composition opposes the utopian argument that computers can revolutionize teaching, that they can create what Nancy Kaplan (1991) ironically terms a "brave new electronic world" (p. 21). This scholarship has engendered repeated, legitimate warnings about the educational uses of technology. Here are two examples:

- Cynthia L. Selfe (1992) cautions us that "[w]hen technology, as an artifact of our culture, is employed by teachers who lack a critical understanding of its nature or a conscious plan for its use, and when these teachers must function within an educational system that is itself an artifact of the political, social, economic forces shaping our culture, the natural tendency of instruction is to support the status quo." (p. 30)
- More recently, Jeffrey R. Galin and Joan Latchaw (1998b) write, "Experimenting with theory and technology may hold potential value; so may exploring new ways of reading, engaging, or challenging students. However, to do either just because it is possible, without much forethought or preparation, is not likely to improve pedagogies or serve students." (pp. 43–44)

In short, as Galin and Latchaw boldly assert, "pedagogy should precede technology" (p. 45).

Based on these warnings, educators must prepare themselves to integrate technology into their pedagogies by realizing a number of truths. These include:

- Computers do not represent merely a better way of doing the same thing. Technology is not "transparent" (Haas & Neuwirth, 1994, p. 321); it changes classroom experience.
- Computers are not "free of ideological constructions" (Kaplan, 1991, p. 27; see also Ohmann, 1985, p. 680). Rather, they "exist within larger social and institutional frameworks and, thus, are at least partially influenced by them" (Howard, 1997, p. 114).
- Computers are not a panacea for classroom problems of power; they cannot completely elide differences in gender, economics, or social classes and may actually exacerbate these differences (Carlacio, this volume; Burkhalter, 1999, p. 62; Frenkel, 1990; Lockyard, 1996, p. 227; O'Brien, 1999, p. 100; Tabbi, 1996, p. 245). Moreover, they cannot automatically decenter classrooms (Kirkley, Savery, & Grabner-Hagen, 1998, p. 231).

If educators are to place pedagogy before technology, applying these salient truths to specific classrooms will require extensive study of existing scholarship, as well as exhaustive review and revision of extant teaching philosophies, course methods, and course content. This effort is, of course, compounded by the need to learn to use the technology itself. As Galin and Latchaw (1998b) report, "teachers should plan to spend a minimum of forty preparatory hours familiarizing themselves with any applications that students will be expected to use" (p. 64).

Given the workload required for preparing to integrate technology into the English or composition classroom, it seems idealistic to expect any teacher to implement technology with pedagogy already in place. First, most teachers don't have the time to educate themselves thoroughly on how best to integrate technology into the classroom or even on all aspects of the technology they plan to use before they use it. Expecting them to do so ignores the fact that technologies that work in one course or with a certain group of students may not work in or with another (Galin & Latchaw, 1998b, p. 51), which means that implementation of technology must always adapt to context. Such an expectation also ignores that technology constantly changes, so time must also be devoted to keeping up with those changes.

Second, the expectation that one can fully lay the ground work for classroom technology prior to use is at odds with the urgency for involving students with computers. The integration of technology into the working world requires instruction of and by technology. This is particularly true of disciplines such as English and composition that emphasize the study of communication. Communication disciplines must begin addressing electronic communication immediately so that students can learn to function in electronic environments and to think critically about them; as Fred Kemp (1998) writes, "Computer technology is not superfluous gimmickry that can be imposed or excluded based on individual likes and dislikes. . . . societal changes in information access and communication will require profound changes in the classroom and in the way most people learn" (p. 149).

Students whose education neglects this area will find themselves handicapped later in life.

Perhaps most important, idealistic expectations of preparing pedagogy before integrating computers into the classroom seem to run contrary to the concept of teaching as a creative, recursive, problem-solving, dialogic process where both teachers and students are learning. Yes, it is vital to prepare for the complexity of teaching with technology, especially to address relevant issues of power and voice in technologically enhanced classrooms, but teaching is not a spectator sport; teachers have to get onto the field and throw the ball around to test why they do what they do. Development of a critical pedagogy occurs not only before but also during and after the active process of teaching.

Such recursive teaching is perhaps more obvious among new teachers, particularly graduate students teaching 1st-year college courses, who may have little or no training in teaching. Lacking background in theoretical and applied pedagogy, graduate-level TAs run their own classrooms without much practice to revise or interpret. Those TAs who have collected pedagogical strategies, like any learners, can mentally juggle only so many before they need to apply them in the classroom. Thus, the classroom becomes their laboratory, where they learn actively by testing what they know against their experience. When they first start, everything they do is an experiment, and even the best laid lesson plans can founder against real students. Yet, as Thomas E. Recchio (1992) argues about graduate students teaching 1st-year composition, this experimental behavior of novice teachers may be an asset: "The transitional status of graduate student compels an experientially immediate sensitivity to the process of learning, a sensitivity that all too easily diminishes with age and experience" (p. 58). This sensitivity to learning may benefit not only students in such courses but also the new teachers themselves because they cannot help but see themselves as learners in their own classrooms. The teaching-assistant (TA) teaching experience may not have much else to recommend it; these teachers, after all, are merely developing the skills from which good teaching may one day emerge. However, their willingness to take risks is worth emulating, particularly in the complex and necessary field of humanities computing. The work of TAs reminds us that, just like our students, we must do our best with the preparation we have, read our failures critically to learn from them, and try again.

To illustrate this argument, we will tell the tale of our collaboration as TAs to teach several sections of a 1st-year English course at the University of Connecticut (UConn) in 1998. When we started, we knew we wanted to work together, to explore the use of technology in teaching literature and writing, and to encourage our students to be responsible, active learners. In attempting to accomplish these goals, we learned that the use of technology in our classrooms created a collaborative environment where technology represented not only a means and an opportunity for collaboration among our students but also the possibility and the need for collaboration between our students and ourselves. Through technology, both we and our students learned about technology, literature, and writing, and the technology, all new teaching tools to us—specifically our Web page, our e-mail discussion list, and a special videoconference with students in South Dakota became the vehicle for teaching and collaboration both in our hands and in our students'.

Looking back on our work as TAs, we see the naivete of our initially noncritical approach. But without taking this risk, we might have continued to avoid technology, relegating the responsibility for teaching it to other disciplines, long after our graduate work was complete. Instead, today, we continue to integrate technology into our classrooms in increasingly exciting and interesting ways. Having learned so much from our early experiences, we would encourage teachers who have not yet tried to work with technology to pick a comfortable, inexpensive application to start with and become a novice teacher again as they learn to incorporate it into their pedagogy and learn from the process of teaching with it.

COLLABORATION MEETS TECHNOLOGY: THEORIES AND RESEARCH

The serendipity involved in teaching collaboratively with technology instructed us through experience about the value of collaboration and particularly the contributions of computers to collaboration. We did not decide to collaborate because of research or theory; instead, our collaboration arose naturally from the informal support we had been drawing on for almost a year and then developed as technology made collaborative dialogues possible and necessary. Even as we were focused more on our specific needs and the specific needs of our classrooms, scholarship in our discipline increasingly reflected the value of collaboration for teachers and students, as well as around and through computers (Crook, 1991, p. 168).

The academic world still clearly and consistently merits the efforts of the individual (Barlowe & Hottell, 1998, p. 270; Bruffee, 1993, pp. 1–2; Leiby & Henson, 1998, p. 176; Nesbitt & Thomas, 1998, p. 32; Reich, 1991, p. 233; Singley & Sweeney, 1998, p. 73; Stahl, 1995b, p. 1; Winkelmann, 1995, p. 445). For example, most educational institutions regularly assign the responsibility for a course to one teacher, and most classrooms evaluate the work of one student on a universal scale (A–F) separate from his or her classroom context. Our students arrive in our classrooms trained to devalue collaboration by their previous educational experience; they often resist collaborative work, and when required to do it, they may not know how to proceed. In Wendy's most recent class (English 249s, Spring 2001), for example, students had the opportunity to collaborate on a final project involving the scholarly annotation of a text on the World Wide Web (Hennequin, 2001). When one student inquired on the e-mail discussion list whether anyone was going to collaborate, Katy Swetz (2001) replied, "I'm not, but then again, I haven't found anyone with whom I could collaborate. What are the benefits?" Katy's question is indicative of our students' experience and attitudes towards collaboration. Students are not eager to collaborate; Katy and several of Wendy's students who were willing to try to collaborate on the annotated hypertext project were unable to find partners. Students like Katy do not know what the benefits are, and, in fact, another student, Jeff Matthieu (2001), doubted that there could be benefits at all; in reply to Katy, he writes:

> That's a really good question! I'd guess maybe just that you can each gather information from sources, but then it seems like it'd be a near-nightmare with two or three different people HTML-ing something and trying to put it together. . . . it seems like more trouble than it's worth. That puts me in the "alone" column, I guess.

Although the mere logistics of collaboration frighten some students like Jeff, others are actively discouraged by group projects gone horribly wrong in past classes. Because of these bad experiences and the problems of organizing collaborative projects, most students see little value in collaborating and actively avoid collaboration when possible, opting instead to work alone. This educational solitary confinement does not occur in our scholarly work, where we share writing with each other informally by trading papers for feedback or more formally at conference and by publication, nor does it occur in business environments where collaboration among employees is integral to success.

However, social constructivist pedagogical models (Bartholomae, 1985, p. 144; Boiarsky, 1990, p. 47; Bruffee, 1984, p. 639; Fish, 1980, p. 318; Flower, 1993, p. 11; Ohmann, 1985, p. 685; Roskelly, 1994, p. 141; Taylor, 1992, p. 135; Winkelmann, 1995, p. 432), which consider knowledge as a social construct, have made collaborative learning a more widely accepted means to encourage active, authentic, student-centered learning. Such an approach is particularly valuable in communications courses like English and composition because, as Russian educator and psychologist Lev Vygotsky (1986) argues, children absorb language from the outside in (pp. 34–35; see also Bakhtin, 1981, p. 259), which means that language is always social. Therefore, collaborative dialogue represents a powerful tool for language acquisition. As Robert J. Stahl (1995b) argues, "Instead of studying about communicating with others, students must learn to communicate effectively within language arts classrooms through the actual practice of communicative behaviors, attitudes, and abilities" (p. 7).

According to the research, collaboration has other academic advantages besides assisting language acquisition. Because thought is connected to language (Bruffee, 1984; Vygotsky, 1986, p. 94), learning, too, regardless of discipline, must be communal and is therefore properly stimulated through collaboration among students, between students and teachers, and especially among teachers interested in continuing to learn about and improve what they do. Specifically, by encouraging students to work together to solve problems, collaboration provides a venue for students to take authority in assisting one another, which can be a model for active, responsible learning. Moreover, as Vygotsky (1978) argues, sometimes a "more capable peer" can be a more effective teacher than the actual teacher because that peer may better understand his or her classmate's existing knowledge (p. 86). Even as the student receiving assistance learns from his or her peer, the student providing assistance learns to better manipulate and express his or her knowledge, creating a mutual learning environment. This environment also includes the diversity of perspectives necessary to advance learning: "The presence of more than one leaner increases the likelihood of alternative hypotheses or interpretations arising, and constructive criticism and conflict is likely to occur" (Anderson, Mayes, & Kibby, 1991, p. 25; see also Craig, Harris, & Smith, 1999, p. 143).

Success of this model is demonstrated through classroom research: "Cooperative learning results in significantly higher achievement and retention than do competitive and individualist learning." In particular, "cooperative, compared with competitive or individualist[,] learning tends to result in more higher-level reasoning, more frequent generation of new ideas and solutions (i.e., process gain), and greater transfer from one situation to another (i.e., group to individual transfer)" (Johnson, Johnson, & Holubec, 1995, p. 53; see also Johnson, 1970, pp. 163–165). Because of the significance of collaboration to learning, the increasing interest in it is justified.

Besides academic benefits directly related to composition and literature, collaboration has other benefits. First, inability to work with others is the primary reason talented workers lose jobs (Roy, 1995, p. 18). Because employment has become students' most important reason to attend college, learning to build interpersonal skills through collaboration at the college level, if not earlier, is vital to the impact college can have on a student's ability to succeed after graduating. Second, because poor interpersonal skills can ruin otherwise successful communication, inability to collaborate can have more impact than the loss of a single job; during a treaty negotiation or a space shuttle launch, for instance, the inability to collaborate can cost lives. Teaching students to work together, even in classes devoted to other subjects, such as English, can change the ability of real-world communities to communicate successfully. Third, collaboration values diversity because it values diverse skills and experience, if not actual social, economic, or cultural backgrounds, and it does so in an environment of equality; members of a team may have different skills but equal responsibility for group success. So, through example, collaboration teaches a democratic philosophy of behavior that respects differences among people (Johnson, Johnson, & Holubec, 1995, p. 56). Students brave enough to try collaboration quickly discover this last benefit and eagerly take advantage of it; one of Wendy's students who recently collaborated on the annotated hypertext projects writes, "Jared [her partner] had a pretty good understanding, better than me, o[f] user friendly formats, backgrounds, fonts, etc. And for this reason, Jared did most of the [HTML] coding" (Williams, 2001, p. 1).

Not only does current research taut the advantages of collaborative learning, but it also increasingly incorporates information technology into collaborative models. Although, as we have been repeatedly warned, technology does not automatically revolutionize curriculum, scholars have begun to identify ways computers can be tools for doing so. On the logistical level, computers can enhance collaboration by enabling students to work together when they cannot meet physically (Crook, 1991, p. 179). Computers can even broaden collaborative work by putting students in contact with experts and information outside the university, making the work more authentic (O'Malley, 1991b, p. 292).

Perhaps more important, however, computers can help to create discourse communities by enabling groups to display knowledge and achievement through immediate, electronic publication. This shared experience and understanding can enhance a sense of participation (Crook, 1991, pp. 179–180). As Tharon W. Howard (1997) writes, "Simply by making the writing process virtually coincident with publication allows the increased interaction and negotiation of competing subject positions in electronic forums" (p. 167). Thus, computers become not just catalysts for collaboration but participants in mediating "intersubjectivity and new, shared meanings" (O'Malley, 1991b, p. 291). For English and composition classrooms, such discourse communities are particularly important because they are formed textually (O'Malley, 1991b, p. 292), which can advance students' writing skills and ability to understand texts as social rather than individual constructions.

This growing accumulation of research in the area of humanities computing is very important to the success of integrating technology into English and composition classrooms. However, if teachers new to technology attempt to master all of it before they begin, they may never begin. We suggest reading not only before beginning to integrate technology into the classroom but also during and after. Make the development of pedagogy related to technology an ongoing experimental and reflective process. But when research knowledge and pedagogical planning reach a critical mass, or time runs out, don't be afraid of beginning to work with technology. Our experience as graduate TAs attests to the value of "just doing it."

OUR FANTASY CLASSROOM

When we began to collaborate on our course, we had no training except a voluntary workshop on using the Internet provided by fellow graduate student David Salomon. We had no departmental or institutional encouragement or rewards for doing what we did, no computer labs for our students to use, no financial remuneration, and no teaching assistants to help us. Any assistance we received we sought and cultivated ourselves. We simply envisioned the course and worked toward bringing that vision to life.

However, our collaboration through technology would not have been possible without the freedom UConn allows its teachers to design their courses, without the willingness of the freshman English coordinator to meet our special needs, or without the physical facilities available to all UConn teachers and students. Nor would it have been possible without inspiration and serendipity to augment the planning we put into our courses. As we will discuss, a great deal of our teaching, both with and without technology, emerged as most teaching does: from our own thoughts, from our reactions to each other's ideas and those of other instructors, and from the lucky accidents that made learning opportunities available. The following details our fantasy classroom that we created through collaboration, technology, and inspiration—fantasy in the sense that we taught fantasy literature and fantasy in that the collaborative structure was, for us, the ideal way to teach.

In the spring of 1998, we requested to teach English 109, Literature and Composition. According to the English Department's home page,

> English 109 is the second semester of the required two semester expository writing sequence. Like English 105 [the first semester's required course], the goals are to develop critical reading, thinking, and writing skills. . . . In English 109, most of the reading is literary while the required writing is expository. (English Department, University of Connecticut, 1998)

Instructors enjoyed, however, considerable freedom to decide what literature should be taught, to allow for different interests and areas of expertise:

> Instructors of English 109 are free to organize the literary component . . . in the course in any way they find congenial and productive, although attention will be given to the formal qualities of narrative, poetry, and drama as necessary components of understanding how literary texts are, in short, literary. (English Department, University of Connecticut, 1998)

During the previous spring, Wendy had presented her students with five possible reading lists for English 109, and they overwhelmingly preferred a list entitled "Fantasy Literature." When Nancy expressed interest in auditing the course Wendy planned, Wendy suggested that Nancy teach it as well, and the collaboration was born. Since we had often in the past mined each other's knowledge and experience of teaching, writing, and classroom management, teaching collaboratively meant simply a logical step from the informal office dialogue to a more official one. As we began to plan and institute this course, we realized that through collaboration not only would we build on one another's knowledge, but also we would provide another, larger audience for the students' writing and thinking in the classrooms and give the students the advantage of more than one classroom authority, thereby modeling scholarly dialogue and reducing the teacher-centered quality of the classroom.

Stephanie Roach, the freshman English coordinator, performed an impressive series of scheduling gymnastics to synchronize our classes—Monday, Wednesday, and Friday at 8 and 9 a.m. We each covered our own classrooms and sections, as usual. Each class would have a maximum of 20 students—the departmental limit for this course—giving us a total of 40 possible students each and 80 for the collaboration. Accordingly, Nancy's classes were scheduled in a room with an approximately 40-person capacity to allow for combined classwork.

Again, thanks to Stephanie, three of our four classrooms were mediated, that is, equipped with many high-tech teaching devices, such as a projection apparatus with hook-ups to a PC computer, a videocassette recorder, a document camera, cable television, academic and foreign television networks, and long-distance lectures. The rooms also came with a slide projector and hook-ups for notebook computers. At the time, the PC computers had a Windows 95 operating system, connected to the school's World Wide Web and e-mail servers, had CD and floppy drives, and boasted the usual programs—Microsoft Office suite, Netscape's Web browser, which had not at that time been displaced by Internet Explorer, and an emulator program for accessing the university's mainframe.

Physically, the mediated classrooms were arranged differently than traditional classrooms. There were no moveable desks; long tables with attached, lightly padded chairs rested on gradually rising tiers. These were positioned around the front of the room, either in a bow shape (in the larger classroom) or a U shape (in the smaller, approximately 30-person classroom). Sliding wet board panels covered the traditional blackboard. Instead of the usual blinds, windows were covered with heavy, pastel drapes. The rooms were painted white with pastel, sound-absorbing boards on the walls in a building where most of the classrooms were beige or yellow, and the sections where the students sat were carpeted instead of tiled.

The students benefiting from this technological and decorative opulence were generally 1st-year students, either 18 or 19 years old, although because of transfers or scheduling problems we also had several older and more experienced students. Our students came from several different ethnic and economic backgrounds, with the majority being White and middle class. The male-female ratio was about 4 to 5 when we combined sections.

Although our students had all been required to take English 105, the first-semester composition course, or the equivalent, their level of writing expertise and experience varied greatly. So, too, did their experience and expertise with computers and the software necessary to this course: Netscape (for perusal of our Web page and several texts available only online) and UConn's mainframe e-mail system.

All students, however, had access to computers; according to the *Yahoo! Internet Life*'s America's Most Wired Colleges survey, UConn was at the time ranked 18th among universities in the United States for computer access, which placed the university 5th among public postsecondary schools (Roy, 1998, p. 1). Students living in dormitories each had a computer port; the university provided state-wide local numbers for off-campus students to dial in to the computer system. Computers were also available in computer labs located in the library, certain academic buildings, and the dormitories for those students who did not own their own computers. UConn's server provided both World Wide Web access and e-mail, accessible to all campus computers as well as the personal ones.

THE EFFECTS OF TECHNOLOGY ON LEARNING
AND COLLABORATION

The following sections detail the impact of this technology on the collaboration that occurred in our particular classrooms. Although one classroom example cannot speak for technology in general, our experience taught us that technology has the potential for necessitating collaboration in using technology and for encouraging collaboration in other areas of classroom learning.

World Wide Web

Introduced to our students on the first day of classes, our fantasy classroom's home page (Knowles and Hennequin) was produced on the UConn mainframe computer during several weeks prior to the beginning of classes, with additions and adjustments throughout the semester. Given the time constraint and differences in HTML proficiency, the technology necessitated a collaborative effort; initially, Wendy was only learning HTML, the programming language with which many World Wide Web sites are constructed, so Nancy constructed the bulk of the programming while Wendy researched links, artwork, backgrounds, and other resources to include on the page. Whoever had opportunity or inspiration contributed original text of assignments and other articles.

The Web page actually consisted of several pages. Two contained our separate class policies, another our mutual schedule of classes and assignments. One page listed the readings for the course and provided links to those texts available online and to the Electronic Course Reserve (ECR) service provided by the university library. We also made all writing assignments, including both assignment requirements and students' ideas for topics of exploration, available through our home page. Other sections of the Web page included links to sites related to texts or authors included in the course, connections to reference materials, and some of the handouts given in hard copy during the class.

Unsatisfied with the number of quality links we were able to locate during the period of time we had to create the Web page, we decided to collaborate with our students in locating links. The technology itself, specifically the vast number of Web sites available online, prompted our decision. We also wanted to ensure all students were capable of using this medium, which even today is not guaranteed. Because technology is best introduced in the context of particular work rather than in a workshop regarding the technology itself (Fitzsimmons-Hunter & Moran, 1998, p. 168), we hoped to provide an authentic need for novice Internet users to learn the technology.

During the two classes held in computer labs early in the semester, we assigned an exercise based on one presented by Tom Lavazzi at the 1998 convention of the Northeast Modern Language Association (NEMLA), which asked students to locate relevant and interesting sites using search engines and then to e-mail their discoveries to the discussion list. Students worked individually or with others to complete the assignment, the more computer-literate students often assisting those lacking such experience, voluntarily taking the more competent peer role without prompting from the instructors. After viewing the Web sites chosen by our students, we selected the most useful, detailed, scholarly, and amusing ones to add as links and mentioned the name of the student(s) who located the site. We hoped that including our students in the construction of the Web page and honoring their choices among our links would convey a sense of investment in the site and of classroom community where student research on the World Wide Web is as valuable as that accomplished by their teachers. This sense of classroom community is vital not only for developing the trust necessary for dialogues that can lead to learning but also for modeling how knowledge is constructed socially through interdependence (Bruffee, 1993, p. 1).

An even more imperative need for teachers and students to collaborate occurred in printing texts from the ECR. Such printing required the computers at which the students worked to have Adobe Acrobat Reader, and if they did not, the students needed to download that software from the ECR page. Downloading proved more complicated than expected, resulting in an inability to complete assigned work and a quick change of lesson plans. But our students came to the rescue; several, upon successfully downloading the program, provided the necessary instructions to their classmates (and us) through the discussion list. Here, too, the difficulty of the technology and the impossibility of the teachers' being prepared for every eventuality forced a collaboration between teachers and students in which students became teachers, not only of their peers but also of their teachers. Such a reversal of traditional classroom roles not only satisfies practical needs—teaching necessary skills—but also allows students to experience learning in a less hierarchical environment (Selfe, 1992, pp. 28–29).

Both informally and on anonymous student surveys, student response to our Web page was generally positive. "It's easy to get to the reading list, to print out the texts," one student remarked on the survey (personal communication, May 5, 1998). Michael Spitzer (1990) notes that online resources give students better and more convenient access to information (p. 65), an experience corroborated by another survey respondent: "If I had a question, it was a good chance that the web site could answer it." When asked for suggestions on the survey for additions to the Web page, most of our students were stumped: "I like it the way it is," one of our students protested, and another agreed, "It's great the way it is :) ." The most common complaints against the Web page—generally expressed informally or brought up during class time—consisted of problems with our university's server or with software or pages beyond our control. In one case, we, admittedly, failed to double-check a link beforehand, and several of our students experienced technical difficulties when trying to print its text. Some, however, disliked the inconvenience of going to a computer lab (located in dormitories) or to the library if they did not have access to a personal computer in their own rooms.

Our experience taught us that the use of the Web page benefited our students in three ways. First—and, considering the goals of a composition and literature course, foremost—by requiring our students to learn how to "surf" the World Wide Web, we connected them to literary texts and secondary resources available online. Many of our students used the resources on the Web to research their term papers, their experience again corroborating Spitzer's research (1990), which shows that online resources make research "an integral part of the writing process" (p. 65). Literally all the students who answered our survey were enthusiastic about the substitution of online texts for copied or bound ones. Second, the activities required by the course forced the students to learn how to navigate the Web and use search engines—skills that may help them research other matters, both academic and real-world topics. Finally—and perhaps most important in the long term—use of the Web insisted that the classroom community, rather than only the instructor, be a source of knowledge. As Bruffee (1993) writes, rather than envisioning knowledge as "an entity that we transfer from one head to another . . . [c]ollaborative learning assumes instead that knowledge is a consensus among the members of a community of knowledgeable peers" (p. 3). Such collaboration makes individuals' participation valuable, has the potential to encourage and improve future collaboration, and instills confidence in the individual.

Those teachers wishing to take advantage of the usefulness and convenience of having materials online no longer need to worry about learning HTML. Since we began teaching with technology, several fine HTML editors have been developed to write the HTML program language as the user composes the page, much as he or she would compose on a word processing program. Some of these programs, such as Netscape Composer, are even free; novices need not be afraid of complexity or expense, and because most of us are already familiar with word processors, users will not have to invest much time in learning to use the HTML editor programs. On the World Wide Web, technical skills and programming knowledge are not as necessary or as important as good content and good proofreading.

Given the ever-increasing number of computer literate students, a World Wide Web page, even if it were to include nothing but a syllabus, would prove useful even in classes where the teacher does not teach the students how to access it. Many already know how to navigate the Web and may conveniently access the information at anytime. Such class sites benefit not only the students but also other teachers who use the Web to find ideas for their own courses. We would further encourage teachers to instruct students themselves to publish on the Web. As Gail E. Hawisher (2000) argues, students should be empowered not only as critical readers of the Internet but also as writers.

Discussion List

Today, many students bring e-mail accounts to college with them. Even those first exposed to e-mail at college quickly become comfortable with it. However, when we offered our fantasy literature course in 1998, we had to use a few classes to teach the students how to use e-mail. After some initial hesitation, the discussion list grew quite active. At the time, we believed that e-mail created the opportunity for egalitarian interaction among all participants in the course, including ourselves. We still believe it can be a tool for collaboration, yet today we are more cautious in claiming it undoes classroom power structures. Reinterpreting our data alongside current research in humanities technology, we acknowledge the need for more critical examination of electronic hierarchy.

With the advent of the Internet and e-mail, the general public and educators especially have worried that a vital part of communication would be lost, namely, face-to-face contact. However, our experience teaching with technology taught us that, bolstered by regular class meetings and one-on-one conferencing, electronic dialogue increases rather than decreases opportunity for communication, the technology itself providing the means for collaboration. As Colette Daiute (1985) argues, "computer talk has extended communication rather than replacing it" (p. 22). It has also made that communication more flexible and effective by erasing time and location boundaries. Teachers being available to students via e-mail can enable individualized conferencing about ideas and questions without having to arrange meetings and can therefore allow a student to begin work sooner. More significantly, the e-mail dialogue became a site for the reciprocity in assistance necessary to build communal ties (Wellman & Gulia, 1999, p. 178).

During the first few weeks, we occasionally needed to supply a question to begin a discussion, but by the end of the semester the students were asking their own questions, answering them, and conducting their own critical debate. Although we required a minimum number of messages sent to the list, many students used the list even more frequently, several communicating often enough to spark face-to-face meetings and others fulfilling class participation requirements almost entirely through the electronic medium, which they found more comfortable than speaking in class. The advantage of e-mail as a class-participation medium for those preferring writing to speaking continues to be replicated in our experience; one of Wendy's recent students, for instance, did not once participate in a classroom discussion but participated so actively on the class discussion list that she received an A+ in class participation.

Even more important than the frequency of participation is that participation occurs in writing. Because this written communication feels like speech (because students' audience consists mainly of their peers), it acts as a comfortable bridge for students attempting to move academic thought from speech into writing. Because "they are totally immersed in writing" (Hawisher, 1992, p. 84), they are learning to write as part of the class's discourse community, without the burden of formality involved in a word-processed assignment. Moreover, electronic conferences enjoy "the high degree of personal involvement . . . more common in face-to-face communication than in written discourse" (Hawisher, 1992, p. 87). Such writing, spurred by peer comments and questions, can lead successfully to more formal writing, as one student, Chris Beaudry, found in moving from an idea about Edgar Allan Poe's "The Raven" to a final exam essay. In a posting dated May 1, 1998, Chris connected the poem with the tale of Aesclipius from his classic mythology class: "Perhaps Poe is making a reference to this story involving Aesclipius and using the raven to imply the sin that the man has committed." When asked why the raven might indicate the man's sin and not Lenore's, Chris responded on May 3rd:

> since it didn't appear as if the Raven was telling the man anything about anyone being unfaithfull [sic] to him and since it seems through the poem that the Raven is trying to punish him in some sort of way, I came to think that the raven simply represents the idea of an adulteress union and that it was the man that had committed the sin not Lenore.

This argument, developed with examples from the poem and the myth, proved an effective basis for an exam essay.

Michael Spitzer (1990) notes that "networks can foster more systematic collaboration. Students using networks can pool their insights and ideas, engaging in collaborative brainstorming in writing, with results available to all participants" (p. 59). Although Spitzer is referring to networked papers, we found that our students began to develop ideas collaboratively through the use of the discussion list. In the previous example, Chris was later able to develop his ideas into a more complex argument through the interest of and interaction with his online peers. Furthermore, several times during the semester, students remarked to us that their paper topics were influenced by or were in fact based on other people's remarks on the discussion list. The flurry of debate about possible lesbian undertones and possible allusions to drug abuse in Christina Rosetti's "Goblin Market" involved students in building ideas among themselves, with merely peer-level guidance from the teachers. For example, when one student introduced the question of lesbian connotations in the poem, another student, Richard Hodge (1998) replied, "a case can be made that there are overtones, but i believe them to be a little wrong. misrepresented is a better word." Debbie Dews (1998) answered the same question: "I think this story did in fact have sexual overtones but more along those lines of woman against man, not necessarily woman to woman." This spontaneous electronic collaboration sparked a number of interesting papers arguing for and against the presence of such connotations.

As Boiarsky (1990) notes, when students are encouraged to use each other as resources of information, "They change from being passive recipients of a teacher's judgment to active seekers of constructive criticism. They give and receive assistance in the same way professional writers request and provide feedback in their own community" (p. 59). At one point during the semester, a student asked the discussion community at large if anyone knew whether to underline or quote the title of Edgar Allan Poe's "The Masque of the Red Death." The student received three answers, all correct, and all from other students, within 4 hours of the initial e-mail. It is important to note that this transfer of authority was only possible for the students over the discussion list; when such simple questions arose in the classroom environment, students deferred to the teachers' authority. The technology, then, makes possible this sort of interdependence among students—a fundamental part of collaboration.

For the instructor new to supervising an e-mail discussion list, there are some negative aspects to be prepared for. One problem with use of the medium is that answering e-mail takes time; reading a discussion list takes more time. Moreover, students, as well as consumers and other e-mail correspondents, often expect quick responses via e-mail, as Sandra Harrison (1998) points out:

> Because we can respond very rapidly by e-mail, e-mail sets up the expectation of a rapid response. A recent German survey complained that half of the companies contacted failed to respond to e-mail enquiries from customers within four days. This was portrayed as a major failing, yet a four-day response time for surface mail would undoubtedly have been perceived as good.

We have often been confronted by students wishing, and even sometimes demanding, to have instant answers to e-mail questions.

More troublesome aspects infuse power relationships on discussion lists. What we initially saw as the advantage of electronic communication—its sense of equality among participants, including, we thought, facilitating instructors—is something that needs to be evaluated as new teachers continue to work with the medium. This issue can be demonstrated by two interpretations of our experience with e-mail flaming.

Early in the semester, the students began to realize that our course requirements were not exactly the same. Our collaborative teaching emphasized the differences in teaching philosophies and practice—namely, grading—that already existed. Yet, such differences, seen so immediately, shocked some of our students and made them feel like victims of an unfair system. Given this righteous sentiment, the discussion list provided a powerful forum that some of the students used to highlight our pedagogical differences and to muster support for complaint. One student wrote a long, verbally charged protest that included a complaint about his low grade. He had completed the length requirement and therefore expected more recognition of that accomplishment, clearly a "customer is always right" concern. He also resented the difference in course requirements that allowed Nancy's students to revise papers for improved grades, whereas Wendy's could not. In closing, he threatened to continue to complain until the grade and the policy were changed. Throughout the message, this student emphasized his complaints with capital letters, a few obscenities and near-obscenities, and a plethora of punctuation. Hawisher (1992) would define this writing style as flaming, "includ[ing] impoliteness, swearing, charged outbursts, and often a high use of superlatives" (p. 91).

This behavior made us uncomfortable. We wondered whether it would force us to undo the collaborative system we'd constructed. We worried that the conflict would ruin our ability to relate to our students and teach them effectively. The conflict also made our students uncomfortable; some reported at the end of the semester in our anonymous survey: "I didn't like the critical not so nice things said in the beginning though," and "Use the discussion list for good comments about the text we are reading, not to criticize teachers" (personal communication, May 5, 1998). One student, Michelle Gagliardi (1998), even felt strongly enough to criticize the perpetrators publicly on line; she wrote, "i think that if you have a problem with something thats going on then maybe you could be more diplomatic about it and approach one of the professors instead of using email to complain."

After flaming began appearing, we acted quickly to provide guidelines for electronic etiquette and to reaffirm our individual course requirements, at the same time inviting constructive criticism. We asserted our authority not only to quell rebellion but also to reestablish trust in the classroom community as a safe place for self-expression and learning. In doing so, we hoped to solve our problem and to demonstrate a calm, considerate approach to the conflict that always arises in a collaborative environment where "participants . . . care about the mutual goals" (Johnson & Johnson, 1995, p. 357). Afterwards, electronic conversation resumed, albeit with some unfortunate scolding of the perpetrators once we had detailed our position. And the principal perpetrator of the disturbance came to speak with his instructor and continued to make regular, cordial, constructive meetings throughout the semester.

Initially, we interpreted this experience as a misuse of the electronic forum. We faulted the student for both his claims and his language, we validated the pain students expressed both online and in surveys about confronting such vehemently expressed critique of their work environment, and we saw it as our teacherly responsibility to formalize the etiquette we had assumed participants already shared, a formalization that we later found many list owners considered a necessary first step to inviting list participation. We acted quickly to model what we saw as an appropriate, sensitive response to conflict.

Looking back on this encounter now, with more experience using technology in the classroom and more understanding of current research about issues of power in an electronic environment, we can reinterpret the situation in a more problematic way. Although flaming is always unacceptable and we should have indicated that from the beginning, the student's concerns about what he saw as injustice could have been addressed by more mature teachers as an opportunity for discussing how teaching and grading actually occur. Our discomfort at admitting students to inquiry that critically examines their own education led us the following year to align our requirements rather than face a similarly disruptive conflict. In essence, our need to maintain our classroom community and focus on course content erased a discussion that might have enabled more active, critical participation in student learning in the long run.

Furthermore, because we sympathized with the pain expressed by students who scolded the flamer, we failed to examine the ways in which those reactions demonstrated similar power issues. In admiring Michelle's bravery for criticizing the perpetrator publicly, we overlooked her criticism of only his complaining style, not the content of his complaint. Perhaps the complaining student's accusations represented a contingent of silent students who also felt manipulated by a system over which they had no control. To these students, the illusory "abdication of pedagogical authority" in the faceless electronic medium, which Terry Craig, Leslie Harris, and Richard Smith (1999) align with Jeremy Bentham's panopticon (p. 141), may represent a frustrating reification of their powerlessness, rather than a means of celebrating democracy. If e-mail discussion lists operate "as benign dictatorships" (Kollock & Smith, 1999, p. 5), then the students who scolded the flamer may have reacted less from a sense of pain at the conflict than from a need to assert the only power they had, voice to support the status quo.

This reinterpretation of our difficult early experience is not meant to discourage newcomers from using e-mail discussion lists. Such lists remain wonderful tools for student–student collaboration. However, we would encourage teachers, as they gain experience in the medium, to further evaluate teacher–student collaboration in the medium. Is flaming, which Susan Herring's studies show is pervasive (Gumbel, 1999), merely an etiquette problem, or is it a means of expressing a challenge to an authoritarian system in diction that belongs outside that system? Teachers need to consider where and why they draw their lines, as well as how to address the competing values of anarchy and safety in creating effective learning environments. Without our early experience and our subsequent reflection on it, we would not be prepared as we are now to negotiate the complexity of this issue.

Videoconference

Besides collaboration within and among our own classes, we also had the good fortune to collaborate with Dr. Barbara Johnson's class at Presentation College in Aberdeen and Britton, South Dakota, via videoconferencing technology. This is another instance where technology provided the opportunity and means for collaboration and where our sense of experimentation led to a truly wonderful learning moment.

Interested in proposing a videoconference funded by Presentation College between students in South Dakota and Connecticut, Barbara contacted Dr. Tom Recchio, the director of freshman writing at UConn, who announced the opportunity. Luckily, we were developing our collaborative courses at the time, so we responded enthusiastically and began to plan an hour-long electronic meeting with Barbara's students. The three of us selected for discussion two texts Wendy and Nancy were already teaching: Harlan Ellison's *Star Trek* teleplay "The City on the Edge of Forever" and Gene Rodenberry's revision of this teleplay (which eventually aired). Although Barbara had experience teaching with PictureTel, given that her class was divided between two sites and relied on such technology for every class meeting, Wendy and Nancy needed and received training. Barbara coached via e-mail, and L. C. (Max) Maxfield, assistant director of the University Center for Instructional Media & Technology, demonstrated the equipment and allowed us an opportunity to practice. Our institution had not encouraged us to do such work nor had even made us aware of the technology's availability.

When the videoconference took place, the conference room at UConn was filled beyond the 18-seat capacity with student volunteers eager not only to discuss a text that had captured their interest but also to experience the long-distance learning that their South Dakota peers participated in every day. Wendy, directing the camera, had trouble keeping up with the hands raised to reserve camera time for a question or comment on topics ranging from the more literary questions about Spock's character and the nature of the characters called the Guardians to more film-related issues, such as costuming, lenses, and slow motion. This interest in the discussion was pleasantly surprising because we had imagined our students would be shy facing the camera, negotiating where to look and when to talk, bewildered as we had been by our first PictureTel experience, and we had prepared a list of discussion questions in case the students ran out of things to say. Exactly the opposite occurred—instructors were sidelined as students discussed their ideas, a collaboration comparable to the student-driven discussions on e-mail.

All the students attending the videoconference who responded to the anonymous survey found something interesting in the experience. For some, the most interesting aspect of the videoconference was the technology itself. Writes one student, "It was amazing to be talking to individuals across the country while sitting in the classroom" (personal communication, May 5, 1998). Another student aligned the technology with the *Star Trek* script under discussion: "It was something right out of Star Trek." Other students valued the dialogue about the text itself. One student liked "[f]inding that across the country (or at least half way) has similar views on the text as we do here." Others related to the human quality of actually seeing the people to whom they were speaking and seeing the way they learn. One reports enjoying "[b]eing able to not only speak with them, but see how different things are for them. It was cool! :) ."

Our experience with PictureTel taught us that such technology has the opposite effect that pessimistic theorists expected when distance learning was first introduced—rather than sterilizing relationships by separating people, our students' reactions to this experience demonstrates that this equipment has the power to draw people together. This visual communication is particularly important in helping participants make a connection between their life experiences and those of others halfway across the country. Although this collaboration lasted only an hour on 1 day during the semester and resulted in some e-mail exchange among the students, the technology itself has the potential for inspiring more collaboration among a variety of groups of people. And even with the minimal exposure some of our students had, the value that we as instructors placed on such dialogue may have a lasting effect on how they envision interpersonal relationships.

Although this technology is expensive to acquire, many institutions already have invested in it. New teachers could inquire into the expense of using it and research ways of covering those costs. Working with institutions that regularly use the technology, as Presentation College does, may mean the expense will be easily borne on the other end of the phone line.

CONCLUSION

Although we cannot argue that using technology automatically causes collaboration, our classroom experience demonstrates the potential for technology to encourage collaboration by providing a variety of opportunities—such as e-mail and videoconferencing—for communication among equals, and in the instances where teacher technical knowledge falls short of student technical knowledge, technology can necessitate collaboration. Such technologically based collaboration represents an appropriate blurring of disciplinary boundaries because, as social constructivist theorists argue, students participating in classroom dialogue gain content knowledge and language and cognitive skills. In particular, the interpersonal communication skills that are gained in such an environment, which are so valued by the humanities discipline, may enjoy long-term positive effects on global communication. Because of this mutually beneficial combination of technology and collaboration in our classrooms, we are actually seeing, to return to Kaplan (1991) and her term, a "brave new electronic world" (p. 21), albeit one that is sometimes uncomfortable and sometimes even frightening.

However, neither the technology nor the collaboration need be feared. Neither is in fact anything new. We instructors routinely seek advice in teaching matters from our more experienced peers from time to time; an instructor asking for the help of a colleague to integrate technology into the classroom, or to learn the technology, is really no different than the collaboration that already takes place in our offices. Because collaboration represents a primary means of professional work, we believe helping students to succeed means providing opportunities for them to practice. What may be unexpected is the idea that we may now collaborate with our students in the classrooms. Many students are very computer literate and are quite willing to play the more capable peer role in sharing their knowledge with their instructors. In fact, doing so is an authentic way for them to gain a sense of authority over their own knowledge and experience.

As for the technology itself, although we tried to integrate several technological tools into our fantasy classroom simultaneously, no one needs to try everything, and certainly no one needs to try everything at once. Those who already use e-mail regularly might want to try a discussion list; others might take advantage of HTML editors to post syllabi, quiz answers, and assignments on the World Wide Web. And no one needs to know everything about the technology—no one can, a situation that fosters the type of collaboration that we saw in our classrooms.

7

Voices Merged in Collaborated Conversation: The Peer Critiquing Computer Project

Mary E. Fakler
SUNY New Paltz–Mount St. Mary College

Joan E. Perisse
SUNY New Paltz–Marist College

Collaboration has become a buzzword in academic circles; other terms being used interchangeably with collaboration are coauthored writing, cooperation, and peer work. We use the term *collaboration* to mean the action of two or more people working together to create something—a concept, a discussion, an essay, a classroom technique. Collaboration best describes our teaching method.

We have been collaborators for many years. The genesis of our collaborative work took place when we became colleagues at the State University of New York College at New Paltz, both teaching 1st-year writing courses. Since then, we have continued our collaboration in other schools: In addition to New Paltz, Joan teaches at Marist College in Poughkeepsie and Mary taught at Mt. St. Mary College in Newburgh. In the beginning, we discovered that each of us planned to have our students read *To Kill A Mockingbird* and decided to collaborate and share ideas about teaching the novel. We began by sharing our students' free writes between the classes. Later, we added a writing exercise in which students in the three schools exchange personal letters.

As we continued our collaboration, we often commented on the differences among the populations in the three colleges. For example, the private colleges are more homogeneous in student population, whereas the public college enrolls students from diverse cultural backgrounds. Because each school has its own unique balance of students with diverse ethnicities and experiences, we knew that our students had much to offer each other, and to be exposed to different ways of thinking and different opinions would broaden their educational horizons. We wondered how we could get the students from one college to share their experiences and reflections with students from other colleges.

Part of the in-class collaborative process of almost every writing instructor is peer critiquing, a process through which students' writing is commented on by their peers. We had engaged in in-class peer critiquing of essays for years. But, as many teachers have remarked, this type of critiquing is often done haphazardly, without much student investment in the work because of students' resistance to the process. Kenneth Bruffee (1993) argues:

> students will not critique each other honestly because students' first reaction to being asked to comment on another student's work is almost invariably to interpret it as an invitation to rat on a friend: mutual criticism as a form of treason . . . the alternative reaction goes to the opposite extreme: almost vile excoriation. At first students refuse to admit that they see anything wrong with a fellow student's work. Then they refuse to admit that there is anything of value in it at all. (p. 26)

We wondered if we could put something of value into the process for them. Unsatisfied with the quality of in-class critiquing, we asked ourselves what would happen if peer critiquers were strangers to each other. We set out to find an answer.

Based on a theory that real learning takes place in a setting where ideas and thoughts are exchanged, internalized, externalized, and exchanged again (Bruffee, 1993, pp. 15–27), we set up a technological arena in which continuous conversation and experience takes place among strangers. In our desire to broaden our students' experiences and bring a realistic view of the world to them, we searched for ways to help them learn something of value to enhance their academic careers and to teach them skills they could value in their everyday lives.

BEGINNING THE PEER CRITIQUING
COMPUTER PROJECT

In this age of information technology and online conferencing, educators and others are finding that boundaries are expanding at rates almost too great to measure. As teachers, we are aware that students need technical skills for academic and future success, and we strive to introduce our students to different avenues of learning. In our search to improve the quality of in-class critiquing, we turned to technology and found our answer. In the process, we instituted The Peer Critiquing Computer Project.

Overview of the Critiquing Process

Using an Internet program called WebBoard, we engage our students in a unique and challenging intellectual writing exercise via asynchronous exchange. We bring together 150 students from eight composition classes in the three colleges where we teach in a collaborated conversation sharing ideas, opinions, and writing. Groups are intentionally created to be diverse, each consisting of three students—one from each college and, if possible, from different socioeconomic and ethnic backgrounds. Members import their essays (by posting essays into the WebBoard conference) in their respective groups, which are then read and critiqued by their group peers. Each student must critique the essays of the other two members of the group, so each receives two critiques of his or her essay.

Critiquing is done within a limited time frame: There is a due date by which the essay must be posted in the groups. Then, students have 1 week to critique their peers' essays and another week to revise their own essays. Critiquing is done according to a particular format, which requires students to answer all questions listed on a critiquing guideline sheet that we formulated. Each mode of writing has its own critiquing guidelines, as shown in the appendix. We do not grade the students' essays until they are revised. We require three essays to be posted on the WebBoard: an expository essay, an analysis essay, and a research/argument paper. Two grades are given: one for the completed essay itself, based on the criteria used in any composition class, and another (counted as part of the Project grade) for the quality and timeliness of the critiques. Because each student's essay requires a critique, each student does two critiques per essay assignment, totaling six critiquing grades for the entire critiquing project.

Technical Skills

Investigating various technologies has been a time-consuming challenge. At first we were limited to working within the confines of each school's computer technology and devising a method that was fairly simple and easy for students in all three schools to use. We needed a common denominator so that all the schools could participate. When we began, we decided e-mail would be the best way to connect the schools because all of them had e-mail access, with which most students are familiar. Eventually we moved from less-flexible e-mail to Hypernews and finally to our current use of the WebBoard. Because our process of peer critiquing has gone from the simple use of e-mail to our more complex Internet WebBoard, we realize that any institution with even limited technological resources can engage in this kind of collaboration.

In using the WebBoard, students gain entry from our individual home pages through any Internet access program. The Project is protected by a password and only our students can enter the groups. At the beginning of each semester, a list of all participants is given to our computer consultant at SUNY New Paltz, Linda Smith, who sets up the groups. Students are allowed access only to their own groups and any other select conferences we set up for them. Linda's expertise is invaluable to us because she has the task of monitoring the hardware and acts as the eyes and ears of all that passes across the WebBoard. She also posts messages concerning technical problems with the system and is available on the WebBoard to help students who are experiencing technical difficulties.

INITIAL STUDENT RESISTANCE

When we announce the use of The Project at the start of the semester, students often react with a startled look: They expressed such sentiments as

> I can't do it . . . I can't write myself; who am I to talk about someone else's writing? . . . it's a writing course, why do I have to use computers? . . . I don't have a computer at home; how am I gonna get all this work done? . . . Computers take away from my creativity and spontaneity. [And our personal favorite:] Computers are evil.

We hear many excuses in the beginning. Students do not want to be embarrassed by being wrong; they worry about giving a peer incorrect information. The Project is seen as an extra, not an integral part of a writing class. In addition, and this is true of most of us, it's hard to have work critiqued. We remind our students that they are, after all, readers and thinkers, and that basis gives them the authority and ability to critique what they read. After the first attempts, when they realize they can do it and that their comments will be accepted, they settle into the process and succeed at it. New Paltz student Lauren Sabo responds this way:

> When the words "peer critiquing project" entered my mind I thought of nothing but negative things. But as the famous quote says, "first impressions may be deceiving". At first, I looked at critiquing another student's paper as extra work for myself, and, besides, why would any one care what I thought of their paper? Well, I learned that people do care and that some comments really help a person out. (Sabo, Student Survey, 1998)

Student Stephen Buffel adds:

> When first introduced to the idea of having a [critiquing] group over the computer, I was reluctant. I am really not a strong writer, so the idea of having total strangers critique my work did not sound appealing . . . [But] after having gone through the procedure, I really began to see that the other students involved were being genuine in their responses and that their comments were actually really helpful. At certain times, they pointed out things that I had thought all along were correct, such as grammatical mistakes, and also gave advice on how to make the paper work smoother. (Buffel, Student Survey, 1998)

In the end, despite the initial reaction to The Project, most students are able to see the benefit of a collaborative computer project.

Goals and Aims

When we began The Project, our goals were many. Using the critiquing exercise, we hoped that by putting our students into anonymous groups they would feel less intimidated commenting on the work of a stranger. We theorized that the act of critiquing via computer would create a more removed and distant atmosphere, thus helping students to become more objective. Because the computer work is completed outside of class, students are not confined to a limited amount of time, as usually happens with in-class critiquing. Because they had more time, we expected more in-depth analysis from them. This also created extra time in our class schedules because in-class critiquing (usually a time-consuming process) is no longer needed. Technology allows students to communicate outside of prescribed class times and allows students to participate from various locations.

Our goals were to provide our students with a real and valid audience other than just ourselves and to expose them to other styles of writing and points of view. Jo Ellen Winters (1997) notes that what she terms *computer-augmented writing* "encourages a far stronger sense of audience than comparable writing produced by traditional methods. . . . Students interact with their computers, with their writing, and with each other" (p. 16). We hoped the act of typing on the computer would encourage the students to write more, and knowing that their essays and critiquing remarks were read by others in the group they would commit more to their writing.

Another overall goal was to help students become an integral part of a group, learning to interact with others in an academic exchange. We want them to realize their own potential of learning from others and to empower them to give valuable input that can help others. In reality, each student becomes a teacher and a learner. Student Seth McBean comments, "[I] enjoyed reading the essays of the people in my group. They helped me to see some of the mistakes that I was making as far as writing for an audience other than the voices in my head" (McBean, Student Survey, 1998). In collaboration, the participants have access to more ideas and avenues of help and feedback. It was our intention for them to see the value in this situation.

OUR FINDINGS

What did we discover? The Project enables us as teachers to see what our students are learning from our lectures and discussions on writing. Because the essays are posted in the groups within a particular time frame, with critiques due within the following week, and we have access to all of the groups, we can track students' progress as they write and revise. We can see how what we have discussed in class is present in their writing and how each writing concept, skill, and element we discuss is applied in subsequent essays. Unlike paper text, which must be collected, scanned, and returned, the computer allows us a surveillance of students' work in progress. After implementation of the Peer Critiquing Computer Project, we saw a significant change in the way our students approached writing and how they incorporated what we tried to teach them in theory in the classroom by applying it practically in the form of essays, discussions, and papers. Significant findings that impact our students and our teaching because of this collaborative project include students having more time to reflect on the writing process, learning rights and responsibilities through peer collaboration, understanding the concept of honesty, and gaining improved writing skills.

Reflection

The Project reveals much about our students' writing and about what they know and understand. Because they have more time to reflect, read, and respond to a peer essay outside of the classroom, sufficient time is spent on the process of thinking about reading and writing. This time frame allows students to reflect on and reconsider what they have seen and heard in the classroom and to develop their own responses and theories concerning this newly acquired information. The pressure of critiquing and responding on the spot, as happens within the classroom setting, is eliminated when more time is allowed. Their critiques on the computer were far more involved and comprehensive than anything we had done previously in in-class critiquing. The appendixes include the directions for critiquing and a critique of a student's essay concerning an argument about cochlear implants and deaf parents who refuse to allow their deaf daughter to undergo the procedure. Students were asked to take a position on this debate in their essays. Jennifer Barckley's critique in the appendix demonstrates how the critiquing guideline is followed and the thoroughness of the students' comments concerning the text, the structure, the support, and the rationale used in arguing this difficult issue.

As you can see, Barckley's critique is meticulously thorough and detailed. Students are encouraged to do their best to be as accurate as possible, and although they are not always correct in their estimation of an essay, students receiving the critique understand they may accept or reject any part of the critique they do not agree with.

Rights and Responsibilities

Students rely on the critiques of their peers for their revisions, and they are given ample time to rewrite and complete the revision process. Although we are members of each student group so that we can monitor each student's work, we do not actually participate in the activities of the groups, preferring to keep the dialogue between the students. Students come to depend on their peers' comments and suggestions rather than our comments to improve their work. We want the students to work out their own problems and negotiate among themselves because one of the aims of the Project is to help our students learn how to work together in collaboration.

By critiquing their peers, students begin to take responsibility for helping each other improve their writing. They teach their peers what they have learned and share advice and personal experiences among themselves. Unlike face-to-face critiquing, which, as Bruffee and others have noted, is insufficient at best, the computer-augmented course (Winters, 1997, p. 13) allows students the freedom to make comments without fear of offending a friend or making an enemy. Student Caitlin Olsen feels comfortable giving one of her peers advice concerning his essay on the violations of women's human rights in third-world countries:

> The tone you tend to use is similar to a tone one would use when talking to a friend. You use a lot of clichés and slang that most people do not understand. It's not persuasive at all. It is not very logical. Some of your sentences and slang words are very confusing. The essay sounds as if it was not thoroughly planned out. Your tone is not very effective. You never try to persuade the reader to agree with your view on the topic. You never actually explain your view on the argument. You simply state it in your thesis statement. I thought your tone was very unprofessional and should not be used in a college level research paper. I wouldn't use this tone in my papers. I try to be more logical and precise with my opinions and ideas. (Olsen, Peer Critique, 2001)

And as student Fred Fetter explains to the author of the essay he is critiquing, "Christine, this is a poor essay. Look at mine, or Tracy's essay [their other group member]. Do you see how they both follow a similar structure? I think maybe you should go to your teacher for some assistance. This essay will not pass as is" (Fetter, Peer Critique, 2001).

Students understand that they do not have to accept all advice given by their peers, just as they do not have to accept our advice. They have the right to question all comments. If a student feels something is not correct or does not agree with the peer, he or she can reject any part of the critique. Often, a student will e-mail the critiquer, and the two will engage in a side conversation about the comments. Face-to-face critiquing generally limits this kind of honest discussion, not only because students are reluctant to disagree with a critiquer one on one but also because of the time constraints of a class session. Although students are free to accept or reject a peer's comments, we encourage them to give careful consideration to each critique; the critique should give them pause to rethink their own papers.

We are intrigued by the cyberspace writing communities, which are formed during the course of the semester. We feel like theorist Kristina Hooper Woolsey (1996):

> [We are] enthralled with the communities of learners made possible with distributed on-line worlds where more information is available to more people . . . their significance is in the new social structures that can be created. You don't have to be isolated with your own ideas; you can find someone else to discuss them with. (p. 86)

Using the Project, students are no longer isolated thinkers; they can now test their opinions, theories, and ideas on objective listeners, who give feedback with their own new opinions, theories, and ideas.

Because they are giving advice to each other, and everyone in the group will be reading the advice, students want to make sure the advice is valid. They take responsibility for their roles as critiquers and begin to be concerned about giving correct advice. After all, the peer might take the advice and use it. What if it is not correct? Critiquing via computer, as opposed to face-to-face critiquing, is particularly valid in this instance. Whereas face-to-face critiquing is more immediate, public, and potentially embarrassing, electronic critiquing affords the students time to check sources for clarity and correctness.

Students become competent at disciplining one another, e-mailing reminders to each other to get their essays on the computer if a deadline is missed or to re-send an essay that has not been sent correctly. When we began the Project, students would occasionally complain directly to us if there was a problem with a student in the group, especially if a critique was poorly done. Instead of directly trying to handle problems themselves, they depended on our authority to take care of behavior problems for them. Part of what we want them to learn, however, is how to handle group members who might not be pulling their weight, so students are expected to deal with each other when problems arise. While in a classroom together, students might expect the teacher to facilitate, especially if there are behavior problems. Facing only a computer screen, students must learn to take care of group problems themselves.

Honesty

More of a surprise to us is how open and honest the dialogue becomes among students. Students are aware of our presence as observers of the groups, but this awareness does not stop them from speaking their minds. They depend on each other for help in catching errors not only in grammar and structure but also in content and practicality of the arguments and explanations being put forth. At times, our students do a better job at catching discrepancies than we do. Student Ramon Espada advises his peer that his essay does not hold enough original thinking: "You give a bunch of information but you do not use your own knowledge . . . Understand? . . . it is like you are referring to too many things. You should try and use your own voice and ideas" (Espada, Peer Critique, 2001).

The students also understand that, as long as the paper is still in rough draft form and not a final draft, any mistakes, including plagiarism, are correctable. Students are not afraid to make strong comments about plagiarism because they know that anything caught prior to a final draft can be corrected without consequences. Student Seth Joseph warned his peer:

Also, you need to cite the sentence "More than one in ten women who reported violence in a current marriage have at some point felt their lives in danger." There is no way that you would know this statistic unless you read it somewhere, in which case you need to cite it so that you are not accused of plagiarism. (Joseph, Peer Critique, 2001)

This honesty comes about because of the unknown writer on the other end of the computer. Face-to-face critiquing has always stifled this kind of forthright scrutiny. About being part of a group, student Marie DesRosiers writes the following:

[I was part of] an excellent group of people, who actually helped to make my papers better. No one was timid about giving criticism because it was good criticism to help a paper. At first, it was hard to tell people I did not know that there is something wrong with their paper, but after a while I knew it was the best thing to do, so they can get a better grade. (DesRosiers, Student Survey, 1998)

At the outset, many students have concerns about hurting their peers' feelings through an honest evaluation of their writing. But this does not keep them from speaking the truth and giving honest opinions of what they really think of the writing. We help the students understand that although we all want to hear "this is really great," those kinds of comments do not help us to improve. Honest, helpful critiques are what will improve our writing. We attribute some of the ease of speaking their minds to the students being virtual strangers. Student Amanda Van Alstyne puts it succinctly:

I had no idea how much better it is to critique a peer when you don't have to deal with them in class, whether it's a bad or good critique! I fully support this method . . . I have always disliked having to look over a classmate's work, especially if I have a relationship with them . . . I find I can be so much more honest and evaluate better when I know I don't have to see that person. Not only that, but it is more accessible when the work is on the computer and not with the author. This gives the critiquer a chance to do the critique at a time when he or she can concentrate fully on the piece before him or her. (Alstyne, Student Survey, 1998)

They also feel that they are doing this work in a spirit of good will, giving their peers the opportunity to revise their essays before Profs. Perisse or Fakler get their hands on the essays.

After students receive the critiques, they are asked to respond to each of the critiquers and to evaluate the critique itself, giving honest opinions about how effective and helpful, or ineffective and unhelpful, the critique was. We have been surprised at the responses, principally because of the honesty with which the students respond. Students show their honest appreciation of a good critique and discuss what advice has been helpful and was not, and why. They comment not only on the advice given but also on how the critiquer gave the comments. They respond frankly to the critiques that were not helpful, pointing out why they were not helpful and how the critiquer could improve them in the future. They learn how to make constructive criticism and understand its value. We have discovered that we empower the students, giving them the understanding that they can make a change, not only in themselves but also in and for others. Whereas classroom critiquing allows students only limited power and authority, computer-enhanced learning gives them almost total control because it creates a student-centered environment.

Teamwork

We know that often in a business situation people work in teams or with partners. Not only do we want students to learn the benefits of collaborative thinking, but we also want them to learn how to be responsible and productive members of a group. In addition, they need to learn how to facilitate having each person in a group do his or her fair share of the work. Students need to be able to stick up for themselves and not be afraid to voice opinions or complaints if things are not working to everyone's advantage. They need to learn to be able to rectify problems within a group. The Project gives them the opportunity to accomplish these goals. After all, it is easier to reprimand a stranger on the other end of a computer than a peer with whom one shares a classroom experience all semester.

Writing Skills

The students begin to learn and remember more about writing simply through the act of teaching and explaining writing to each other. More students question us in class on specific writing styles, grammar questions, structure, and documentation than ever before. Because they are responsible for another student's success, the critiquers are more conscious about the correctness of their comments to the author of the essay. Because we do not critique the writing until it is handed in as a final draft to be graded, the students need to depend on one another for advice and feedback on their writing. They create an academic conversation, discussing essay requirements, ideas, sources, what to change, how to change it, and so on. Ordinarily, without this collaboration students would write in isolation, not reaping the benefits of an audience with a vested interest in giving them feedback and advice on their writing.

PROBLEMS, SOLUTIONS, AND BENEFITS

When we started the Project, we did not grade the critiques; only the final essays were graded. But this did not work as well as we had hoped. Some students will give a minimal amount of effort. Once we began to grade the critiques, investment in the work seemed to improve. For some, there still seems to be a need to do hard work only if there is some reward at the end. However, they do like helping, and, as they continue to critique, they become better and more responsible workers. The computer allows a permanent record of comments, which can be continually referred to, and they can see the process of their own progress.

The grades on the critiques give them an idea of what level their critiques are achieving. We show the students examples of good critiques and how and what other students are writing to help them improve their own critiques. Because the students write a total of six critiques during the course of the semester and each critique is graded and often discussed in class, the students learn how to improve the tone, content, and depth of future critiques. Students often improve as they go along, especially when they see how well other students critique.

We also noticed how others in a group affect an individual. Initial placement in the groups is done randomly. We put together a list of students as soon as possible at the beginning of the semester. We know little of our students and their writing skills, and we cannot take advantage of selectively placing certain students within a group where they will get the maximum benefits. But when we become aware of a student who writes at a lower lever (e.g., an English as second language student), we might move him or her into a group with a more advanced writer who could give that student some much needed assistance. We find that when a group has two competent writers who do their work on time, the third student rises to the level of the group. If he or she has not been very conscientious, the student becomes more so, and his or her writing abilities improve.

On the other hand, we have also discovered that, in a group with lower level writers or with students who are late meeting deadlines or whose critiques are poorly done, the level of the entire group seems to drop, or they might maintain the same level throughout the semester, but there is little improvement. Because each group only has access to its own members, students are unaware of how much work other groups are doing. They do not strive to improve, and they seem to accept the level at which they are working. There is no one in the group to set an example of a higher standard.

To address this situation, we created a public group that any student from any group can access. There we post work that other students are doing that exemplifies good critiques. Because all students have access to the WebBoard, all have access to this public group. This is certainly not possible in a regular class of about 20 students. The public group exposes work to a potential audience of 150 students and exposes readers to many more examples of student writing than a regular classroom could ever offer. When a student receives a poor grade, we direct him to go to this conference group and review other critiques written by his or her peers.

We often have an in-class student tutor (a former student who finds the Project so valuable he or she continues with it!), who will enter the public group and critique the essays, showing the writers how good critiquing is accomplished, and how valuable and helpful the critiques are. We refrain from doing a critique for the students because we do not want them dependent on our advice instead of that of their peers. We also do not want other students to feel that one group has an unfair advantage over another. In some rare situations, we might move a student from one group to another. We do not like to do this because our students do not want us to disrupt their groups. They seem to form a bond and a feeling of unity very quickly and would rather continue to work together than to have someone moved. In a classroom setting, there are generally about only 20 students to pair up. Electronically, we have the choice of pairing up a weak writer with 150 others.

Another benefit of the Project is the students' willingness to give a peer a second chance and their demonstrations of understanding and patience. When we send a message to a group concerning one of the members, it is for only one of two reasons: to respond to missed work and deadlines or to the breaking of one of the few rules that govern the project. The electronic format allows our responses to be immediate, as opposed to waiting for the next class session, and because of the anonymity of the group, this causes less embarrassment.

If a student misses a deadline, the group is informed that the writer is not entitled to receive a critique. We leave it up to the other members of the group whether or not they want to critique the delinquent student's work, but they are not required to do so. In class, this kind of decision making would be almost impossible; could we have some students agreeing to critique while others sit and wait for them? In almost every instance, the late writer gets his or her work critiqued. We see the members of groups protecting their peers against us—the outsiders—as often as we see them gang up on slackers and read them their rights! This behavior does not surprise us within the classroom itself, but to find it among students who have never met fascinates us.

One of the greatest benefits we have found involves our ESL (English as second language) students, most of whom are either Asian or Latino. In addition to being exposed to spoken English in their classrooms and written English in the texts they read, the Peer Critiquing Computer Project exposes them to written texts that are created by other students. This encourages them to write cogently in English when they write their critiques. In addition, the native speakers are exposed to the writing, philosophies, and styles of foreign students, while being able to help them improve their grammar and structure. All this takes place without us, as the students come to accept the responsibility to and for each other's writing improvement.

DISCUSSION GROUPS

Considering Nel Noddings's assertion (1991) that "the capacity of moral agents to talk appreciatively with each other regardless of fundamental difference is crucial in friendship, marriage, politics, business and world peace" (p. 157), we felt compelled to expand the Project to include her assertions into the classroom through our collaborative discussion groups. In addition to the formal critiquing aspect of our project, students engage in a collaborated conversation, sharing ideas in a less formal atmosphere. Each semester we share a common thematic syllabus featuring the same novels, reading, and writing assignments. Because the electronic groups are anonymous, Noddings's differences are limited and lack importance because students are unaware of them. Because there are no physical appearances, and rarely are there descriptions of economic and social standing, all students have to relate to are words on a computer screen.

So, along with the critiquing portion of our project, several of the smaller groups are combined to form larger discussion groups. During the course of the semester, four questions reflective of the theme of the class are posted. Students engage in a threaded conversation, first answering the posted question and then responding to their peers' responses to the same question. We have discovered that through these discussions our students evolve into a community where they can share their experiences, try out their own ideas, and explore other ways of thinking, and do so in a less formal way than they are doing with essay writing and critiquing. Synthesizing from the readings, essays, and their own personal experiences, they form conclusions that can be shared with their peers.

At the onset, either one of us would pose a question to our students concerning a topic/theme from the course. But we found that the less we were involved with the project the more it became the domain of our students and the more freely they communicated with one another. We soon invited guest speakers to become a part of the collaborative discussion groups, which was met with great success. Our current and most beloved contributor to the discussions is Mike Riso, a cousin of Joan's, who is serving in the U.S. Navy and has taken our students with him virtually as he tours the world aboard the USS *Shilo*. We have, via the WebBoard, experienced with Mike the world in peace and in conflict. Through Mike, our students have been exposed to a way of life and shared experiences far removed from their own lives and limited experiences. Mike is close to their own ages, and they look up to him as he honestly relates his perceptions of the world and the situations in which he finds himself. As he relates or shares an experience or idea with them, he poses a question (usually somewhat controversial) to which the students can respond. Our students have never met Mike, but many e-mail him privately and have become his friends, and at times they confide in him about their own struggles with school or life.

By expanding on the collaborative process, we engage our students in a collaborated conversation, a back-and-forth exchange of ideas, opinions, questions, and conclusions. They enter the conversation with enthusiasm, happy to be able to speak out without having to work at it. In a traditional classroom setting, discussions are limited by time and number of students present. Electronic discussions do not have these limitations. Students have plenty of time to consider their responses and opinions, and the written text allows them to review peer responses as often as they would like.

Those students who would not ordinarily voice an opinion because of shyness or gender issues, for example, feel competent and comfortable giving their opinions. Student Brian Plourde says that the discussions "gave us a chance to hear the different views that people had on certain subjects. It also allows us to comment on their thoughts and feelings without their taking offense to our responses" (Plourde, Student Survey, 1998). More important, the WebBoard becomes a community to the students who, as they begin to learn about one another, come together to form a cohesive whole, creating alliances and friendships. They use the WebBoard not only as an academic tool but also as a place to be heard and reach others connecting on all levels. This semester they even collectively helped one of their peers choose a name for his first baby, with suggestions and explanations of names and why they felt the names they suggested were significant.

OVERALL BENEFITS

One of the most positive aspects of the Project is the possibility for multidisciplinary and cross-disciplinary applications. Because of the asynchronous nature of the Project, the technology enables anyone anywhere to participate in a collaborated conversation. Charles Fisher (1996) notes that "digital technologies are exceptional in their capacity to create, manipulate, transport and store artifacts of learning activities" (p. 122). The ability to adapt the Project to any learning environment is one of its greatest assets. Either component, the critiquing or collaborated conversation, can be used by any instructor in any discipline. In coming semesters, for example, we plan to add a philosophy course, political science course, or history course to the project. Sharing a common text, our composition students will critique the papers of the students from these courses from a writer's standpoint, while the students in the other disciplines will critique the contents of the essays. Elementary and high schools can exchange work between classes or between schools. Older students can critique the work of, and share ideas and knowledge with, younger students. Sociology, art, music, dance, theatre, mathematics, biology—the possibilities for the exchange of ideas and writing styles are bounded only by the number of courses offered. And collaborative projects can be made as simple or as detailed as the instructor chooses, as students ease into them, learning step-by-step how to use the technology and how to interact with their peers.

The benefits of both aspects of the project are many, for both students and teachers. The Project does far more than teach students how to write well. Our students learn skills that are required in the professional arena. Every part of the Project teaches knowledge, skills, and social interaction that can be applied to future academic or professional experiences. As Janis Forman (1994) notes, "as a microcosm of the broader society, student groups can embody the tensions, born of differences in age, class, gender, and race, that themselves reflect cultural, political, and socio-economic inequalities" (p. 130). Students learn the importance of meeting deadlines, how to work closely with others, and what their responsibilities are to those with whom they work. They learn to accept and offer criticism, follow instructions, and work out anticipated and unanticipated problems. They become familiar with and comfortable around computers, while learning computer skills and problem trouble-shooting approaches. Finally, they gain confidence from the knowledge and skills they acquire, realizing that their voices count and will be heard.

Student Kelly Sullivan writes:

> This program has truly helped me in time management. Everyone knows that as a college student the one thing that is essential to surviving in academia is time management. This project has helped, in one word—"dead lines." The tremendous amount of deadlines from this project has helped me as a student because I have learned to manage my time better. I have noticed how, in other classes, I tend to get things done much faster than in the past, and I do believe that this project contributed to my time management progress. (Sullivan, Student Survey, 1998)

And student Brian Plourde comments:

> This project has helped to prepare me for the real, working world. It has taught me how to communicate online and the importance of meeting deadlines. I thought that this was going to be a worthless, annoying project when we started, but it actually turned out to be extremely useful. I might not have met all the deadlines on time, but I now definitely understand the importance of deadlines and have been working harder in an attempt to meet all of them. (Plourde, Student Survey, 1998)

For us, the benefits include being able to see what our students are learning, and having the critiquing done outside the classroom gives us more time in the classroom for other work. Woolsey (1996) asserts that "we need to figure out how technologies can lessen the burdens of teachers' every day activities as well as encourage some innovative teaching" (p. 80). We find we do less grading because the students critique the rough drafts—a job we once did. The grading is spread out evenly during the semester. More important, though, we can watch our students develop and grow as writers and as individuals. We watch as they gain respect from one another and accept the responsibilities of their roles as peer tutors. They find their voices and learn to risk opinions, in the knowledge that what they have to say counts; their ideas are valuable and meaningful. The Project is more than a learning tool for writing; it allows us to see the fruits of our labors.

Introducing our students to a new process of collaborative critiquing and discussion is an intellectual expansion that they have not previously encountered. Learning to use technology as a means of sharing experiences and opinions opens new ways of thinking for them. Collaboration between diverse students expands the arena of social awareness and reinforces lessons about life and living. Because of this diversity, the intellectual expansion of ideas has dimensions of sameness and difference, which help our students to see things from perspectives they might not otherwise have gained.

Technology allows for a great number of students to be part of this diversity and new knowledge. Students often feel that there is only one way to read, to analyze, to write. But each comes to the work with her or his own perspectives. Being exposed to the ideas of so many others helps students to expand their own ideas while finding similarities with others'. Finding common threads and ideas helps them to discover that although our differences can be vast in some cases, we are, truly, more alike than different.

Collaborative theory has held sway for a long time. As Selfe (1989) notes, "indeed, we know that literate individuals collaborate, in a sense, to make meaning, engaging in a unique kind of asynchronous, two-sided interpretation of personal experience, even when one party never meets the other" (p. 4). But the Peer Critiquing Computer Project takes the theories of collaboration into new realms: the accessibility of the technology makes the conversation encompass vast numbers; the asynchronous quality allows for greater thinking and reading potential; the anonymous nature of the groups encourages honesty and clarity not found in traditional classrooms; and ease of use of the technology and its communicative values allows for easy conversation and exchange of ideas.

This collaborated journey of cyber conversation and discovery helps students to realize that as we hurl ourselves into this new century while heading towards newer and better technology, because they are the future, they will be responsible to and for each other. The collaborated conversation—the Peer Critiquing Computer Project—helps to prepare them to do just that.

APPENDIX A: ARGUMENT ESSAY CRITIQUE

When writing argument the writer seeks to convince a reader to agree with him/ her concerning a topic open to debate. A written argument states and supports one position about the debatable topic. Support for that position depends on evidence, reasons, and examples chosen for their direct relation to the point being argued. One section of the writing will attempt to refute opposing view points on the topic, but the central thrust of the essay is to argue convincingly for one point of view.

EXPLORING WHO? About the Author

What do you understand about the author from reading the essay?

How is the argument being presented to the reader through the author's tone/ voice? Is he/she angry, offensive, someone you can trust, calm, logical, and does he/she exhibit good thinking skills?

Is this tone effective? Why or why not.

How do you respond to the voice/tone of the essay?

Would you use this voice/tone in your essay? Why or why not?

EXPLORING WHAT? Thesis

Is the overall topic/argument suitable for college writing?

What is the primary idea of the essay? (Locate the thesis and include it in your response.) Did the author take a firm position on one side of the argument or the other?

What are the secondary ideas the author is trying to convey?

Have you had similar ideas, thoughts, feelings, or experiences? Did you relate to the author's ideas?

If the thesis cannot be located, explain why it cannot be located and then explain how to create a thesis statement.

EXPLORING WHY? Purpose and Goal

What seem to be the author's goals or purposes? Are his/her intentions to win for the sake of winning or just to prove he/she is right?

Did the author suggest reasons why this is an important argument? What are they?

Have the writer's goals been achieved with you as the reader?

EXPLORING HOW? Structure and Organization

How is the essay structured?

Discuss the structure of the body of the essay. How is it set up and does it have two parts? (a pro and a con).

How is the essay organized?

Discuss how the introduction is set up: how effective is the general opening; is the argument developed and explained in the background information; is there a clear thesis taking a position; what are the three pro points?

Discuss how reasonable and effective the author's argument seems to be? Does it make any sense?

Does the author use specific details, facts, examples, illustrations, quotes, and statistics to support the points of his/her argument? What are they?

How well is the conclusion developed and how effectively does it close the essay? Does the author suggest possible solutions to resolve this argument? What are they?

What are the strong parts of the essay?

What are the weak parts of the essay?

Did the author win you over to his/her side of thinking? Why or why not?

APPENDIX B: JENNIFER BARCKLEY'S
ARGUMENT CRITIQUE

Hello Katherine!

I hope you are enjoying these wonderful snow days. I know I certainly am. We have midterms this week, so it is a stressful time for me. I've definitely needed these "free" days. Thank you for posting your essay, and I hope I am able to help you in some way! Good luck, and have a great week!

EXPLORING WHO? ABOUT THE AUTHOR

From this essay, I understand that you are a person who is accepting of others. You do not desire to change others to make them more like yourself, which is an admirable trait. Your argument, however, is not as strong as I believe it can be. You need to support your ideas and develop your thoughts and ideas in greater detail. Furthermore, you tend to give the impression that it is easier to never try something new because you can then avoid putting the time and effort into it.

While this may not be your intention, the overall tone of the paper is somewhat pessimistic. This is nothing against you or your opinions—it is simply a matter of rephrasing your words to transmit a more positive and convincing undertone. Overall, while you introduce some good ideas, you do not provide your own opinions. It seems as if you are merely stating facts without evidence and your own voice. I think your tone is a little too passive, as you did not persuade me to change my opinion on this controversial issue. Personally, I would not choose to use this tone in my essay, as I prefer to be more direct and aggressive when arguing. Furthermore, I became quite passionate about this topic and I felt it was imperative to express my strong emotions throughout my paper creating a more prominent and aggressive undertone.

EXPLORING WHAT? THESIS

The overall topic/argument is suitable for college writing. You addressed the issue in a mature manner, and you did not use any derogatory references or phrases. You applied knowledge and facts to your paper, giving it viability. Your primary idea in your paper seems to be that a cochlear implant may not be worth the ramifications that can be produced from this piece of technology. I believe your thesis statement to be "Getting a Cochlear Implant may not be the right choice". The thesis statement needs to follow background information, and this sentence was the first sentence of your entire paper. You need to address "why" a cochlear implant may not be the "right choice", rather than offering a blank statement. For instance, you could write something on the lines of, "A cochlear implant may not be the best choice for a deaf individual because there are many complications involved with such a device". I am not recommending that you write these words, as I feel that you should express your own ideas, however, hopefully this will help put you on the right track. You did take a firm position on one side of the argument, however, after addressing the situation at hand, you continually stated that the "parents need to understand . . .". In actuality, based upon the situation, it is the child that needs to understand the ramifications that can ensue from a cochlear implant.

Your secondary ideas or points are a bit ambiguous. You state, "Three of these include oral language development in all deaf children, even with implants, is a very slow, intense process, there are many psychological and social effects, and the realization that many implants, is very slow, intense process, that has many psychological and social effects, and the realization that many implanted children get little or no benefit even long auditory therapy." This sentence needs to be broken up into three separate sentences and rephrased to make it more clear and understandable. Furthermore, each of these "three" points, seem to blend into one point. Each point is redundant of the other. In your second to last paragraph you discuss the physical effects of a cochlear implant, which can be made into one. of your three points. Your first point seems to blend with your third point, (discussed in the 2nd, 5th and 6th paragraphs). Your second point can incorporate both paragraphs three and four. Overall, it seems that you essentially only have two points. Thus, you need to develop a third point and combine your ideas in the other paragraphs as I discussed above to create your two other points.

I can relate to your ideas as I considered them myself when conducting research for this paper. However, your ideas can be more firmly expressed by using outside support in the form of quotes. I definitely agree with many of the points you made in terms of stress inflicted on the child through the rehabilitation and the misconception that a deaf child needs to be "fixed"/cured of his/her "problem".

EXPLORING WHY? PURPOSE AND GOAL

Your goal or purpose, as an author, seems to be to persuade the reader of the idea that cochlear implants are insignificant and even detrimental to those who are deaf. You seem to believe in your point of view, as you are attempting to dissuade the audience from supporting cochlear implants. Your reason as to why this is an important argument is not blatantly stated. While I do not feel that you supported your opinion to your greatest ability, you did give some indication through your tone that this argument did have some meaning to you. Your reasons pervade through your point, in which you discuss the psychological and social effects of a cochlear implant on a child and the effectiveness of the implant. This however, tends to transmit the idea that gaining the ability to hear is not worth the long journey taken to attain this ultimate goal. While you may feel this way, perhaps you could rephrase your ideas and include support so they have more medical viability. I accepted your ideas and facts as truth and considered them, however, my preconceived opinion on this matter could not be budged by your argument.

EXPLORING HOW? STRUCTURE AND ORGANIZATION

Your essay's structure is a bit confusing, as, previously mentioned, you have scattered your points into various paragraphs. Thus, these points need to be condensed and modified into single paragraphs. Your essay lacks a rebuttal (or con). You need to show the reader that you understand some of the arguments in support of cochlear implants, and then undermine these opposing views by demonstrating your "stronger evidence" or rebuttal. You need to offer two opposing view points in two separate paragraphs with a transition between the two. You also need a transitional paragraph leading from your main body, where you have argued your [pro] points, to the "opposing view" segment of your body. Think of the body of your essay as two parts, A and B, where A expresses and argues your point of view, and part B discusses the opposing view point.

Your essay lacks an initial "attention grabber" as your opening sentence is your thesis statement. Instead of beginning with your opinion, you need general information relating to hearing or being deaf. Furthermore, you describe the hypothetical situation as "Two parents have a prelinqual deaf child (deaf before language acquisition) are faced with the dilemmas which includes a choice between the 6 year old daughter having a cochlear implant or not." This sentence is a little confusing, and the situation could be more accurately portrayed. The fact is the parents are deaf and they are at "battle" with their six year-old daughter who desires the implant. As previously mentioned, your three points are a bit ambiguous. Overall, there really appears to be only one point. You also need to provide background information to help develop the argument. You begin discussing your opinion on cochlear implants without ever telling the reader what a cochlear implant is. It is difficult to agree or even disagree with you without fully understanding what it is that is being debated.

Thus, I would recommend including a general opening followed by background information in one paragraph. This paragraph could then be followed by your thesis statement and three pro points (each one in a separate sentence). You also need to include transitional sentences between paragraphs so the reader clearly understands the connection between your key ideas. Your argument is reasonable, however you do not introduce it enough to make it clear to the outside reader. The fluency of your sentences is unclear. The fifth sentence discussing modern technology is most likely deadwood, as it does not provide relevant background information or support your point. By clarifying your ideas your argument will be much stronger. Just remember the point you are trying to convey throughout your paper, and this should help direct your organization and sentence structure.

Throughout your essay, after each paragraph of your main body, you cite a source. However, it is confusing as to what this source is and how it was used. You need to directly list the source, rather than an article name. Furthermore, it is not necessary to paraphrase and/or summarize. Instead, you do want to use quotes from scholarly people within the medical field to support your points. The way your paper is "cited" now leads the reader to believe that everything you have written has come from the source in parentheses. If this is indeed the case, you need to incorporate some of your own ideas without using words such as "I" or "us", and use quotes to strengthen your ideas and arguments. I was especially confused by the "(cochlear implant opinion)". I did not know if this was your opinion or an excerpt from an article. All of this needs to be made clear to the reader so he/she knows that your information is viable and legitimate.

Your conclusion should probably include more detail. It seems to leave the reader hanging. While you do not want to introduce any new material in this paragraph, you do want to reinforce the ideas you have already discussed to affirm and fully convince the reader to accept your opinion. You do not need to start this paragraph with the transition, in conclusion, because the last sentence of your previous paragraph should make it clear that you are preparing to conclude your paper. In addition, this first sentence contradicts with your previous explanation of the situation between the parents and their daughter. Your mentioning of "language acquisition age" in the second sentence of this paragraph is confusing. This argument is unclear and somewhat weak, as the child can still develop language and speech comprehension long after this point with continuous therapy and a concerted effort. I believe your ideas are once again good, but can be expanded to end on a stronger note. Your discussion of cochlear implants as "not being a miracle cure" is a good point, however, your final words addressing sign language seem to be somewhat haphazardly placed. You have not previously mentioned sign language at any point in your essay, so you almost seem to be putting the reader on a new topic in your concluding sentence.

Your information and data is the strong part of your essay. You seem to understand the points you want to argue, yet you have difficulty expressing them clearly in words. However, you do need to expand on this data by providing quotes from experts to create strong and saleable points. Your essay could be much stronger and convincing if it was more structurally sound. As I mentioned, your essay is organized in a random and inefficient manner. Your ideas are scattered throughout the paper, taking away the reader's ability to immerse him/herself in an aspect of your argument and come out convinced. Instead, the reader is left to determine what message is being conveyed at each point in the essay. Overall, because your ideas can be organized and phrased in a more effective manner, you were not able to thoroughly convince me to take your point of view. There is some ambiguity surrounding your ideas and the way in which they are expressed. I do, however, understand where you are coming from and think that you have introduced some good ideas that can simply be expanded.

Good luck Katherine! I hope you have not taken offense to anything I have written. I know that is difficult to write a paper, especially one of this nature for the first time. My goal is only to help you. Every thing I have written is my attempt to help you learn something new and earn the grade you desire. Talk to you later!

8

Reentry Women Students' Online Collaboration Patterns: Synchronous Conferencing in a Basic Writing Class

Alice Trupe
Bridgewater College

Early adopters of computers in the writing classroom radiated an excitement that seemed disproportionate for talking about a writing technology. The adoption of classroom networks and hypertext authoring programs suggested almost revolutionary possibilities for writing instruction. Concurrently with the technology's emergence came the trend to social constructivism in educational theories and the use of critical pedagogies that privileged student interaction, active language use, and collaboration in the classroom. Lester Faigley (1992) captured the spirit expressed by early adopters in a book chapter exuberantly titled "The Achieved Utopia of the Networked Classroom." Others, like Faigley, celebrated the juncture of theory, practice, and technology in the emerging virtual classroom, while the possibilities generated by the technology bred new classroom practices, reached new populations, and in turn spawned new theories.

Yet early adoption brought critical questioning as well, resulting in calls for more research into the impact of computers on learning (see, especially, Hawisher, 1992; Hawisher & Selfe, 1989, 1991a; Selfe & Hilligoss, 1994). Was computer technology equally utopian for all students? Feminist and critical pedagogues were already challenging language patterns and classroom hierarchies that reproduced existing power structures. They raised questions that resonate at the beginning of a new century, as more portions of our lives move onto the Internet: Is the brave new, utopian world created by computer environments a world that embodies White males' biases and power structures, subordinating women's and minorities' concerns and interactions in much the same ways that society at large frequently does? If it is, indeed, a masculinist Westerner's world, we must consider the implications for equity in integrating computer technology into classrooms where many women and minority students find themselves second-class citizens. Furthermore, as integration of computer technology into classrooms progresses at an astonishing rate, and political leaders mandate computer literacy as a basic component of education, we must ask whether differences anchored in race, socioeconomic class, and age, as well as gender, will be exacerbated by different learning styles and unequal access to computer technology (see, especially, Gomez, 1991; Jessup, 1991; Selfe, 1998; Selfe, 1999a; Selfe & Selfe, 1994).

These questions, along with personal experience, led me to conduct a classroom ethnographic study early in 1996 in a colleague's basic writing class of reentry women students at an urban community college. I had made my first forays into computer classrooms as a 40-year-old graduate student and single mother supporting two teen-aged daughters as an adjunct composition instructor. My own experience of the classroom as a student, after 9 years' part-time teaching experience, sharpened my empathy for reentry women students. Like most reentry women students, I was driven by high standards for my academic performance and fear of failure, while trying to meet conflicting demands on my time growing out of existing responsibilities to my family, friends, and students. Previously somewhat computer-phobic, I plunged into computer environments after my first summer as a graduate student, learning simultaneously to use word processing, the university's network programs, and the LAN-based classroom software in my basic writing class. My own enthusiasm for the pedagogical potential of networked environments was tempered by the moments of frustration and panic I felt as both teacher and student in those environments.

Thus, I found myself asking research questions that reflected the critical questions articulated by Hawisher and Selfe (1991a, 1991b) and a small but growing body of research on reentry women students, a rapidly expanding population on college campuses. Especially important for my study was the literature on women and technology (Cherny & Weise, 1996; Flores, 1990; Haraway, 1985; Hawisher, 1996; Jessup, 1991; Kramarae, 1988; Kramer & Lehman, 1990; Spender, 1995, Sullivan, 1996; Takayoshi, 1994; Turkle, 1995; Wahlstrom, 1994). The literature I reviewed in developing the study varied and included theory and research on women in the classroom, reentry women students, women and language, women and technology, computers and writing, composition and basic writing instruction, and feminist frameworks for research.

The theoretical basis for my study came primarily from social constructivist theory, especially the dialogism of Bakhtin's (1981) and Vygotsky's (1986) work on the development of language through social interaction. Bruffee's (1984, 1986, 1993) work on collaborative learning and its importance for composition instruction had shown that the moment was right for study of collaborative electronic environments. The outcome was a turn toward articulation of the possibilities and research by composition specialists using computer networks (Barker & Kemp, 1990; Batson, 1989, 1993; Burns, 1992; Cooper & Selfe, 1990; Duin &

Hansen, 1994; Eldred, 1989; Forman, 1991, 1992, 1994; Hawisher, 1992; Schriner & Rice, 1989; Taylor, 1992).

I was particularly interested in community college populations like the ones I was used to teaching and in entry-level composition classes, where students would be least likely to have prior experience with computer networks, so that I could observe the impact of the network on a classroom population with whose challenges and interactions I was familiar. I planned to address several questions in my semester-long study:

- Does the electronic environment itself pose additional obstacles to reentry women students' adjustment to college writing instruction, or does it facilitate their adjustment?
- What means do reentry women students develop to cope with the demands this environment makes on them (e.g., collaboration in ways not initiated by the teacher; creation of informal support systems)?
- What role do personal attributes other than sex and age (e.g., race, class) play in the findings of this study?
- What teacher and student behaviors seem to create positive educational results for reentry women students in developmental writing classes conducted in the computer classroom?

To obtain the information that I hoped would answer these questions, I offered my services as a technical assistant, a role that facilitated my status as participant-observer in my colleague's classroom.

THE STUDY

I conducted the study at the urban campus of a large multicampus community college. Most of the 18 women enrolled in the class were participants in a program designed to support underprepared reentry women students' transition into college. All of the women had placed into the lower level of two levels of basic writing. All but two of the women were 1st-semester students. About two thirds of the women were African American or biracial, two of the White women had biracial children, and the instructor and I were White. Two of the women were younger than the minimum age that the college defined as reentry women students (age 23); the rest were in their 20s, 30s, or 40s, with the oldest being 49 years of age. During the semester, attrition would bring the students' number to 14, 10 of whom passed the course.

The Reentry Women's Program provided support for students in the form of blocking their classes (scheduling the group together) and keeping 1 day a week open for home and family responsibilities, so most of them shared several classes. The program also promoted informal and formal networking for mutual personal and academic support and introduced the students to the college's range of support services. The instructor in my study frequently referred to principles of women's education articulated by Belenky, Clinchy, Goldberger, and Tarule (1986) and critical pedagogy anchored in the work of Freire (1993). Thus, the atmosphere fostered by the program and the instructor was intended to facilitate women's development of their voices in a collaborative learning environment.

The class was scheduled in a locally networked lab equipped with Daedalus Integrated Writing Environment (DIWE), and the instructor was an experienced basic writing teacher who had taught writing in computer environments for several years. Departmental requirements for all basic writing students included an exit review of in-class writing and a portfolio of reading-based essays. The instructor's use of the network focused primarily on Interchange sessions as a vehicle for discussion of readings on which essay assignments were based. (Essay topics were suggested by subtitles of the text, Anderson's [1991] anthology *Sisters of the Earth: Women's Prose and Poetry about Nature.*)

Given this combination of classroom environment, course requirements, and pedagogy, I had access to transcripts of Interchange conferences that I analyzed for patterns of discussion and collaboration and for the relationship to essays the students subsequently wrote for their portfolios. This chapter focuses on the two Interchange conferences that were devoted to this purpose during the semester, each covering multiple class periods.

Patterns of Classroom Interaction

Like other classes using Daedalus that I have taught or observed, this group began with mainly teacher-directed discourse and quickly moved to student interactions with other individuals or posts addressed to the whole class. The instructor's initial messages asked students to respond three times and ask at least three questions, instructions designed to promote interaction.

A look at the numbers of students participating and the length of their conference posts not only illustrates the nature of the group's online discourse but also reflects other patterns of interaction and, not too surprisingly, the pattern of success in the course. The first conference included 15 participants. Instructor comments accounted for 45% of the total word count, nine students posted 42%, and six students posted 13%. The nine students who were substantially involved in this conference were the nine participants in the study who would complete the class, and eight of them would pass the class with a C or higher grade. Their patterns of contribution ranged from one message of 165 words to seven messages totaling 526 words; one student posted four messages totaling only 70 words, and another posted five messages totaling only 72 words. Generally, their responses conformed to the instructor's expectations and seemed to help them think about the writing topic—women's kinship with nature.

Given the optimism about interactive computer technology that composition instructors, myself included, have voiced, it would be rewarding and self-vindicating to focus on the success stories. However, it is instructive to look at the participation patterns of those students who posted very little to the conference, most of whom would fail the course as well. The instructor's opening posts explicitly directed students to participate first by introducing themselves and then by discussing the assigned readings. Nonetheless, six students (40% of the student posters to this Interchange conference) accounted for only 13% of the conference transcript. Some of the nonparticipation might be explained as being rooted in difficulty with the computer environment: One post contained only 10 hard returns, indicated by the ¶ symbol. Another obstacle to participation might have been a misunderstanding of what the Interchange conference was for: One student's post consisted of only the standard heading (name, course number, date, etc.) for an essay, suggesting that the student believed she was supposed to begin drafting her essay.

This first Interchange conference demonstrated the shift from an initial pattern of students' responses to teacher questions to a pattern of true interaction among students. The trend was initiated by a group of three women whose success as students would be supported through their friendship and collaboration outside the classroom as well as in it.

The new pattern of communication was signaled when one student, Marie, asked a question, addressing it to another student rather than to the instructor. (Pseudonyms used in this chapter were chosen by the students.) Marie followed this by directing another question to Malaysia, who was one of her friends and would become another leader in the class. Marie's third question was addressed to Gail, her other close friend in this threesome. Gail was the first student to respond to Marie's questions.

Marie didn't wait for answers to her questions, though, before continuing to post. Her next Interchange post addressed a question the instructor had posed to the class at large. The instructor (known as Marcella in Interchange transcripts), seeing the students' difficulty with the topic women's kinship with nature, which the writing assignment was based on, had asked how they felt about weather as one aspect of nature. Marie responded to the question about weather and then expanded the discussion to address values:

> Marcella I am like you I myself loves when it rain and thunderstorm. It does clear the air and has a sense of peacefulness. Me myself am a country girl. My fathers' mother grew up in Macon, GA. She has taught me the real way of life. People from down South have a greater sense of morals and values and I take that to heart. My grandmother have taught me how to fish, hike, and how to catch crayfish and how to cook them, although I have to amit I forget how to cook different typesof recipes (Marie, Interchange Session 1).

Following this post, other students followed Marie's lead. Of the remaining 32 messages in this Interchange conference, 10 questions or comments were addressed specifically to students by students, and half of them came from Malaysia. Malaysia also made the closing comment of the conference, which was addressed to all the students and drew on information she had gleaned from headnotes to the readings under discussion:

> Hello ladies, the poems that you are reading, and have read. There are two woman that are deceased. Carrighar, and Swenson try to enjoy the readings of our exposure, because the ladies mentioned will not be writing anymore. Its' just my opinion Malaysia (Malaysia, Interchange Session 2).

What struck me about this closing comment is that Malaysia assumed authority and took initiative here. "Ladies" was the term the instructor frequently used when she addressed the entire class, and Malaysia's use of it suggested that she felt authorized to communicate information to the whole class. Additionally, the information she volunteered was not a response to any teacher question, though the theme of death in one of the assigned poems had been touched upon during the conference. Malaysia's final qualifying statement ("Its' just my opinion") was followed with her signature, an unnecessary measure because Interchange messages were displayed under posters' names in the conference environment. Although she seemed to disclaim the significance of her opinion, she also claimed it with her signature. This posting seemed to me to show the full move from teacher-vested authority to student authority, a characteristic of computer conferences that is often identified by proponents of classroom computer conferencing.

In the second conference, posts were fewer and longer. Only 13 students participated in this conference: 3 very active students contributed 26 of the messages and 33% of the total word count; 10 students contributed 18 messages and 27% of the word count; and the remaining 13 messages and 40% of the word count were posted by the instructor. The lead that Malaysia had taken in the previous Interchange conference showed more markedly in this second discussion, in which she posted 13 messages.

Clearly, just producing words is not a measure of quality, although it is a measure of class participation. To show how this class participation affected the more traditional, required texts produced for the final portfolio, I would like to show, first, the evolution of one student's thinking through the conference and the relationship between that and her essay on the assigned topic; second, the types of posts produced by two study participants who would have difficulty passing the course; and, third, the single post contributed by one student who tended to work alone rather than in collaboration throughout the semester.

The Relationship Between Interchange Participation and Writing for the Portfolio

Case 1, Gail: "Something Inside of Me Is Wild Like a Flame That Burns Eternally."   The topic under discussion was the "wild and untamed in women," based on several textbook readings and Gary Snyder's (1990) essay,

"The Woman Who Married a Bear." Gail, like other students, seemed baffled by Snyder's essay, which retells a Native American tale:

Msg #7
I really can't relate to this text. I'll try to choose another topic. Maybe someone can help me get started (Gail, Interchange Session 2).

After 12 more posts, Gail was drawn into the conference. She commented on the text and related it to her knowledge of people and her own feelings:

Msg #20
The little girl looked at the bear and seen something that she was interested in. I think it must have been that time in her life when her hormones start to get a fire burning. Just like when teenagers start feeling wild. That is how the wild side of me starts to come out. When I see a nice looking man on T.V. and I say "he looks like ice cream and cake". The bear must of looked like ice cream and cake to her (Gail, Interchange Session 2).

At this point, it looked as though Gail had moved toward better understanding of the text and was prepared to write about it. What had happened between Message #7 and Message #20?

Half of the intervening messages were posted by Malaysia and Marie, who had commented on this text and other assigned readings, addressing their messages to specific individuals. Malaysia wrote:

Msg #9
The man was gorgious, something like a GQ model, but the story is she chose to go with the nature of the man/bear, and live and learn in his domane. . . . To Marie (Malaysia, Interchange Session 2).

and:

Msg #13
When I die I want you to take my head, and my tale. Burn the both of them, and sing this song. To Marie (Malaysia, Interchange Session 2).

Malaysia was referring to the portion of the text that describes the girl's mourning after her brothers have killed her bear-husband. Snyder's (1990) words are "The woman burned the head and the tail, then she sang the song, until all was ashes" (p. 160). To this request, Marie responded:

Msg #19
Malaysia, no problemo! I will perform it to a tee. Just you make sure that you burn my tail and my head. Also, if you could try to make it alittle more exotic (Marie, Interchange Session 2).

These playful messages, as well as longer, more serious ones written by her friends, drew Gail into this Interchange conversation. The students have drawn upon Snyder's text to find a metaphor for their own growing friendship, but it Malaysia's message about the "GQ model" was the direct inspiration for Gail's "ice cream and cake" comparison.

Gail's thinking continued to develop through the remaining two messages she posted to the conference. She reflected on the requirements of being ladylike rather than wild in a message that Marie disagreed with, and then she revised her thinking in response to Marie's objections.

Building on what she had said in Message #20 about "hormones" and the "ice cream and cake" man, Gail posted this message about society's response to assertiveness in women:

Msg #33
When [women] show feelings and they speak out it is considered as wild, untamed. But I think that the bear was wild and being untamed even as a bear/man. Women have to be wild in a classy way to get respected as a lady (Gail, Interchange Session 2).

Marie's response, coming substantially later in the conference, repudiated the idea that respect is gained through ladylike behavior:

Msg #51
Every woman, has a wild and untamed side of her. She must in order to survive in this ruthless world of today. Society has changed. Women don't have the time to be ladies, we must challenge the men of our world to gain respect from them. To let them (MEN) know that we will not tolerate being walked over (Marie, Interchange Session 2).

Marie's comment had a clear impact on Gail, who began working toward balance in her next message, posted almost immediately after Marie's. Here, Gail focused on the idea of control, acknowledging the wild feelings and her desire to act on them but emphasizing her personal control over these feelings and over her life. She moved from this assertion to acknowledge Marie's point, identifying men's position in the world as one of control over, presumably, women, or perhaps women and children.

Msg #53
I am on the wild side just alittle bit. [That] keeps my life style interesting. I do have to make sure I only take it to a limit. With everything you do even if its being wild there must be some kind of control to the wild side. I have a lot of fun with the wild side [in] me. You can gain some control of your life with being wild. Wildness in women is not always a bad side, it can help protect their self-esteem. Also most women need to be wild to get that backbone that is needed to make it in this world with the men in control (Gail, Interchange Session 2).

Gail thus linked the wildness with independence and "self-esteem" in this comment that seems to be a rethinking of the idea of "respect" she had mentioned earlier. Watching Gail's thinking emerge through these posts, we can see that the Interchange environment required her to clarify her ideas, to define her terms, and to respond to others' objections. The thinking and writing processes begun here in the conference showed clearly in the essay that she wrote later. In her essay, Gail continued to explore the need for control through balancing feelings and desires with others' expectations.

Gail wrote about her "good girl image" in her introduction, using a phrase the instructor had not only used but emphasized: "women are often taught to be GOOD GIRLS, to not assert themselves, to give up their personal power to others (most of the time men, but not always)." In contrast, the central metaphor of her essay apparently was suggested by another assigned text, Judith Minty's (1991) poem, "Why Do You Keep Those Cats?" in which the cats are described as lying in front of the fire, waiting "for the flame of yellow moons" (p. 106).

Gail's essay began as follows:

Going back into my mind with thoughts about what makes me wild and untamed, I would say something inside of me is wild like a flame that burns eternally. The flame is bluish, and it is so beautiful. The flame can be wild and out of control at times, even though the flame reaches new heights and is full of beauty. One drop of water and it all can be washed away. The blue-like color comes from the heavens. The beauty comes from the heart.

All of life's ups and downs are the heights of the flames. Strong and bold as I am, with one drop of water my flame can go out. The wind can blow out my wild flame easily if I do not use a cover to protect it. I have learned to control my wild and untamed side. That is one of the harest things for me to do, to keep the flame at a low burn. I always have to maintain a very low key good-girl image when really I want to be some what wild and untamed. I am married to a baptist minister. In the baptist religion a "minister's wife" must appear to be tamed (Gail, Essay).

Gail sustained this metaphor of a carefully protected flame of wildness throughout her essay. She seemed, ultimately, to define the flame as a capacity for sexual love, building on the Interchange comments on "hormones," although the flame is also a spiritual gift that might be identified with individuality and a zest for life. Her essay showed the same concern with control that her final conference post revealed. In discussing Snyder's (1990) essay, she explicitly identified the flame with passion:

The anticipation of seeing the bear/man was like a dream come true. That was like the flame in my life; out of control. Reaching out to touch the beauty of the fantasy of going off with the bear/man was a wild and untamed dream. . . . When she decided to let her wild side out and become the woman that she wanted to be, that is what made her untamed. The bear/man was sort of a blessing, but at the end of the story the drop of water came and washed her blessing away. It put out the flame because now the wild and untamed side of the bear/man came out. The little girl was not that innocent good girl any more (Gail, Essay).

There is a strong sense of retribution, of punishment for impurity here: The life-flame of wildness was put out because the sexual wildness of the bear/man had destroyed the girl's innocence. Gail pointed out that the girl's mother had tried to protect her innocence, but those adolescent hormones (alluded to in her Interchange Message #20) drew the girl to her fate:

> Being a wild and untamed woman cannot always add up to what you think it should be. The little girl disobeyed her mother when that flame inside of her started to burn out of control (Gail, Essay).

The connection to the Interchange session became even clearer in the next sentences, where Gail incorporated the ice-cream-and-cake metaphor of Message #20:

> When she saw the bear/man he looked like ice cream and cake to the little girl, and that made her go off and be wild and untamed. After she had a taste of that ice cream and cake, she could see that it was not as good as she thought it would be (Gail, Essay).

Here, Gail suggested the conclusion she would reach, that the flame would be doused with maturity, and this seemed to limit the meaning of her flame metaphor to the ways in which physical passion overrules reason in young people. Her essay would end by rejecting the positive and beautiful images of the flame's burning high that she included in her opening paragraphs.

This essay was a very successful one for Gail. It was about 1,000 words long, twice the required length, and it showed a markedly lower level of deviation from standard written dialect than most of her other essays. It also showed how the Interchange conference could foster development from initial frustration and bewilderment to an interweaving of metaphor, reflection on others' views, and multiple textual references. And, significantly, Gail's essay sounded very different from the conversation in Interchange, showing her ability to make a transition from the informal dialogue to the more formal discourse requirements of the final portfolio.

This case might make a strong argument for the advantages that synchronous conferencing gives basic writers to develop their ideas and sense of audience. However, this story shows only one aspect of the networking shared by Gail with Malaysia and Marie, who sat together in classes, ate lunch together, talked over personal problems and assignments on the phone at home, and gave each other feedback and help on assignments for this class and others. It would be a gross oversimplification, then, to attribute Gail's writing success to the use of Interchange. Yet the conference transcript helped Gail remember and tie together her developing ideas and keep her audience in mind. The conference transcript also showed me the connections between electronic classroom discussion and formal writing, and this was useful for analyzing students' failure and near-failure as well as students' success.

Case 2, Kimba "I didn't get a chance to read the story" and Peaches "Two time honored essays surface." Like Gail, Kimba and Peaches networked with other students, forming a group whose members who sat together in classes, ate lunch together, talked over problems on the phone at home, and helped each other in class. However, this group's behaviors showed strong resistance to the class, typical of many basic writers. They came in late together, as much as 20 to 25 minutes late in a 50-minute class on more than one occasion, when they were already on campus for an earlier reading class. If they were late, one or more of the women would ask questions that required the instructor to repeat information they had missed by coming late. Their loud talking angered other class members, and interviews with other students revealed that their attitudes were offensive to class members in other ways as well.

Peaches was the only member of this group who would pass the class. Sitting between Kimba (not a reentry woman student but a recent high school graduate) and another friend, Peaches sometimes found it difficult to stay on task. One or the other of them might talk to her, and she was cajoled into typing for her friends, whose keyboarding skills were inferior to hers. Kimba was frequently off-task, working on a different essay from the rest of the group or on something altogether unrelated to the basic writing class. She sometimes left the room for brief periods, and she asked off-topic questions that showed she was missing instructions that were keeping other students oriented. Urged by the instructor to meet regularly with her for individual instruction on grammatical problems, Kimba missed appointments and failed to reschedule them.

Kimba failed the course because she failed to turn in a portfolio. She negotiated a grade of Incomplete with the instructor but did not meet the extended deadline either. Some of her difficulties can be explained with a look at her two posts to the second Interchange conference, which accounted for less than 2% of the total word count of the conference. For comparison, note that Malaysia's messages accounted for 16% of the word count, Marie's messages amounted to more than 10% of the word count, and Gail's messages totaled almost 7%. Both of Kimba's messages were addressed to a student identified here as V. Her first message was on task:

Msg #11
V, I wonder what made the little girl walk away from her home and family. and what did the bear do to make her feel so comfortable with him, also what was she thinking at the time (Kimba, Interchange Session 2).

Her message was, however, focused on the opening paragraphs of Snyder's (1990) essay, and her next post would make it clear that she was unprepared for class:

Msg #18
V, I didn't get a chance to read the story so can you tell me a little more about what you read so that I can get an clear idea about what your saying (Kimba, Interchange Session 2).

On the basis of these brief posts, it is easy to understand Kimba's failure to profit from the collaborative opportunities afforded in this class. If she used the computer conference as a substitute for doing the reading for the class, there was little chance that her personal network could empower her as a student.

Peaches, in contrast, was prepared for class, and she posted two messages totaling just under 4% of the conference word count. Her first post showed her engagement with the text as she attempted to answer Kimba's questions:

Msg #23
The little girl is scared and also afraid. She should have listen to her parents maybe she would not be there right now. She did what the man/bear told her to do, he also would make circle around her head so she wouldn't think about home or no one else. I think if she try to go home the bear would do something to her. Kimba this should teach use not to go crazy about gorgious men (Peaches, Interchange Session 2).

Here, Peaches explained the character's actions, as Kimba had asked, but then she went on to ground her explanation in the text, and then she connected the text to her personal experience. Furthermore, in echoing Malaysia's wording and spelling of Message #9 ("The man was gorgious"), Peaches showed interaction with the entire group.

Peaches's post also makes for an interesting contrast to what Gail would write in her essay on this topic. She suggests the girl was afraid but subservient to the bear. Not only had the bear undermined parental authority, but also the girl faced violence if she left him. The girl is passive, subordinate either to her parents or to an attractive but controlling and potentially violent man. Gail, in contrast, saw herself and the girl as agents, capable of making the choice to follow passion rather than reason. These two positions reflected their classroom behaviors as well. Gail, Marie, and Malaysia interacted with other students and initiated topics in Interchange; Peaches moved from responding to Kimba to answering a question posed by the instructor, a move that showed her concern about authority.

Peaches's second post followed up on a thread initiated by the instructor and expanded by Malaysia. The instructor (Marcella) had asked, "Why does the beauty stay with the beast and love him even though her family thinks she's nuts?" Peaches responded:

Msg #40
Marcella, I can understand what your saying about the little girl and Beauty and the Beast. Everyone thought that the beast was evil but, when the girl stay for awhile it also brought the beauty out. So the little girl seen someone with alot of beauty but,

deep inside he was the beast. And she didn't realized it until it was to late (Peaches, Interchange Session 2).

Here, Peaches reversed the inner-outer dichotomy of Beauty's beast (the beast was threatening on the outside but had inner beauty) and asserted that the bear/man had an attractive surface that belied his inner violence.

The kinds of writing Peaches undertook in this Interchange session resembled the conference activity of Malaysia, Marie, and Gail, though her participation was more limited. Peaches responded to others and developed a theme in successive posts. However, Peaches's essay on the wild and untamed did not incorporate any conference material. Her text attempted a formal tone and was heavily reliant on summary of the two readings she chose. It began:

> Two time honored essays surface from the works of Annie Dillard and China Galland on what does it mean to be wild and untamed as a woman? Also what is it about the wild and untamed in nature that women relate to. In searching for a valid method of getting to know one's wild side I will respond to these two presentations and determine which philosophy would be most comfortable to follow (Peaches, Essay 2).

Instructor comments on this draft would question, "Whose words are these, Peaches? They don't sound like the rest of your paper." As a matter of fact, Peaches told me in a casual in-class conversation that her brother-in-law had helped her extensively with the essay. He was probably unaware that a transcript of the class discussion even existed. It is likely that Peaches's concern with producing the kind of text she thought the instructor wanted led her to collaborate outside of class rather than in class, relying on a relative whose writing skills seemed to her strong enough to impress the instructor favorably.

Thus, Peaches made a contribution to the computer conversation, but it did not transfer to her production of a formal text for her portfolio, which probably means that she did not see the relationship between the two activities. The instructor had presented the Interchange environment as a place for students to discuss the readings, and she provided them with copied transcripts of the sessions. It would be difficult, though, to fault Peaches for failing to connect the writing she did in Interchange with the writing she did for her final portfolio because she rightly deduced that what counted (i.e., what her grade would be based on was the essay, regardless of her classroom participation in computer conferences).

In marked contrast to Peaches's limited connection with the classroom conversation is minimal conference participation by Mo, one of the best students in the class, who worked alone for the most part. Like Peaches's second post, Mo's would focus on obedience to authority under threat of violence.

Case 3, Mo: "Sharing Your Thoughts and Feelings with Other People Is What We Need More Of." In the first Interchange conference, Mo posted four messages for a total of 70 words. Each post was a single sentence or phrase that appeared to be an attempt to answer the instructor's questions to the whole group. Her participation in the second Interchange conference consisted of one message of 51 words. This post was the last message in the conference, a position that made it likely most students never read it. It sheds light on the background and values Mo brought to her college experience, and it reveals what made Mo a solitary rather than collaborative student:

Msg #57
I wasn't allowed to be wild. My Mother was afraid of everything. I got lots of beatings when I was growing up, me and my sisters weren't allowed off of the front porch. If you got your clothes dirty, that was another beating. Going off the porch was a wild thing to do (Mo, Interchange Session 2).

The essay that Mo wrote after this conference echoes the theme of a restricted childhood. She wrote:

My own feelings of feeling wild and untamed have more to do with feeling free, the freedom to be yourself, and to do and be what you want to be. That can be a way for me to feel wild, no more you can't do this or you can't do that.

If Mo was still thinking about childhood restrictions, the ideas she expressed in an essay were an extension of what she had posted in Interchange, in contrast to Gail, whose thinking evolved through interaction with other students. Mo's essay focused on her fears, rather than on the wild and untamed. Responding to China Galland's (1991) description of white-water rafting, Mo wrote:

I like the feeling of feeling free and wild, like [Galland] feels when she confronts nature. She knows what she wants to do, and she doesn't let fear get in her way; she overcomes fear. My fear has to do with the fear of trying new things, the fear of failure. To me, overcoming those things is like climbing a mountain or going whitewater rafting (Mo, Essay 2).

This passage makes it clear that Mo was facing big challenges as a reentry woman student.

Indeed, Mo was one of the students most baffled by the computer environment, and the risk taking involved in writing was for her a constant exercise of great courage. She was often frustrated but doggedly persistent over the course of the semester. She repeatedly had problems with saving and retrieving files, she talked to me about a long-standing preference for reading over writing, and she recalled a high school typing teacher's criticism more than 30 years in the past. It is likely that all of these factors contributed to her limited participation in the Interchange conferences.

Early in the semester, Mo had to retype one essay in its entirety twice because she twice lost her file. In an interview, she told me, "I probably ended up shortening my essay a little bit because I was just tired of typing it in." Frustrated with her inability to keep her files, she employed a strategy of sitting patiently in front of the computer at the beginning and end of each class period, waiting for someone else to help her open or save and close whatever file she was working on; she continued this for several weeks.

In our interview, she also told me:

> I definitely need some writing, because I hate writing, and this is like a challenge to me. I mean, I love reading. I can sit and read and read, for days and hours, but to sit down and put it in writing, I just can't do it. . . . This is new, this English, it's a challenge. And those machines . . . I mean, I've got pages of stuff written, notebooks of stuff, but just getting it into the computer.

Although Mo said she had difficulty with writing, her comment that she had "pages of stuff written, notebooks of stuff" shows that it was the computer that was the problem and that she had no difficulty in generating text.

Keyboarding was not a problem for Mo, though, as it is for many students new to the computer. Mo had taken typing in high school. However, her classroom experience with typing had been negative, as she told me in our interview:

> I had typing in high school many years ago. And I had a male teacher. And he would get so upset with me, and then he would say, "You better think about doing something else." I was so slow! My problem was I was too busy reading this material we were supposed to be typing, and I would just be sitting there reading.

What Mo brought to this class, then, was the kind of negative experience of writing and the keyboard typical of many basic writers. She also brought the high anxiety level and fear typical of many reentry women students.

Mo's minimal participation in the Interchange conference could be attributed largely to these fears and negative experiences, but she remained isolated in the classroom in other ways. She worked 40 hours a week and was enrolled in only two classes, unlike other students in the Reentry Women's Program. These circumstances certainly limited her networking opportunities. Typically, if she talked with another student in the class, that student was Jordan, another good writer and usually solitary worker.

As the semester progressed, Mo did become more comfortable with the computer environment, and the collaborative atmosphere of the classroom and program attracted her, despite her fears. In her final essay, written on an old typewriter at home, Mo reflected:

> I think to live life to it's fullest you have to particapate with other people, and to join in with the universe. Getting to know yourself and your fellowman is one way of getting in touch with the universe. Sharing your thoughts and feelings with other people is what we need more of (Mo, Essay 4).

Mo's determination to be a good student manifested itself in a plan for taking an introductory class in computer use before enrolling in her next writing class. It seems to me likely that if the class had incorporated Interchange sessions more frequently and extensively throughout the semester, Mo might have taken more risks. The demands of the departmental exit review, however, put a premium on the writing and revising of essays, and the instructor made scheduling choices that focused more on revising matters than class discussion and inventing.

TOWARD SOME CONCLUSIONS

Did Interchange fulfill its potential as technological utopia in these students' classroom experiences? It appears that online communities reproduce dysfunctional group behaviors as well as functional ones, codependencies as well as mentoring, conflicts based on race and class as well as appreciation of multicultural diversity, exclusion as well as collaboration, when the classroom maintains its traditional structure and dynamics. Although the classroom readings and pedagogy were anchored in feminist understandings, and the students were supported by a program that fostered networking and collaboration, these collaborative feminist practices were subverted by two factors: real-life demands on the reentry women's time and resources outside the classroom, which necessarily limited their access to technology, and grading based on departmental standards of assessment of traditional essays.

In the classroom I observed, the narrative of success—the use of a networked environment to produce individual portfolios of essays that could earn grades of B or A—was counterbalanced by the narrative of failure—resistance and nonparticipation in the same environment leading to Fs for most of the resisters. It is possible that fear of the technology posed barriers to some of the women's learning, exacerbating the anxiety they brought with them and resulting in their resistance to participating in the study and their attempts to circumvent classroom requirements for producing passing texts.

Ultimately, departmentally mandated requirements for the final portfolio drove the course, creating more pressure on students and instructors to allocate class time to drafting and revising of the essays that would count for assessment. Thus, assessment requirements functioned to devalue the collaborative conversation either as a goal in itself or as a means to producing better essays.

However, a careful look at Peaches's and Mo's experiences suggests the potential of the networked environment to foster academic success, bolstering the more obvious success stories shown in Gail's texts and the assumption of authority shown in Malaysia's Interchange posts. Peaches networked with students who failed the course, but she also made connections with the students who passed. Handing in her portfolio at the last minute, she earned a C in the course. Mo's experience of technology as an obstacle may have exacerbated the isolation she already experienced: she had a full-time job and was a part-time student, not fully part of the cohort that took a full schedule of developmental coursework together, and she was one of only three White women who stayed the whole semester, so racial difference may account for some of her isolation. Mo's actions, however, indicated a powerful commitment to succeed in the classroom, typical of reentry women students, and her Interchange posting suggested her desire to join this community of her peers.

Growth in the texts of writers like Gail should encourage optimism about the power of a supportive online conversational environment for generating traditional text. Peaches's use of Interchange suggests the powerful possibility for movement of basic writers from codependent group behaviors into connection with a success-oriented peer group. Timid group members like Mo may be able to interact more fully with peers in a supportive online conversational environment.

We are realizing daily the potential of interactive computer environments for connecting beginning writers with their peers and others as a real audience for their writing, for making rhetorical concepts concrete, and for showing writing as a dynamic activity. Synchronous conferencing places the emphasis on the process of making meaning, and its use is likely to be limited or seen as irrelevant when assessment rests exclusively on a set of standard writing products. Because of the potential of computer conferences to foster successful communication through writing, we should value the texts produced in such conferences and include them as texts to be assessed in final portfolios. If final grades had rested, at least in part, on class members' ability to contribute meaningfully to classroom conversation, marginal and isolated students would have felt a stronger incentive to participate. They might have written more and developed their ideas more fully in online conversation. Such meaning making authorizes students as successful participants in academic environments and should not be undervalued.

Our students figure out what we value, and they choose to adopt, or not to adopt, the values our classrooms embody. We can use conferencing as a novel means to a traditional end, or we can rethink the end itself. The truly transformative impact of interactive environments may finally be a reconsideration of what genres count as academic discourse.

9

Using a Virtual Museum for Collaborative Teaching, Research, and Service

Jo B. Paoletti
University of Maryland, College Park

Mary Corbin Sies
University of Maryland, College Park

Virginia Jenkins
Chesapeake Maritime Museum

For some of us, our first glimpse of the Internet was love at first sight. Perhaps it was the power to access and project primary source collections of images and manuscripts, or the Web's ability to visualize—through the miracle of hyper-links—interdisciplinary connections between different facets of our American studies scholarship. But the most enduring advantage to us (so far!) has been the ability to create—and connect—interdisciplinary scholarly communities. The Internet offers a way to achieve in practice what we have taught in theory for so many years: that American Studies fosters the connecting frame of mind (Mechling, 1979). We bring scholars, students, and resources together in dynamic combinations to tackle complex historical and cultural ideas.

Most of us learned our craft within the paradigm of the lone humanities scholar whose work focused on a single period or text or person. If he or she had help, it existed within a structure of a hierarchical pyramid, with the scholar at the pinnacle and students below. Interdisciplinary scholars following this model found themselves stretching to master several disconnected fields and working alone on projects that cried out for collaboration. The authors of this article are scholars with different areas of expertise within the larger fields of material culture and American studies. We all study different kinds of *things*—food, clothing, houses—to understand and critique American society and cultures. Our one common activity was not research but teaching; we have all taught the same introductory American studies course, Material Aspects of American Life. Based on what we have observed and experienced in the classroom, we will argue that one of the most important potential effects of the Internet will be its transformation of humanities scholarship from a mainly solitary pursuit to one that will depend increasingly on collaboration. Furthermore, we believe that this collaboration will gradually break down the traditional hierarchical paradigm and blur the divisions between scholars, students, and the interested public.

The project we describe below, Virtual Greenbelt—or VG, as we call it—has involved not only the three of us but also four graduate students, two alumnae, and half a dozen undergraduate assistants during the past 5 years.[1] We have created this virtual museum and pedagogical Web site, working in partnership with each other and with volunteers and staff at a nearby museum. There have been several happy results: We have built not only a Web site but also a web of relationships, and we have discovered solutions to many of the pedagogical problems that led us to the Web in the first place. These include the following:

- Developing more effective learn-by-doing techniques for teaching material culture studies by having students learn fieldwork methods and apply them to research projects of their own, which more closely models how scholars specializing in material culture do their work
- Teaching visual literacy
- Fostering collaborative learning and modeling the collaborative scholarship that occurs in the museum world, where much research in material culture studies is generated and presented to the public
- Teaching critical thinking
- Teaching interdisciplinary thinking and problem solving
- Improving students' writing and research presentation techniques, which in material culture studies includes developing individual and collaborative exhibitions

In addition to these achievements, our work on VG has enabled us to realize an important academic ideal: a project that integrates teaching, research, and service. Much to our surprise—and delight—VG has involved us in truly interdisciplinary collaborative scholarship and helped our American studies department reach out to a local community with an innovative and mutually beneficial community-based learning project. But with this more complicated mode of scholarly work, we also find ourselves pondering new challenges, many of them familiar to other online educators. Among these are the following:

- Issues of intellectual property and copyright: Who owns VG?

---

[1]The graduate students are Ann Denkler, David Silver, Sandor Vegh, and Psyche Williams-Forson; the alumnae are Katie Scott-Childress, current curator of the Greenbelt Museum, and Dr. Joan Zenzen; and the undergraduates are Arthur B. Hobson II, Adam Fegely, Tom Klancer, John Morsa, and Jason Schlauch.

- Issues of creative control: Who determines the appearance and content of VG? The faculty directors? The graduate students who have labored long and hard on the site? The museum's professional curators? The Friends of the Greenbelt Museum (FOGM)?
- Issues of strategic planning: How can we plan and coordinate the rational development of the site? How do we anticipate turnover in personnel? To what extent does a collaborative project need to become institutionalized to survive?
- Issues of leadership: Who is in charge? How are decisions made? Who maintains the site? Who fields questions from Internet users about the site?

We are a long way from being able to supply definitive answers to these questions, but perhaps this article is a beginning.

BACKGROUND

The focus of this chapter is Virtual Greenbelt (http://otal.umd.edu/~vg), an online museum that has been used in three of our undergraduate classes since 1995. VG serves three separate but related purposes. The site is an electronic version of an existing museum in a nearby town, the Greenbelt Museum. Like many other museum sites, it serves to publicize the museum and to make its exhibits and collections available to a larger audience than would otherwise be possible. Virtual Greenbelt is also a pedagogical platform, providing resources and project space for a variety of educational purposes connected with three different American studies courses. The third purpose has evolved along with the site and was not a development we foresaw. VG has helped those of us who have designed it and used it in our teaching create an effective interdisciplinary collaborative community-based project.

VG owes its existence to Jo Paoletti's frustration with the temporal and spatial limits of the traditional classroom. She was teaching a course in material culture studies: the analysis of objects to understand and interpret the human experience. The final project for her students was a term paper and an exhibition based on the paper—a display, actually. The students had no advance time to set up their exhibits because there was a class using the room right before them. The 25 or 30 displays overflowed the space. It was difficult to evaluate all of the exhibits in the span of one 75-min class period or for the students to visit their classmates' projects and give them feedback in any thoughtful way. Jo expressed her frustration to a colleague in computer science, outlining her dream: a digital exhibit gallery, where students could create and display their projects over the span of an entire semester, with time for feedback and revision. He said, "Have you seen Mosaic?" And Virtual Greenbelt was born.

FIG. 9.1.   The real Greenbelt Museum occupies a small one-bedroom townhouse built in the late 1930s. Courtesy: Greenbelt Museum.

Rather than create an online museum from scratch, we approached the Greenbelt Museum, a historic house museum in a nearby town, 4 miles from campus (see Fig. 9.1). It is very small, housed in one of the original structures in one of three "green towns" built during the Great Depression of the 1930s by Franklin Roosevelt's Resettlement Administration, part of the New Deal. The museum is supported by the City of Greenbelt and FOGM. Like many small museums, it operates on a modest budget; at the time we began the project, the museum had just hired its first 2-day-per-week curator.[2] It is open to the public 4 hours a week, on Sunday afternoons, and depends on volunteers to provide tours to visitors and school groups on Sundays and at other times by special arrangement. Despite its small size and modest appearance, the Greenbelt Museum interprets an important moment in planning history and in the evolution of the American suburb. In addition to its historic International Style building, set in an innovative planned community, the museum's collection includes furniture, household items, toys, and clothing typical of a White lower middle-class—but upwardly mobile—American family of the late 1930s and 1940s. The museum is rare among historic house museums in presenting the lives of ordinary American families during two of the seminal events of the 20th century, the Great Depression and World War II. The museum collection is supplemented with hundreds of photographs of town and community life, planning and construction documents, oral histories, and the town archives, located in the public library across the street.

---

[2]The City of Greenbelt agreed to support a full-time curator position, effective January 2001.

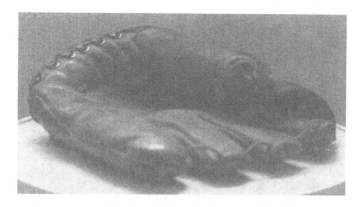

FIG. 9.2.   Objects such as this well-used baseball glove formed the nucleus of the original Virtual Greenbelt collection. Courtesy: Greenbelt Museum.

Much of the work on the first 2 years of the development of VG was devoted to photographing and scanning as much of the museum's collection as was practical (see Figs. 9.2 and 9.3.) The materials cost was low—about $200 for film and processing. But it was, as many of our readers are aware, a very labor-intensive process and one that was accomplished with volunteer labor, mainly that of Mary, Gina and Jo, Ann Denkler, the original curator of the museum, and two graduate students, David Silver and Psyche Williams-Forson. But creating an online museum was only the beginning of the process. The next challenge was to incorporate those objects, photographs, and documents into classroom learning and activities. To help with this process, we applied for and received a grant from the Center for Teaching Excellence, the College of Arts and Humanities, and the Vice-President for Academic Affairs Office of the University of Maryland campus.

Each class uses VG for a different purpose, but in none of the classes is Internet technology a central concern. For example, AMST 205 (Material Aspects of American Life) is an introduction to material culture theory and methods; Mary and Gina's AMST 205 classes interpreted objects and living spaces from the Greenbelt Museum singly and in thematically related groups. Mary's HONR 259J (American Suburbia) was more concerned with how the community interacted with the new suburban landscape; students did primary research on early features of Greenbelt's original architecture, planning, and social organization, using plans, photographs, and other archival materials housed in the Greenbelt Public Library or on VG. Jo's course, AMST 418D (Growing Up American) combined ethnography and material culture with projects that featured interviews with Greenbelters about their experiences as teenagers during World War II, with special attention to rationing and clothing restrictions.[3] The students had visited

---

[3]See http://www.otal.umd.edu/~vg/community/oralhistory/oralhistory.html

FIG. 9.3.   In addition to images of museum objects, students have access to drawings, oral histories, and photographs, such as this scene from a community baseball
game. Courtesy: Library of Congress.

Greenbelt, toured the museum, and read some of the AMST 205 projects online
before the interviews, which helped them place the informants in context from the
beginning.

Though their subject content differs, the courses that use VG share some common goals and teaching strategies. All are taught by faculty who specialize in material culture. Just as a literary scholar may use many different works to teach
students the discipline of literary criticism, we use our various courses to teach
students the fieldwork and presentation techniques used in material culture research. For example, students learn to analyze artifacts systematically, so they can
understand their features, functions, how well they work, and what they mean to
the people who crafted and use them. Because our courses fulfill general education requirements for the campus, we also share a common mandate to hone basic
humanities skills in our students. We foster critical thinking, which we define as
considering a problem from many perspectives and drawing careful conclusions
through a continuous process of information gathering, synthesis, criticism, revision, new knowledge, rethinking, and revision. We also help our students improve
their writing and presentation skills. In addition, general education courses at

Maryland expose students to the modus operandi of the offering department's discipline. We use the Web and its hypertext capability as a means to help students see and develop interdisciplinary connections in their coursework: to link points within their projects, between a current and previous project, between their project and work a classmate has done, or a related site on the Web. Finally, we want our students to acquire basic skills in visual literacy, to be able to evaluate critically exhibitions and other visual presentations. VG plays an important role in achieving these learning objectives, by allowing students to simulate the research and exhibition publication processes in an intellectually authentic way.

COLLABORATION AND ITS DIMENSIONS

The notion of collaboration for learners, teachers, and scholars is hardly new. After all, group projects, team teaching, and coauthoring have been around for a long time. Still, some definitions and a bit of history will help put our work in context.

Collaborative Learning

According to Neil Davidson (1999), not all group work is collaborative learning. He prefers the term *cooperative/collaborative learning*, or C/CL, defined as "an instructional approach in which students work together in small groups to accomplish a common learning goal." Although students are still individually responsible for learning and participation, in true collaborative learning, the collaborative element is a necessary and deliberate part of the activity. The value of collaborative learning is well documented, in studies dating back to 1949. Numerous reports have established repeatedly that cooperation "promote[s] greater intrinsic motivation to learn, more frequent use of cognitive processes such as reconceptualization, higher-level reasoning, metacognition, cognitive elaboration, and networking, and greater long-term maintenance of the skills learned" (Johnson & Johnson, 1999).

It also provides students with a more realistic experience of how research on material culture is developed and presented in the museum environment. The objects and other materials from the Greenbelt Museum offer our students authentic primary sources with which to practice fieldwork techniques. They can use images of museum artifacts to create online exhibitions and illustrated papers—some of them remarkably professional looking—and participate in a more fully realized research presentation process than our temporary classroom displays could ever simulate. Part of that realism is collaboration because most actual museum exhibits are created by teams, not by individuals. We have identified several levels of collaborative learning our students experience while using VG.

The most basic level is collaborative discovery. Students work in teams to learn and practice fieldwork or visual literacy techniques. For example, we often teach fieldwork analysis by asking students to apply a technique in class to a group of artifacts from contemporary life (e.g., beverage containers). By working together, team members are able to provide a more accurate description of their artifact, understand more of its historical context and a greater range of its possible social, cultural, economic, or ideological uses and meanings. This exercise enables students to glimpse the potential information an artifact tells about the people who made it or used it. We also employ collaborative learning to teach visual literacy. In HONR 259J (American Suburbia), students analyze photographs together and learn about the production processes—image composition, printing techniques, cropping, digital manipulation, and so on—that shape the final image. For example, students can study Resettlement Administration and Farm Security Administration photographs of Greenbelt from the VG image bank to evaluate how government photographers represented the model New Deal community. They can examine how those images have been used and reused differently over time and compare the images of life in Greenbelt projected in the photographs with the actual experiences of residents captured in oral histories mounted on the VG site.

A second level of collaborative learning is peer editing. As with collaborative learning in general, the value of collaboration in the writing process is widely recognized. Advantages include reducing students' writing anxiety, reinforcing the importance of addressing a particular audience, encouraging multiple revisions and drafts, and establishing a habit of critical self-evaluation (Living Lab Collaborative, 1999). The AMST 205 (Material Aspects of American Life) classes are taught in a teaching theater with 24 terminals. Each student creates a personal Web page for the course, where all the assignments for the class are posted during the semester, easily viewed by the instructor and their classmates, for critical review and revision. In Mary's classes, students post drafts and complete a formal critique of another student's project, providing feedback and specific suggestions for revision prior to the final due date of the project. This has resulted not only in better projects but also in more revisions with less coercion. In fact, some of the students have continued to revise and maintain their sites after the term has ended.

In addition to peer editing, students work in pairs or larger teams to complete complex research tasks. A very successful exercise in AMST 205 has been the examination of a super artifact, such as the radio or refrigerator, with each student in the class engaging in a short research project about one aspect of the object (see Fig. 9.4). These aspects include technology, manufacturing, advertising, entertainment, family dynamics, gender issues, and communication, for example. Each person contributes a piece of the research, and in pulling the separate components together for the final Web site, students explore the multiple relationships between factors that shaped the artifact's production and use. This assignment requires a heavy time commitment from the instructor, but fosters much more interaction between teacher and students, and between students and students. It has generated a sense of collegiality that is often lacking in

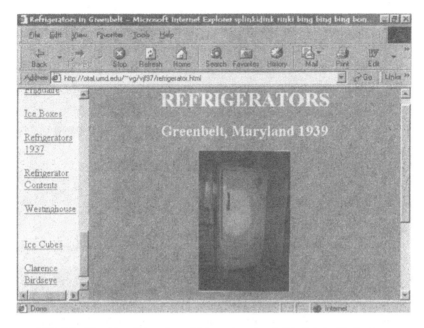

FIG. 9.4.   For this collaborative exhibit, each student in Virginia Jenkins's class researched a different aspect of the refrigerator in the Greenbelt Museum kitchen. Courtesy: Virtual Greenbelt.

the regular classroom situation. The device of having to work out technical problems to produce the final Web site has engaged students in more intense discussion of content issues as well.

The final project in AMST 205 takes collaboration to another level, working with people beyond the classroom. This project asks students to choose an object from the museum and to design an exhibit for Virtual Greenbelt. Students meet with the curator and are expected to visit the museum at least once to select and begin study of their object. They are asked to describe, analyze, and interpret the object in the context of Greenbelt in the period 1937–1945 as well as today. Students may work singly or in collaboration with others, but in either case they must link their exhibit with at least two or three others so that a visitor might grasp the connections between them and find additional topics of interest at the site (see Fig. 9.5). In pursuing their research, students worked with the curators to gain additional resources or corresponded with former residents of Greenbelt by e-mail or in face-to-face interviews to obtain pertinent primary material. The informants, in some cases, have helped with the final revisions of the projects based on their interviews; similarly, student research projects have spurred informants to undertake additional research or think about Greenbelt's history.

One of the reasons this community-based learning project succeeds is that students can see that their research is important to the museum. Paper copies of their

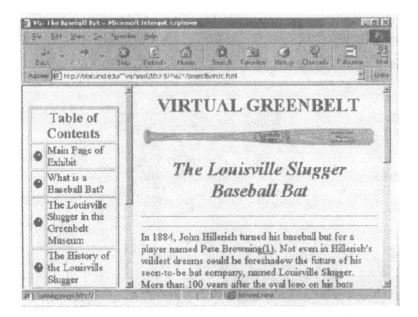

FIG. 9.5. This project on a baseball bat in the Greenbelt collection was one of three interconnected projects about baseball in early Greenbelt. The other students chose the fielder's mitt and the town ballfields (see Figs. 9.2 and 9.3) as their artifacts, and the three students collaborated on a single exhibit. Courtesy: Virtual Greenbelt.

projects are kept in an archive to be read by volunteer docents and used by the curator to support future exhibits. The best projects are chosen for permanent inclusion in the public museum Web site, available to viewers throughout the world. Students occasionally receive e-mails from Greenbelters or other visitors to VG, praising their exhibitions and asking for more information.

To these familiar forms of collaborative learning, we would like to add a new variety, self-collaboration. At the end of the semester, we ask students to prepare a Web-based portfolio of their work.[4] Web-based final portfolios have become standard features in many of our courses using VG, not only because they encourage thoughtful revision but also because they afford each student the opportunity to engage in a dialog with his or her earlier self. Perhaps some would disagree that this represents collaboration, and perhaps it doesn't, in the strictest sense. But as humanities scholars, we want to encourage self-reflexive learning, by giving students the opportunity to assess their own intellectual growth.

The final portfolio generates self-reflexive learning in at least three ways. Students revise each major project for the portfolio. Because projects have been

---

[4]For an example of a portfolio assignment, see http://www.otal.umd.edu/~vg/msf99/homework/portfolio.html

posted and discussed all semester long, students can draw from a range of sources—peer feedback, the instructor's comments, and their own new knowledge and rethinking—to stimulate the critical thinking needed for their revisions. A second component of the portfolio requires students to craft links to the projects of classmates. For example, a student revising a project on depression glass might link to a project on dining during the Depression era, another on food rationing during World War II, and a third project on depression glass or fiesta ware that argues a contrary point. In each case, the student revises his or her own text to engage with, build upon, or argue against the linked project, affording another opportunity for reflection and rethinking. A third opportunity for self-reflexive learning occurs when students write a brief essay summarizing what they have learned from the course and integrating each feature of the portfolio. In American Suburbia (HONR 259J), for example, students complete three projects that showcase different American studies' research techniques: an artifact analysis; a study of representations of suburbia from film, journalism, television, music, or literature; and a primary research project on the history of Greenbelt. In their summary essays, students delight in recognizing genuine changes in their thinking during the course of the semester, and we delight in watching them forge those insights by integrating different disciplinary perspectives developed in their own projects.

Collaborative Scholarship

By now, the benefits to students of Web-based projects are well known. We have heard repeatedly how the medium encourages active learning, recasts students as producers of knowledge, facilitates group efforts, and more closely models the ways professionals in various fields do their business. To these familiar benefits, we would like to add one from the faculty perspective: the integration of research, teaching, and service into collaborative scholarship.

Research, teaching, and service are the Holy Trinity of professional advancement in American universities. For those of us who have always chafed under their implicit separation, the discovery that projects such as VG can provide a venue for all three is particularly exciting. In the first years of its existence, VG erased the boundary between teaching and service, particularly as we selected the best of the student exhibits to create a virtual museum serving this important historical planned community. This year, Jo will use VG to conduct ethnographic research on home sewing during the Great Depression and World War II, using forms and follow-up online interviews to gather information. Undergraduate students will assist with the interviews and the resulting VG exhibit. Mary's students will produce a series of research projects ruminating on how to create better residential communities. We plan to mount a new series of oral histories the museum curators have obtained from Greenbelt's first generation of children and link to the Greenbelt Public Library's Web site which provides the complete run of the *Greenbelt Cooperator*, the community's weekly newspaper. It has become impossible to characterize the VG project as research, teaching, or service; it is scholarship, broadly defined, with a variety of potential outcomes for different audiences.

Moreover, the faculty and graduate students in our department have been drawn into ever-increasing collaboration, with VG as the center of our efforts. Before VG, the American Studies department was a fairly typical set of faculty, each focusing on his or her own specialization, interacting mainly over theses and dissertations. Because VG required a tremendous amount of work and could be used in different classes, it fostered the evolution of a working group—the Virtual Greenbelt Collaborative—composed of faculty and students, which stayed active for 6 years. Other projects and working groups have spun off from the Virtual Greenbelt Collaborative, but there is considerable overlap among the projects. We have copresented papers and coauthored articles. When graduate or undergraduate students have assisted with the class, the result has been team teaching. Our relationship with the staff and volunteers at the Greenbelt Museum has grown beyond the original quid pro quo arrangement (we get to photograph their objects; they get to use our students' research) to collaboration on exhibits and grant proposals and involvement in policy considerations.[5]

THE FUTURE

Virtual Greenbelt has evolved from a display space furnished with a few photographs of museum objects to an ongoing project that is part classroom, part museum and—with the combination of image bank, oral histories, primary documents, and the online ethnography project—part research lab. This transformation has affected our students, the Greenbelt Museum community, and ourselves, not only in the kind of work we do but also in how we work and the ways we relate to each other. It has been an exciting new adventure and mostly positive.

However, as we all know too well, these advances bring with them new concerns and controversies. We face several ongoing challenges at this time. The first concerns issues of intellectual property and copyright. Who owns VG? How can both the university and the museum protect their intellectual property? How much can or should student work be edited? By whom? How do we attribute credit for the resulting exhibit? At present we ask students to sign permission forms granting VG the right to edit and exhibit their virtual exhibition course work. We pledge, in turn, that we will not alter their point of view during the editing process; student exhibitions will appear under their own bylines. With the Greenbelt Museum, we maintain a gentleman's agreement that the museum curators will grant permission for use of images or texts of primary

---

[5]As this chapter is going to press, a second member of the Virtual Greenbelt Collaborative, Psyche Williams-Forson, has been elected to the board of directors for the Friends of the Greenbelt Museum.

sources from their collection and retain the right to review student exhibitions for inclusion in the public face of VG. The university maintains control over all virtual exhibitions created under its aegis.

Underlying these intellectual property issues is the awareness that if a project of this scope and complex nature merits copyright protection, it should also merit recognition as scholarly publication. We are long past the point where we should have to argue for the acceptance of electronic works as scholarship; in fact, large-scale electronic works are often more intellectually demanding than traditional research and publication. In our opinion, a close reading of the intellectual property discourse reveals concern about value and status, not remuneration.

Closely related to intellectual property issues is the problem of creative control. How much control should FOGM and the museum's curators have over the appearance and content of VG? Both FOGM and the Greenbelt Museum have stories they want to tell about Greenbelt. Do we tell only those, or as scholars do we have a responsibility to offer alternative voices and other perspectives? This tension has always existed between the academy and practitioners in the field, but as long as we were guest-curating on the latters' turf, the lines of authority and ownership were clear. If a student creates a high-quality exhibit that is unflattering to Greenbelt, should FOGM or the museum be able to prevent it from appearing on VG? Our gentleman's agreement with the Greenbelt Museum gives curators the right to review student exhibitions for inclusion on the permanent site. During our 6 years of collaboration we have not yet had a serious disagreement over content issues. Both the curators and the VG collaborative have rejected student exhibitions for quality reasons, however.

More tension has been generated by aesthetic issues than content issues to date. Who should control the look of VG when graduate students labor long and hard to provide the technical formatting for the Web site, but museum curators feel a proprietary interest in the appearance of what has become the most widely distributed source of information about the museum? This has boiled down to a resource issue for both VG and the museum because neither entity has the budget to employ a professional designer. Currently we are compromising on a graphic redesign. Curator Katie Scott-Childress has mocked up a new look for VG based on historical materials from the 1930s, and our web master, graduate student Sandor Vegh, will implement the new look gradually, as he finds time. In these circumstances, it is important to maintain open, honest communication with the museum and the friends' group, so that everyone understands what a redesign entails and no one is ever taken by surprise by changes in the appearance or content of VG. It has helped tremendously to formalize these relationships; the VG collaborative feels very gratified that FOGM has selected one of its faculty members as vice-president and has elected a second VG member, a graduate student, to the board of directors this year.

A third set of issues we struggle with, perhaps less successfully, concerns strategic planning. With so many collaborators, and no firm managerial structure, how can we plan the rational development of the site? It is hard enough for a lone researcher to discern when it is time to redirect or close out a research project. How do we accommodate the inevitable ebbs and flows of interest and energy, not to mention turnover in personnel? To what extent does a collaborative project need to become institutionalized to survive? We struggled with these issues this past year, when none of us used VG extensively in any class. We took advantage of the hiatus to do some long-overdue site work, with the help of a paid graduate student, and in preparation for two national conference presentations on the VG project. We feel significant pressure to apply for external grants to institutionalize the project and to garner the financial and administrative resources it would take to elevate VG to a place among the nation's leading teaching and research Web sites. At the same time, our resources for maintaining the site are stretched so tightly that it is challenging to find time to prepare a competitive grant application. Alternatively, if the day comes that VG is no longer in use for our classes, and no more primary sources or projects are being added, we will need to have the energy and motivation to write its final chapter. What, then, will be our obligation to FOGM and the Greenbelt Museum?

Logically connected with issues of strategic planning are issues of leadership. Who is in charge? Who maintains the site? Who manages the correspondence connected with VG? Who gets credit for work on the site? The energy generated during the first few years was enough to drive the project forward, and the particular classes using VG in a given semester largely determined its direction. The casual structure of VG has been enjoyable, but at some point, we need to formalize a leadership structure, and we should appoint an external advisory board. This is particularly vital if we are to compete successfully for funding. The most likely candidate of the three of us to assume leadership of the collaborative is also the one with the least amount of free time (of course), leaving the leadership issue unresolved as this article was written. If collaborative electronic scholarship has an Achilles' heel, it is its ability to provide for its own perpetuation. This issue is complicated by our observation that, to date, our work on electronic projects has not garnered the same recognition in the academic reward system as our more conventional research efforts.

CONCLUSION

In the last 6 years, VG has been a vital part of our scholarship service, and teaching in American material culture. The collaborative nature of the project has transformed the way we define and practice that scholarship, in ways that we have only now begun to grasp. The ultimate question we face at this moment is not "Is it worth doing?" but "Can we go further?" Having experienced the value of collaboration to our students and ourselves, any step back to the days of being solitary practitioners seems unthinkable. If VG—and the many, many other similar academic sites now available—represents an improvement in research, teaching, learning, and public service, why would we even consider returning to the old ways?

But clearly we have come as far as we can with existing resources and within the existing paradigm of humanities scholarship. The demand for funding of electronic projects in the humanities quickly outstripped available grants and foundation support, and competition will only intensify in the future. One answer to the resource issue would seem to be more collaborative projects with multiple uses. The main obstacle to moving in this direction remains the traditional reward system in humanities scholarship, which privileges the solitary scholar. How do we encourage, support, and reward collaboration? How can we involve graduate students and adjunct instructors in this innovative scholarship and at the same time socialize them about the traditional expectations surrounding merit? Both continued innovation and maintenance of existing projects like VG require energy, commitment, and, most of all, time. In the face of barriers to teamwork, or the threat of being denied promotion because one produced a collaborative Web site instead of a monograph, the path of least resistance would lead back to the solitary carrel in the library.

But we cannot countenance that solution. For the first time in our professional careers we feel we are modeling the interdisciplinary potential of our field through a creative fusion of research, teaching, and service. We are poised between the forward momentum of innovation and exploration and the retreat to the old conventions with all their limitations. Should we go further? Absolutely. Can we go further without resolving the thorny issues of ownership, collaboration, and academic reward? Probably not. Humanities scholars involved in creative electronic collaborations clearly have their work cut out for them, changing the conventions of their fields and disciplines to enable innovative scholarship to flourish. But there is strength in numbers. One encouraging sign is the establishment of centers or institutes for information technologies in the humanities (e.g., the Maryland Institute of Technology in the Humanities, founded at the University of Maryland campus with an NEH Challenge Grant in 1999).[6] Such centers can help garner support for needed changes in the profession by nurturing these emerging modes of collaborative scholarship, sponsoring problem-solving forums, and directing the efforts to update outmoded academic policies. Although we have serious issues to resolve as we continue to develop VG, we are confident that we are moving in the direction that will best serve innovative work in American studies. Ironically, it turns out that the toughest challenges of online scholarship are not technological.

---

[6]The MITH Web site can be found at http://www.mith.umd.edu/ Jo Paoletti has held a MITH Fellowship and Mary Corbin Sies is presently a member of MITH's Internal Advisory Board.

10

Across the Cyber Divide: Connecting Freshman Composition Students to the 21st Century

Dagmar Stuehrk Corrigan
University of Houston-Downtown

Simone M. Gers
Pima Community College

In the science fiction novel *Neuromancer,* William Gibson (1984) coins the term *cyberspace,* sharing with readers his vision of a virtual frontier:

> Cyberspace. A consensual hallucination experienced daily by billions of legitimate operators, in every nation, . . . A graphical representation of data abstracted from the banks of every computer in the human system. Unthinkable complexity. Lines of light ranged in the nonspace of the mind, clusters and constellations of data. Like city lights, receding. (p. 51)

Although other metaphors for the spaces of Internet life have come forward—the information superhighway, for instance—none seems to have the aesthetic signif- icance of Gibson's, a feature too often ignored in analyses. Indeed, *Neuromancer* is often considered the most extreme of future visions; other possibilities, like that outlined in William Mitchell's *City of Bits* (1996), do not go so far in imagining such a richly and holistically technological world. We begin with *Neuromancer,* in part, because we find the term cyberspace most appropriate for this chapter but also because we ourselves fascinated by the romantic appreciation with which readers appear to view the Gibsonian cyberpunk and the prominence of a guide, or necromancer, in the novel. Like the necromancer for Case, we seek often to be guides for our students, and we too enjoy imagining the different ways that cyberspace can be a home for innovative teaching and learning.

The pedagogical foundations for those who teach in electronic environments are primarily based on social constructionism, and a relationship between the postmodern notion of the self and social constructionism can be drawn. According to Kenneth Bruffee, "A social constructionist position in any discipline assumes that entities we normally call reality, knowledge, thought, facts, selves, and so on are constructs generated by communities of like-minded peers" (quoted in Ward, 1994, p. 50). In the electronic classroom, students have access to a wider range of knowledge and experiences from their peers and participants through the use of e-mail, discussion boards, and other tools that allow collaborative exchanges to take place without having to meet face to face. Students and instructors create a discourse community where boundaries of what used to be known as the classroom shift beyond the physical confines of space. When the traditional, socially constructed classroom community consisting of like-minded peers is transformed by a technologically enriched environment, what results is a not-so likeminded, socially constructed community of participants connected by electronic strings, which, according to Marilyn Cooper (1999) in "Postmodern Possibilities in Electronic Conversations," just creates rhetorically different spaces for postmodern selves to interact. The major difference between the like-minded, socially constructed, traditional classroom environment and the electronic, socially constructed environment Cooper claims is that the electronic environment "sets up a different rhetorical situation and encourages different writing strategies than writing for print technologies does" (p. 141). Students in the electronic classroom must become rhetorically savvy; they must rhetorically multitask because they are limited to conveying meaning only through text and because body language, tone, and placement are limited. But students may, as Sherry Turkle has studied extensively in her works, recreate themselves in the word-likeness in which they desire to be created.

Although some students may enjoy the opportunity for creative transformation of self, some critics of the electronic classroom and postmodernism claim this approach of shifting boundaries and knowledge as a sociopolitical construct tends to fragment society and lead to an array of selves for the individual that may be confusing. Sherry Turkle (1995) claims:

> We must understand the dynamics of virtual experiences both to foresee who might be in danger and to put these experiences to best use. Without a deep understanding of the many selves that we express in the virtual we cannot use our experiences there to enrich the real. If we cultivate our awareness of what stands behind our screen personae, we are more likely to succeed in using virtual experience for personal transformation. (p. 269)

Turkle warns instructors of what could go awry in the virtual, especially for those who do not see these virtual selves as tools for facilitating self-knowledge. From the postmodern perspective, it is human nature to revise self, especially because our actions are met with the placement of others within the socially constructed environment and their body language and intoned responses. These language revisions of self can be transformative, and recreating and revising aspects of self in an electronic classroom environment can be transformative, too.

In "Self-Structure as a Rhetorical Device: Modern Ethos and the Divisiveness of the Self," Marshall W. Alcorn, Jr. (1994) writes about how the self is positioned. Alcorn asserts that rhetoric can be controlled and used to develop an image of the self in addition to social interactions. Alcorn proposes that the self is a stable organized collection of voices rather than a random mixture of chaotic influences. Alcorn further claims that rhetoric and experiences create self. He summarizes Amelie Rorty's idea that character becomes person, and here person is defined by the social expectations with predictable roles and behavior:

> Rorty's work encourages us to acknowledge that different cultures not only imagine and define selves differently but also formulate social and cultural conditions that permit the creation of different selves. Within each social context is a reciprocal relationship between the self that a culture imagines and the shape of self-structure that is lived. (p. 11)

Cooper (1999) agrees with Alcorn and Rorty and about the context of the electronic classroom:

> Intellectual self realization, in the tradition that stretches from Plato to Wordsworth to Peter Elbow, is a process that results in the discovery of the universal forms of truth that defines knowledge, and that relies on developing the thought processes of the individual in line with these universal forms. In electronic conversations, the individual thinker moves in the opposite direction into the multiplicity and diversity of the social world, and in social interaction tries out many roles and positions. (p. 143)

The electronic environment provides an opportunity for experimenting with that which is not like the real self, that which is a revision, a recreation and, yet, a creation of self.

To honor the multiple selves that are created and documented in this textual environment created by the electronic classroom, we agree with feminist scholars who critique the basic tenets of social constructionism in that this socially constructed space perpetuates the current hegemonic distribution of power that is mirrored by our society. We propose revisiting and revising expressivism as defined by Donald Murray as a way for the previously underrepresented and disenfranchised, as represented by student populations found in community colleges and open admissions universities, to position themselves in the academic discourse community. Murray writes: "When we discover what we have said, we discover who we are" (quoted in Ward, 1994, pp. 21–22). Murray believes that as writers find their voices, they discover their identities, and they are motivated to continue exploring "the constellations and the galaxies which lie unseen within us waiting to be mapped with our own words" (quoted in Ward, 1994, pp. 21–22). Even though Murray presents a neo-romantic view of the self, we revise that quest for inner knowledge by proposing that expressivisim is a heuristic through which students may expose the many facets of the postmodern self. In the context of an electronically mediated, socially constructed classroom experience, an expressivist approach is one way through which postmodern selves are exposed and a pinnacle of understanding becomes attainable.

Through expressivist writing assignments, such as the profile and the cyberfriends project, students voice identities that are arising and developing. Personal experiences for these beginning writers cannot be discounted, and instructors who honor students' literacy experiences use that honor as a way to facilitate students' learning to position themselves in the academic discourse community. Students' expression of selves are grounded in direct experience, and, as expressivist writers in electronic environments, they are given agency as to how they choose to represent themselves. Rather than the banking model (transmission model) of education decried by Paulo Freire, in which students are " 'receptacles' to be filled by the teacher" (quoted in Ward, 1994, p. 92), facilitating students' self discovery in electronic environments with expressivist strategies at the beginning of a writing course facilitates their taking control of their learning, risk taking, and developing agency in this new discourse community. Students for whom personal exploration is facilitated take control of their learning and decide what they want to write about and how they want to frame their discussions. They take risks by trying out new selves in the often new and personally uncharted environment of cyberspace. They develop agency as they create in this socially constructed, electronic environment. Though new to academia, academic writing, and electronic environments, students are often willing and excited by the opportunity to recreate or create self. However, because some students do not fully comprehend the socially constructed intercontextual rhetorical context, the established rhetoric of the academy, or the Internet, they do not understand the parameters of context and experience disequilibrium and sometimes rhetorically miscue. These rhetorical miscues, as Ken Goodman (1973) points out in his important work *Miscue Analysis: Applications to Reading Education*, tell us what students know and how they come to know. These rhetorical miscues allow teachers to value what students know and need to know and facilitate teachers' supporting students' creation of agency (Goodman & Marek, 1996, pp. 203–207). These miscues as well as the experience of disequilibrium occur when neophytes are exploring who they are and who they want to be, and they provide Vygotzskyian zones of proximal development.

Critics of expressivisim focus on the disasters that have been analyzed in recent research when students expose themselves to unaware and unintentional audiences; however, in a postmodern electronic environment when emerging, fragile selves are codified online and reflected back, expressivism is a lens through which the writer may view, analyze, and reflect on selves, and through which students become reflexive to these many selves. In this manner, we extend Murray's application of expressivism and Turkle's findings of personal transformation to include how participants portray themselves in an online environment and demonstrate through Ana Floriani's (1994) ideas about contextuality and intercontextuality how expressivism, social constructionism, and radical pedagogy can be synthesized in a manner that facilitates students' development of critical literacy practices and transformation of self while bridging the gap between personal, academic, and technological frontiers.

The Cyberfriends Project expects students to develop new writing and technology skills in the context of the classroom and in the electronic environment of e-mail, using the textual artifacts from the classroom as well as e-mail artifacts from a cyberfriend to produce essays. In this manner, the cyberfriends assignment extends Ana Floriani's project on intertextuality and intercontextuality to include texts produced in a virtual environment. In "Negotiating What Counts: Roles and Relationships, Texts and Context, Content and Meaning," Floriani (1994) demonstrates how students use prior knowledge to negotiate situations in the writing classroom. Floriani demonstrates how constructed texts include what is written, what is spoken, and what oral and visual interaction takes place between members of a situational group (p. 235). She uses a definition of intertextuality Bloome and Bailey (1992) provide to set up her notion of intercontextuality:

> Whenever people engage in a language event, whether it is a conversation, a reading of a book, diary writing, etc., they are engaging in intertextuality. Various conversational and written texts are being juxtaposed. Intertextuality can occur at many levels and in many ways. Juxtaposing texts, at whatever level, is not in itself sufficient for intertextuality. Intertextuality is social construction. (pp. 181–182)

Building on the concept of intertextuality, as previously framed by Bloome and Bailey, Floriani (1994) reveals her notion of intercontextuality: "Given the view of context as meaning with text . . . intercontextuality refers not only to previous texts, but to the social situation in and through which a text was constructed. That is, prior contexts may be interactionally invoked in the local context being constructed" (p. 257). These definitions of intertextuality and intercontextuality also illustrate what happens in electronic writing environments as students instinctively employ their present literacies: Even though cyberfriend partners never meet each other, share the same physical space, or speak to each other, students learn to read contexts and employ intertextuality and intercontextuality in the production of meaningful texts. Although Floriani's study focuses on the oral and written construction of context, the Cyberfriends Project extends her work by adding cyberspace to the arsenal of tools 21st century writers use to create meaning.

THE PROJECT

For the Cyberfriends Project, students are required to write a profile of assigned cyberfriends with whom they begin corresponding through e-mail. Adapted from the profile assignment found in the *St. Martin's Guide to Writing* (Lunsford &

Connors, 1995), in which students choose an interesting person, place, or activity and write about it from a fresh perspective, our Cyberfriends Project requires students to write this profile based solely on e-mail correspondences with cyberfriends. Students would not only draw upon the rhetorical modes of description, narration, and exemplification to write the essay but would also negotiate the complications of communicating in cyberspace.

After introductory e-mails have been exchanged, students have to elicit information from cyberfriends to develop an intriguing profile of them. Ultimately, students are required to develop a fresh perspective about their cyberfriends' e-mails so that readers of the profile will discover a surprising insight about the writer's subject. To ensure that the interviewee's personality is reflected in the piece and to promote reader interest through a lively pace and tone, students must also integrate quotes from correspondence. The Cyberfriends Project promotes the development of critical-thinking skills because it requires students to present themselves in writing in a manner in which they want to be perceived. Students also learn how to draw information from their subjects, how to analyze discussions, and how to develop a focus on a specific aspect of their subjects as opposed to presenting a biographical sketch. To successfully focus on some intriguing aspect of their cyberfriends, students use interview, note-taking, observation, and analysis skills to motivate meaningful correspondence. From these information-gathering exchanges over approximately two thirds of the semester, students should be able to identify a focus for the profile. Developing a focus requires students to think critically about the e-mails they have collected and identify a theme or isolate a specific aspect and develop it from an interesting or fresh perspective. Because they can never actually speak to their cyberfriends to gain insight, students must often think like anthropologists, piecing together the story-profile based on artifact-texts. This is a new and awkward experience for most students—they are unable to negotiate information through dialogue, nonverbal communication, or visualization; they are entirely dependent on their own and cyberfriends' written communication skills. Although occasionally cyberfriends have intriguing lives that make for easy-to-write, exciting profiles, most often the focus of a profile is developed from a recurring theme in many e-mails. Sometimes the focus is found in the undercurrent just below the surface of what the interviewee writes.

THE STUDENTS

Dagmar's students from the University of Houston-Downtown (UHD) were freshman composition students from what in 1997 and 1998 was named the most diverse regional liberal arts college in the western United States by *U.S. News and World Reports'* 11th- and 12th-annual *America's Best Colleges* guidebook. UHD is an open admissions, urban university offering bachelor's and master's degrees and is located literally in the heart of downtown Houston. A commuter university, it serves about 8,000 students per semester. These UHD students were matched with those from Simone's Pima Community College's Desert Vista campus in Tucson, Arizona. An open enrollment, 2-year institution, Pima is the fifth-largest multicampus community college in the nation, serving more than 55,000 residents from Pima and Santa Cruz counties each year. Unique to the Pima system, the Desert Vista campus serves a student population consisting of 70% minorities. On the surface, our students seemed to be very different from each other, coming from distinct regions of the country and with diverse cultural backgrounds. However, what we, and our students, discovered was that the only real difference was that many of these students had never been exposed to e-mail technology before this course and assignment.

THE CYBERFRIENDS PROJECT: SOCIAL AND CULTURAL ISSUES

The Cyberfriends Project was executed during the spring, summer, and fall semesters of 1998, and although it had not changed drastically, there were variables for each semester. During the spring semester, students were matched gender for gender in most cases. Because of the large group (approximately 80 students total), we did not want to invite any potential problems with sexual harassment, especially in view of our feminist concerns about socially constructed space. This situation lead to in-depth planning and discussions between us because the potential for online sexual harassment became an immediate concern. Our concerns for minimizing any potential for sexual harassment were well founded as Floriani's notion of intercontextuality shows that students will undermine instructors' intentions of keeping assignments intertextually based while trying to minimize any potential for sexual harassment.

The most blatant example of students' ability to subvert intertextuality occurred during the spring semester when one pair of students' participation in the Cyberfriends Project had an ulterior motive besides gathering information relevant to the assignment. While Chico was busy trying to write himself in a presentable manner for the assignment, an alternative motivation for corresponding became meeting girls in Houston. Chico and Juan started discussing the women in each other's class, making recommendations and forwarding e-mail addresses to each other. Chico and Juan were drawing on their previously formed schemata of virtual environments as a means for meeting women as seen by online chat rooms and used this information to determine that this assignment provided an opportunity to extend their knowledge of the women in Tucson and Houston. Chico and Juan had these deliberate conversations, even though they copied e-mail to us as a requirement of the assignment. Their discussions then provided fertile ground for class discussion, thereby exponentializing the intercontextuality of the situation. Although their e-conduct surprised and created concern for us, fellow students were not shocked, offended, or surprised, and they willingly engaged in Chico and Juan's pursuits. The classmates' acceptance of the behavior as appropriate well demonstrates the significance Floriani's intercontextuality.

Cross-gender partnerships were tried in the summer semester in which all students were matched with cyberfriends. We decided to allow this group to choose randomly from the pool of cyberfriends to see what kind of results this would produce by comparison. The drastic affect intercontextuality can have in a virtual environment was demonstrated in the case of Linda, who was a returning, nontraditional, female student. Linda's husband was less than understanding because she was paired with a male cyberfriend; in fact, Linda's husband was seriously threatened by his wife's assignment, referring to Linda's cyberfriend as her "cyber-boy-friend." Linda's cybercommunications were, as Floriani (1994) might argue, affected by the "invisible" social histories of her husband's cultural literacies, the history of her marital relationship, and her approach to platonic social relationships (p. 271). Dagmar asked if Linda would like to trade cyberfriends with another student; however, she declined. This choice created several problems for Linda: She could only write about her cyberfriend during class time; she could only access her e-mail from school during the week; and she was under additional stress because her husband relentlessly probed her with jealous questions regarding her "cyber-boy-friend." Yet, this choice also gave Linda academic agency because she was driven by her personal goals for a college education, despite the challenges faced as a result of her overtly hostile and subversive home environment. Interestingly enough, Linda's cyberfriend, Bob, and his wife were both in Simone's class. There was no jealousy in Tucson, but the whole Pima class routinely teased Bob, who took everything tossed his way with grace and humor. Bob dealt with the class activities in an objective manner. His attitude about the assignment, including his e-mail, textual relationship with his cyberfriend, her husband's intercontextually significant jealousy, the classes' and his wife's contextual knowledge of their shared history of the text, were reflective of his personal literacies.

Students' developing notions of academic agency did not limit the influence of intercontextuality. Many of our students' experiences demonstrated that they do not separate their academic and private lives. One Pima student, Don, admitted in response to his cyberfriend's probing questions that he did not really love his wife. Don felt the need to be completely honest with his cyberfriend, so he admitted that which forced him to confront his true feelings. This confession created a paradox in Don's life, causing him to ask Simone and his cyberfriend not to comment on or use part of that confession in e-mails to his personal account because he did not want to hurt his wife, and because, more important, he did not want to be forced into having the conversation with his wife. At the same time, Don was able to demonstrate his developing academic agency as he felt confident in trusting his cyberfriend and Simone with his request for privacy. These potentially intertextual forces had an affect on Don's ability to produce other texts or as Donald Rubin notes, "the writer's text helps to define, even change, the social context" (quoted in Duin & Hansen, 1994, p. 91). In this case, the electronic format provided the veil of a confessional: The very public gesture of e-mail became privatized because it could be deceptively perceived as confidential correspondence between two people. Just as Turkle suggests, Don used his experience in the virtual to revise his experience in private life. Additionally, intercontextuality is illustrated here as Don realized that his disclosure now had direct bearing on his private and academic life, moving the parameters of the assignment beyond the confines of the classroom.

We hoped that as students wrote to get to know their cyberfriends, they would also develop self-awareness. In addition to learning how to use e-mail and compose with computers, students would engage rhetorical strategies that would help them to improve their writing throughout the semester. What we discovered through writing this chapter is that this expressivist, yet postmodernist assignment expanded more than their technological literacies; it built on their cultural literacies, expanded their academic literacies, and developed their critical thinking and writing skills.

INNOVATIVE SOLUTIONS TO PROJECT CHALLENGES

What we hoped would happen and what actually happened had us drawing on our imaginations and our abilities to organize and problem solve to keep the project going. One challenge was what to do with a cyberfriend when the other dropped the course. If this happened early enough in the semester, we were able to reassign cyberfriends, but more often than not, students dropped when it was too late. When the scenario unfolded, the remaining cyberfriend was usually left with just a tidbit of information to write the profile essay. A similar challenge arose when cyberfriends were just not able to write enough information about themselves. We resolved these challenges in ways that complemented our teaching goals and course objectives.

Dagmar's students whose cyberfriends had dropped the course or who perceived that they did not have enough information to write a cyberfriend profile were allowed two options: profiling anyone else, either through e-mail or face-to-face interviewing, or authoring a self-profile. All UHD students who were in this situation chose to profile someone else via face-to-face interviewing. More or less, face-to-face interviews became essential because students did not have the time to develop another e-mail-based cyberfriend relationship. Students who had to start over intercontextualized the e-mail experience. Using interview questions generated during the e-mail process helped these students do a much better job of soliciting information from their new interviewees. Because of their experiences with the failed cyberfriend connection, these students had a better understanding of what was necessary to fulfill the profile essay's basic features.

Simone's students who did not feel they received enough information from their cyberfriends to write a profile, whether that perception was real or perceived, were required to self-profile. Simone offered only the option of self-profiling because when she reviewed what her students sent to their Houston cyberfriends, she found they were receiving approximately the same quantity and quality of information as they were sending. During conferences, these students admitted that they would not be able to write self-profiles based on what they had e-mailed their UHD cyberfriends, even though several class periods were devoted to e-mail correspondence. Simone felt that if these students had more fully participated in the assignment, they would have received enough information to complete the profile. The students in these partnerships were, in a sense, hyperintercontextualizing because they decided that they would give only in proportion to what they received and that they would not extend themselves, doing only what they perceived as absolutely necessary. Because one of the constraints of the assignment was that all information contained in the profile must be documentable through e-mail, students who chose to self-profile had to eventually e-mail a copy of the essay to the cyberfriends they had duped. Amazingly, when some UHD cyberfriends received copies of these self-profiles, they wrote back, demonstrating the learned intercontextual value of reciprocity. Even though reciprocal correspondence was no longer required, these UHD students wrote because they were motivated by their cyberfriends' attempts to complete the assignment and felt more comfortable about revealing more of themselves.

This self-profiling strategy provoked an interesting exchange between Larry and Joseph. Larry continually tried to get Joseph to write but to no avail. After a few brief messages, including information about himself and specific questions for his cyberfriend, went unanswered, Larry chose not to continue including information about himself. Larry sent only questions to his cyberfriend until he e-mailed his self-profile to Joseph. At the top of his profile, Larry wrote, "Here is a copy of the profile I wrote on myself because you didn't respond to my e-mail." Joseph wrote back, "Had you asked me, I would have sent you information." Clearly, both students believed that, in their shared context, the other was not playing the game, or participating in the assignment. It might have seemed to Joseph that there would be no consequences for not writing because, after all, Larry was not in his class—he would not have to physically face or interact with Larry in class. But Joseph's choice not to participate became real when Larry sent Joseph his profile.

The situation between Larry and Joseph typifies the challenges of motivation and control that some students face. As instructors, we perceived the Cyberfriends Project as one in which students have complete control. One aspect of the assignment that promoted student control was that students chose what they wanted to write about themselves, thus controlling how others perceived them. They did not have to worry about any of the visual contexts that exists in a face-to-face interview, including what they looked like, how they were dressed, where they met for an interview, and how they acted. Students had control over global concerns, such as what they wrote and how much they wrote, as well as over local concerns, such as the manner in which they wrote. And because there were seemingly no real classmates or instructors watching their actions and discussion as would happen in a classroom environment, students could choose to behave in cyberspace in ways they might not behave in real time. For instance, Joseph chose not to return Larry's e-mail or respond to his questions because, quite possibly, he perceived Larry's position as nonthreatening. After all, Joseph would not have to see Larry or Larry's instructor. But it was clear that Joseph chose to open and read Larry's e-mail. Had Joseph been deleting the e-mail unopened, he would not have read Larry's final note about why he had written a self-profile. However, in his final e-mail response, Joseph reveals that he was participating as a lurker, watching and reading but not responding. Though Joseph was in control of his and Larry's interaction, Joseph could not control Larry's reaction and judgment, nor could Joseph control Larry's communication of that judgment to both of their instructors. Unlike Joseph, students like Don who choose to be more honest through cyber communication than they were in real life faced a different set of contextual consequences. Don's choice of self-expression to a seemingly anonymous and nonthreatening listener became a force that controlled him, instead of a situation he controlled. So although the assignment was intended to motivate students and to teach them to take control of the writing process, some students became controlled by their levels of motivation or lack there of.

Though students realized that everyone participating was faced with the same dilemmas about what and how much to write about themselves, some students still felt awkward discussing unsolicited information about themselves. Many students did not view anything about their lives as interesting, fascinating, or valuable. These tendencies mark as important cultural literacies, such as modesty or humility, and personal literacies, such as shyness or arrogance, in social situations. To empower students and help them to expand their current literacies, a class period was spent brainstorming problem-solving strategies for soliciting information and motivating cyberfriends to feel comfortable writing about themselves. Both instructors continually reiterated that everyone involved faced the same challenges.

Even when students were able to write to their cyberfriends, they faced the challenge that e-mail communication is much flatter than verbal, face-to-face communication. Just as in meeting a person face-to-face, many students did not know what to say or how to solicit more in-depth information after a perfunctory introduction. And though awkwardness could be present in face-to-face interviews, at least face-to-face interviews offer contextually significant opportunities for meaning making, such as facial expression, tone, nonverbal communication, and the regular chit-chat that occurs when first meeting someone that can be used to stimulate further conversation. In e-mail communication, the students could not draw upon these contextual markers. So instructors facilitated intertextual class discussions about how to motivate partners, and students shared written examples of how to beg or use guilt, kindness, and directness as catalysts for responses. Students demonstrated how they used shared context to inform social interaction by modeling how some of the same motivational strategies used in face-to-face interaction could be used in e-mail.

These class discussions allowed more advanced writers to demonstrate how tone can be affected through humor and through textual symbols, such as smiling faces and frowns that are used in electronic discourse, by sharing their cyberfriend correspondence with the class. Sometimes these strategies worked, but not always. Just as in face-to-face interviews, the tone of e-mail conversation can be misread, as was the case between Larry and Joseph. Sometimes what one student perceives as a friendly question or a smiling face can be perceived as sarcasm. In a face-to-face interview, however, when tone is misread, the opportunity for immediately setting the situation straight exists. In e-mail interviews, this opportunity exists only if the person insulted writes back. If so much damage is done that the person chooses not to write back, it is almost impossible to reopen the lines of communication, effectively silencing the other person. With e-mail communication, students cannot draw on their personal literacies for reading the interactional context available in a live interview. And though they can read the interactional context of the text, most freshman composition students we taught were new to electronic discourse and had not developed a level of literacy in that venue directly correlating to the level of literacy they possessed in face-to-face contexts.

Another aspect of the assignment that the instructors felt promoted student control was that students chose which aspect of their cyberfriends' lives was interesting. But because some students wrote short, perfunctory e-mail that provided little information, finding an idea or theme on which their profilers could focus was difficult. In a face-to-face interview, interviewers could have asked follow-up questions to sustain and develop a topic of discussion. In e-mail, students could write follow-up questions, but that did not ensure that the questions would be answered. In the face-to-face situation, there would have been more social pressure on the interviewee to respond to a waiting interviewer who asked questions in good faith. This social pressure does not exist with e-mail and allowed interviewees the option of silence.

So although students and instructors continually worked within the context of their classes and cyberfriend correspondence to develop problem-solving strategies that would facilitate success, sometimes how a student constructed shared context with a cyberfriend adversely affected the outcome of the assignment. Indeed, the manner in which students acted within teaching and learning contexts affected the cyber-discourse community that was created in each class. The relationships between cyberfriends affected not only the relationship between the partners, but it also affected what was discussed in the individual classes as problem-solving strategies for better e-mail communication were developed. Either way, students were in control of the process whether they chose to write or not.

Whether they wrote a profile based on e-mail correspondence, self-profiled, or wrote about a live subject after participating in e-mail correspondence, students were able to build on their personal literacies and develop critical literacy.

LEARNING OUTCOMES

During the planning process, the following specific learning outcomes were established as a basis for evaluation:

- Transference of skills: audience awareness, purpose, and critical-thinking skills
- The quality of students' writing and the writing process as an effect of technology
- The level of intimacy between students as an effect of technology
- Metacognitive abilities to further assess their own writing process

Of special interest was how students would match up with their assigned cyberfriends and gain insight into their abilities to produce meaningful texts in what was originally viewed as a "self-contained discourse community" (Duin & Hansen, 1994, p. 102). But as we applied Floriani's notion of intertextuality and intercontextuality, we found that these constructions were infused into our specific learning outcomes as well. We learned that although we were searching for our specific learning outcomes, we also saw why they emerged in our shared electronic classroom by looking through Floriani's lens.

Transference of Skills: Audience Awareness, Purpose, and Critical-Thinking Skills

Like most composition instructors, we find that audience awareness, understanding of purpose, and critical-thinking skills are perhaps the biggest challenges in teaching freshman composition students; however, these particular learning outcomes seem easier to achieve in this shared learning situation. With the Cyberfriends Project, students developed transferable, critical-thinking skills by drawing intertextual relations between the model essays discussed in class. Already having the experience of at least one round of correspondence between cyberfriends, the next cyber connection required students to use critical-thinking skills. They were required to e-mail a rhetorical analysis of the model essay and a discussion of how the essay satisfied the features of a profile to their cyberfriends. Electronic and critical literacies were reshaped intercontextually when students read and responded to each others' analyses within their own individual and shared contexts. This dialogic and electronic exchange of analyses provided a framework for some students' profiles, especially students who felt they did not know how to begin writing or did not feel they had enough information to write an intriguing profile of their cyberfriends.

Connected to Floriani's sense of intercontextuality, the idea of an audience was ever present and powerful because students were writing to others, who they realized would judge them by what they wrote in outside contexts. No longer could they rely on looks or behavior; the focus was intensely on what they wrote and how they wrote it. Instead of having the understanding of the instructor as the only real audience, students realized that someone else, someone strange but who was also in a similar position, would not only see but would also come to know them and write about them through their words. Some students excitedly opened their e-mail to find they did not understand what their cyberfriends were saying because of misspelled words, extensive use of slang, and incoherence of ideas. Also, some students were so shocked by how poorly written they perceived the correspondence to be that they read it to the class or made jokes about the correspondence to the whole class. We reminded the students that just as they perceived others based on what was written, so were they being perceived. These incidents brought about yet another opportunity for us to discuss audience. We suggested that students focus on the content of the e-mails rather than the grammar. This idea in turn led to a discussion of how grammar markers could affect opinions formed about an individual's character. So through textual correspondence with their partners and through intertextual class comments, students were able to contextualize their partners and realize the purpose of writing these e-mails. In her work with adolescents developing critical media literacy through after-school discussion groups, Donna Alverman, Jennifer Moon, and Margaret Hagood (1999) found that positioning is not only socially constructed but the audience situates a participant (p. 130). Abelardo, a Pima student, candidly stated in his response journal entry about audience: "Yeah, because I knew my buddy would be reading, and Ms. G., I had to be my good self and not my bad self." Many students wanted to present themselves in their "Sunday writing," if you will. Abelardo realizes that the writing occasion now extends beyond the borders of the classroom context that created the assignment, demonstrating the influence of intercontextuality.

However, this Sunday writing was not of interest to all students. Thankfully, there are rebels in every situation. For instance, audience awareness defined through the lens of intercontextuality can be fueled by cultural heritages and differences, as the intent to impress the audience proved to be a double-edged sword in the relationship between Francisco and David. David continually wrote to Francisco about material goods he possessed. From the Lexus David drove to the chrome Mongoose bike he rode as a child, Francisco perceived David as consistently trying to present himself as "better," and Francisco was not impressed. A Hispanic male from a modest home, Francisco was raised to believe that it is not polite to discuss money or material goods with strangers. He tried to refocus the conversation with David, who continually bragged about yet another thing. Put off by David's obsessions, Francisco quickly realized what his intriguing focus would be: David's desire to impress through material possessions. Unable to disassociate from his cultural literacy, therefore, Francisco could not perceive David in a positive light. David was an international student from South America who physically and emotionally appeared to be younger than his contemporaries. His naiveté and cultural background could therefore explain the content of his e-mail exchanges with Francisco. To effectively gauge their audience in this virtual environment, David and Francisco should have been able to "actively monitor their own and others' actions and language, and interpret these interactions in order to select from their repertoires possible ways of interacting that meet the local demands of the situation" (Floriani, 1994, p. 245). However, David and Francisco's individual and sometimes-conflicting contexts, coupled with the relative anonymity of e-mail, caused them to misinterpret their e-mail correspondences and skewed the sense of audience for both students. David seemed to be unaware of Francisco's attempts to turn the conversations around, and Francisco became less tolerant of David because his attempts failed. This scenario demonstrates that even though students use contextual knowledge to construct their e-mails, confusion occurs when students' sense of audience is misconstrued.

The relationship between context and text is further distorted when students seek to manipulate the audience by creating a persona to present to their cyberfriend. One student, Sergio, who was already well versed in the posturings of e-mail personas through his experiences in chat rooms, decided to enhance his image. Sherry Turkle, in "Virtuality and Its Discontents: Searching for Community in Cyberspace" (1996), reminds us that through personalities created by role-playing games "the fake seems more compelling than the real" (p. 3), which was Sergio's impetus for wanting to experiment with a faux self for the assignment. Sergio created the stereotypical persona of a young, well-to-do male, complete with the material trappings of fast cars, fast times, and fast women. Unconsciously, Sergio probably thought this would be a way of expressing what he felt was lacking from his life and be infinitely more compelling than was real life. Consciously, he wanted to add a bit of fun to the assignment. In the e-mails he copied to Dagmar, Sergio would confess which parts were the faux Sergio and which were real. In his confessions, he felt that he was truly putting one over on his cyberfriend. Dagmar wondered if Sergio's audience could tell that Sergio was not all he seemed to be. She asked Simone to discover if Sergio's cyberfriend, Victor, noticed anything. Sure enough, in class discussion, Victor admitted that his cyberfriend seemed flat with personality traits that were stereotypical of the persona Sergio was creating—there was no "life" to Sergio because there were never any details following general claims, causing most of his e-mail messages to be short. Applying Turkle's view, Sergio's "virtual experience may be so compelling that . . . [he] believe[s] that within it . . . [he's] achieved more than . . . [he has]" (p. 3). Turkle's insight into cyber personalities corroborates Sergio's belief that his faux self was real, both to himself and his cyberfriend. However, in reality, Sergio could neither elaborate on the faux parts of himself nor successfully weave them into his real experiences. Unable to write a profile based on the e-mails Sergio sent, Victor profiled himself. Although Victor was disappointed, he still continued to write to the faux Sergio and provided him with enough material for Sergio to complete the assignment. In this situation, when context is less genuine due to a forced contrast between real and virtual identities, intercontextuality becomes much more complex to observe in terms of its specific influence on meaning making. Even though Victor had to self-profile, both he and Sergio developed critical perspectives. Victor guessed that Sergio was making up information, but he did not seek confirmation. Sergio realized that he did not fully realize and communicate a persona to Victor. Both participants ended the experience with new understanding of and appreciation for critical communication.

Once audience and purpose became real for these students, they spent a lot more time composing their correspondence. They were careful about not only what they said but also how they said it. This led to a discussion on how focusing on the mechanics of writing might inhibit someone from writing back and how the informality of e-mail allows students to compose as if writing a rough draft. As Mark wrote in his evaluation of the project, "You stopped me from commenting on her grammar, which was a good move on your part." Mark was able to concentrate on the content of his cyberfriend's messages and respond to that, rather than the grammar errors that seemed genuinely to annoy him. Students began to expand the textual boundaries of the writing class, receiving feedback from both instructors on formal writing and peers on informal writing. The synergetic relationship between intertextuality and intercontextuality in the transference of either the ability to deal with global issues in an e-mail or local issues is demonstrated when students produce meaningful texts that present themselves in the manner they felt would be well received.

Another transferable skill developed was the understanding of how outside resources can enhance one's writing. Students who used many quotes, letting their subjects speak, realized how the quotes helped fulfill the fast-paced and entertaining aspects of the assignment. As students became more thoughtful and spent more time composing e-mail, their correspondence became more moving to the reader and quotable. When Greta cited from Lisa's e-mail about how her father threw her and her young son out of their house and how they lived under a bridge, readers were moved to tears. Greta was able to move her readers because Lisa had taken the time to develop her story clearly and because they developed the shared context for that correspondence to be meaningful.

It is true that students who write profiles based on face-to-face interviews could experience the same kinds of transference of skills. Any profile writer develops the ability to incorporate quotes into a text from well-spoken and thought-out verbal answers. But in a live interview, context is constructed in helpful ways that do not exist in text-only interviews. In a face-to-face interview, generally the interviewee chooses the subject to be profiled. This selection is usually based on prior knowledge and could eliminate the impetus to write a chronological account of the person's life. A profile based on a face-to-face interview could use visual texts as well as the spoken text provided. In contrast to face-to-face interviews, all participants in the Cyberfriends Project were complete strangers, known only by their writing voices. Furthermore, students who understood how audience and purpose affected a coherent piece of writing demonstrated an understanding of tone. Although tone can be easily recorded and analyzed in verbal communication through inflection or emphasis and through nonverbal cues such as expression and body language, in text-only communication an understanding of tone was difficult to develop.

The Quality of Students' Writing and the Writing Process As an Effect of Technology

One factor that heavily influenced students' perceptions of technology was the e-mail software used. Simone's spring semester students were brought to the computer lab and instructed by computer technicians how to use the free e-mail program provided by Pima. This antiquated program required typing commands to perform functions. Confusion arose when Simone required students to use different software, as both faculty and students learned how to use the system. Dagmar also had her share of frustrating technical difficulties. Even though her classes were conducted in an electronic classroom, the e-mail software available to students caused problems similar to those experienced by Simone's students. Students at UHD were able to obtain VAX accounts, and Dagmar taught them how to use Pine, e-mail software that she had been using for many years. Students learned how to use Pine quickly and, with practice, found it adequate. Allotted space on the server filled quickly for students who did not delete old messages. They found that they could receive new messages but were unable to send anything until they freed up server space. Frustration levels soared as those students who had not maintained their accounts painstakingly composed messages to their cyberfriends only to find that they could not be sent. Because we knew that the majority of our students did not have computer access at home and the current use of school-provided systems caused undue stress on the project's final outcome, we resolved to find better software for our students. During the next semester, we decided to require students to use one of the free, Web-based mail programs, which drastically improved students' technological literacies and overall personal success and enjoyment of the project.

After we resolved the software challenges, some students found that overall the technology conflicted with cultural literacies and developing academic literacies and inhibited their ability to produce meaningful texts for the assignment. We noticed that some Hispanic male students who had weak English skills were not very willing to participate in the assignment. Their attitudes could be, on the one hand, stereotypically interpreted as machismo behavior or, on the other hand, as protecting their dignity. Frank Pajares and Margaret J. Johnson (1996) in "Self-Efficacy Beliefs and the Writing Performance of Entering High School Students" demonstrate that self-perception can affect students' ability to write, and cyberspace adds yet another complication to this scenario by adding technology to English proficiency. Male, Hispanic, freshman students with poor English skills may unconsciously feel emasculated. Adding the fear of technology to poor English skills in an environment in which written English skills are the primary tool for communication in e-mail compounded these students' resistance to the assignment. In this case, context took predominance over text, and the two seemed to work against each other, reinforcing current literacies rather than reconfiguring them.

Level of Intimacy Between Students

For the most part, 75% of our students were conscious of how their written words could effect communication with their cyberfriends. This also affected the level of intimacy that was established. If students connected on a personal level, their essays were perceived as easy and fun to write. Obviously, these were the students who wrote the more interesting profiles and were able to find their cyberfriends' intriguing aspect. Sandra writes: "I think the fact that me and Ann sounded a lot alike helped me to write the profile. If I had had someone that I had nothing in common with it might have been harder to do." Others connected so intensely with their cyberfriends that they exchanged phone numbers, pager numbers, photographs of each other, and physical addresses. Those who had international students as cyberfriends reached a different level of intimacy and interest. Lorynn in Tucson recalls that although she and Xuan were very different, this did not affect her ability to write about her cyberfriend because Lorynn was interested in Xuan's background. Lorynn explains, "I think she [had an] intriguing life, where she comes from, going to another world, and coming here in America for a better future."

Although liberating and fun to some, others found the whole e-mail and profile process to be a large waste of time because of poor partner response. Some were not able to establish a certain level of intimacy required to write the essay. Those who outright disliked their cyberfriends and were not able to get past this were psychologically unable to move to a more professional level of communication for the assignment. Mitchell was prejudiced against his assigned cyberfriend from the moment he received her Hotmail pseudonym, sexeetbear. Based on her self-appointed name, Mitchell made certain assumptions about his cyberfriend's personality. In face-to-face communication Mitchell would have other indications— giggling, body language, dress, eye contact—to tell if sexeetbear was really flirtatious. He had no real evidence of her nature other than that she picked this screen name, yet he perceived her as a very young and immature woman—someone with whom he would not be able to communicate at all. Mitchell's contextual assumptions skewed the way he interpreted and responded to her e-mails. His prejudice also worked against him as it sabotaged his efforts to write the essay.

We, along with our students, discovered what Tom so aptly wrote in his journal: "Getting to know the person in general is just not good enough because what's really required is that you get to know a specific aspect of the person." Even when students did get to know specific aspects of their cyberfriends, however, they had to be genuinely interested in what they found to write enthusiastically. It seems that those who did not connect on a personal level and did not move beyond the context of their individual classrooms to form some sort of bond perceived the essay as merely another teacher-centered writing exercise. This was the case with Kristy, who was a returning, nontraditional student with very good writing skills and who was extremely proficient with the technology. She was one who dutifully wrote her cyberfriend when asked to and responded to her cyberfriend's questions. Kristy never complained about not receiving enough information, only that she found her cyberfriend boring. Kristy chose to function only textually and came into the project with the same literacies as she left. She never perceived herself as a writer who had control over contexts within the project to make it meaningful. Although Kristy was able to incorporate the basic features of the profile essay into her paper, her writing was vapid, reflecting her boredom and her perfunctory relationship.

Metacognitive Abilities

We wanted to see if the cyber aspects of this project contributed to our students' abilities to think about their writing processes. This was accomplished indirectly through the use of the portfolio transmittal letter in Dagmar's class and in an in-class guided writing assignment for Simone's class, which challenged students to revisit their goals and discuss their progress as writers. Students who chose to include the Cyberfriends Project in their portfolio reflected on how much they enjoyed the project but also identified the problems they had with it and articulated how they solved those problems. Paulina, a Jamaican immigrant, wrote:

> Last but not least is personally my favorite, a profile essay. This essay not only helped me with my writing skills, but it also helped me with my computer skills. . . . In writing this essay, I was able to take information that was given to me about a person and convert it into an essay telling you all about the person. This helped me organize my thoughts and set up a paper. . . . English 1301 has helped better my mind, writing skills, and it has given me great experience to taking on harder and more difficult writing tasks.

Dagmar found overall that students who participated in the Cyberfriends Project wrote more thoughtful portfolio letters in which they were able to analyze their portfolio selections. The development in critical-thinking and writing skills to which Paulina points was noted by many of Simone's students. Greta notes,

"Well, I had never sent an e-mail before this class. It was hard to memorize the commands sometimes but I got used to it. It definitely helped my writing skills because I was very aware of making myself clear and explaining myself." In general, students who participated in the profile project developed self-awareness and realized that basic principles such as thesis, focus, development, audience, and purpose are pertinent no matter what one is writing. Janie notes, "I can see the improvement not only in this class but my other classes as well. My attitude about writing has changed forever for the better." Juan, however, captures the reason why we hope to continue developing the project:

> I chose to revise the e-buddy [cyberfriend] essay because this was a great experience. I thought that the e-mail process was very interesting. It gave us another take on what fellow students are going through in their freshman year. It was a great learning experience. I never was into the e-mail activities others were into. I did not even know how to get started. But making it an assignment and part of the writing class gave me the extra push to learn. For this I thank you.

Juan verbalizes development of academic agency and transformation of self. He realizes that these experiences are a product of the intercontextuality of the electronic classroom experience.

CONCLUSION

The impact of students' personal literacies on academic and technological literacies is undeniable. Floriani's (1994) work proves that "[i]n classrooms, events are not isolated occurrences, rather they are often intertextually . . . and intercontextually tied across time and space and serve particular purposes within larger cycles of activity" (p. 271). With the Cyberfriends Project, indeed, classroom events are not isolated occurrences. They are inextricably bound to other classrooms and to students' personal literacies. These classrooms are intertextually tied through students' e-mail correspondences and the constraints of the assignment and intercontextually tied across time and (cyber)space through cultural and social climates of geographically different students. The Cyberfriends Project serves to initiate freshman composition students to interacting electronically in a global society and developing academic agency. This project demonstrates that rather than discarding or replacing personal literacies, students connected to others in cyberspace are able to create meaningful texts and experience transformation of selves by drawing on and then reconfiguring their present literacies to meet the challenges they will face across the cyber divide.

11

Web Writing and Service Learning: A Call for Training as a Final Deliverable[1]

Christina L. Prell
McDaniel College

Since the social turn (Clark, 1996) in composition studies, composition scholars have been emphasizing the social relationships surrounding writers as they create texts. In actualizing these relationships in the classroom, writing teachers have developed many approaches, chief among these being community service, or service learning. Composition and service learning, commonly known as CSL, is a growing movement within composition studies and seeks to develop students' writing skills and civic awareness by exposing students to writing experiences within their local communities. This movement has grown in popularity and has taken a variety of forms, ranging from reflective journals of student experiences to collaborative work environments where students write a useful document that fills some real community need.

Parallel to this development in service learning has been the introduction of electronic or computer-based writing in the composition classroom. Hypertext, the World Wide Web, and multimedia are quickly becoming resources that writing teachers use in the classroom and with which students are asked to compose. Recent projects of note to explore these issues include Ilana Snyder's *Page to Screen: Taking Literacy into the Electronic Era* (1998), Todd Taylor and Irene Ward's *Literacy Theory in the Age of the Internet* (1998), and David Reinking et al.'s *Handbook of Literacy and Technology: Transformations in a Post-Typographic World* (1998). Additionally, electronic publications from Eastgate

---

[1]I would like to thank Jim Zappen for being my mentor throughout this whole project, Lee Odell for his teaching and good advice, and Carol King for her support and insight.

Systems and *Kairos: A Journal of Rhetoric, Technology, and Pedagogy* continue to push the bounds of scholarly texts.

However, although service learning and hypertext scholars have explored the challenges and exciting potentials of these new movements, few have explored the relationship between service learning and hypertext and what such a union might afford composition teachers. And although many may be excited about the possibilities such a union might afford, no one seems to have grappled with the problems that come with it, specifically those of technological knowledge gaps, the uniqueness of the media, and access to these technologies. These problems complicate the more traditional CSL scenario and give rise to more fundamental concerns of ownership and sustainability. In this chapter, I describe a case study in which I attempted to bring about such a union and found myself grappling with many unforeseen technological obstacles. In revealing these difficulties, I offer suggestions to those writing teachers thinking of designing a course that blends CSL with the Web. In particular my chapter addresses an issue that CSL courses generally overlook and one, moreover, that becomes highly problematic once technology is introduced: sustainability. My chapter finishes with a call for teachers to look beyond segmented school calendars and to build ongoing programs that sustain the efforts of individual students from one semester to the next.

SERVICE LEARNING: ACTUALIZING VALUES

CSL classrooms are growing in popularity: The recent publications of the book *Writing the Community* (Adler-Kassner, Crooks, & Waters, 1997) and the journal *The Writing Instructor*'s special issue on service learning attest to the variety and strength of recent initiatives in bringing writing students to the community. In reviewing these different attempts, however, one can generally situate these initiatives in one of two scenarios. The first involves students working with nonprofits to write documents that fill some need of an organization. The second has students tutoring others in literacy skills and asks students to fulfill the writing requirement of the course through a combination of research papers, essays, and reflective journals. Benefits of these two sorts of service learning vary, yet scholars tend to agree that students attain a deeper sense of civic responsibility at the same time that a real community need is being met (e.g., Dorman & Dorman, 1997; Huckin, 1997).

Herzberg (1994), one of the early initiators of CSL, provides a sound model for one sort of service learning approach.[2] He has created a 2-semester course that focuses on sensitizing students to literacy issues within the community. Herzberg spends the 1st semester discussing with students the cultural forces that give rise to conditions where people in our country are left without reading or writing

---

[2]For another CSL example involving tutoring programs, see Schutz and Gere (1998).

skills; this semester also provides basic tutoring skills. In their 2nd semester, Herzberg's students assist local community members gain reading and writing skills, while, at the same time, the students prepare a research paper on a topic related to their service work. Herzberg's course helps students expand their notions of writing to include involvement with the outside world. His approach also gives students an opportunity to rethink fundamental issues surrounding the writing and reading process. What is particularly noteworthy, though, is Herzberg's (nearly) equal attention to the different dimensions of his course: Students learn tutoring and writing skills, spend ample time out in the field, and are given a space for critical reflections pertaining to the social forces shaping literacy.

This attention to time—for preparation, critical thinking, and community involvement—is not a common characteristic of CSL courses. Most, in fact, teach these courses within a single semester. This seems difficult when one considers that, beyond writing papers, CSL students must spend time interacting with community members and somehow gain insight into what it means to be a civic participant. How can one semester achieve so much? What should teachers focus on? In fact, as the next two examples illustrate, 1-semester CSL courses cannot offer the sort of in-depth approach found in Herzberg's course. Instead, teachers have to prioritize and make choices about what to include and exclude. In doing so, a tension develops between time constraints, needs of students, and the needs of outside community members.

Huckin (1997), who teaches a technical writing course geared toward community service, has taken an internship approach toward engaging students with the community. Students are asked to write technical documents for nonprofits with the idea that these organizations act as clients or real-life audiences for the students. To prepare for interaction with these clients, Huckin offers students readings and lessons in interviewing, observing, and teamwork. Such skills are taught alongside lessons in writing and rhetorical strategies. Huckin's priorities, then, are giving students real-life writing experiences as well as giving them the opportunity to help local nonprofits.[3]

Similar to Huckin, Eddy and Carducci (1997) teach a CSL course that asks students to work with community members to write documents. Students meet with community members to discuss potential writing projects, develop a project goal, and create a plan for meeting that goal. Eddy and Carducci also see value in exposing their students to real audiences, yet the value of such exposure seems not so much to prepare students for the real world but rather to prepare students for changing that world. As students work with different communities, say Eddy and Carducci, they learn how to fluctuate and connect with these communities, thus

---

[3]An obvious question that flows from this example is how does Huckin's work differ from that of an internship? In other words, if service learning is to help students gain civic awareness, then do internships with nonprofits meet this goal? Perhaps, but such an approach relies entirely on the influence(s) of the client organization rather than the additional help or influence of the classroom and teacher.

gaining entrance and potentially changing how these communities operate (p. 79). Here, the priorities are writing as a type of social agency.[4] Yet do students actually go out and change the world or is this the hope of the teachers? How would these teachers know that the agency learned in the classroom gets carried out beyond that particular semester?

These examples help illustrate the many directions a CSL course might take, depending on a teacher's prerogatives and priorities. That a CSL course is more complex than a traditional one is obvious. What all of these courses have in common, however, is the question of how successful they are in helping their community or society at large. In all cases, eventually the course has to come to an end, and the question of "what next?" is asked. Isolated semester-bound actions are helpful in the short term. But they remain precisely that: short-term, isolated actions. What, then, can be done to ensure long term success?

More important, in the context of this volume, none of these examples grapples with the complexities of hypertext or Web technology. How would the introduction of a hypertext technology such as the Web influence this already complex scenario? To begin answering this question, the next section looks at how these technologies are currently being used in writing classrooms and attempts to understand to what extent these new technologies can be seen as entirely new or merely as alternative tools for current traditional practices.

THE NATURE OF HYPERTEXT: POSTMODERN MEDIUM OR RHETOR'S INSTRUMENT?

Like service learning, hypertext's popularity in the writing classroom has been steadily rising within the past few years (DeWitt, 1996). Because these new hypertext technologies are different from standard print, they present a variety of problems for writing teachers.

Those in praise of the new media speak about its liberating qualities. This perspective sees hypertext as breaking down traditional roles of writer and reader, thus offering new democratic ways of constructing texts (Bolter, 1991; Joyce, 1988). In a true hypertext, one where very little formal structure is present, the author creates the words, and the reader determines the structure (Bolter, 1991, p. 154). This collaboration gives more authority to the reader and offers the reader the opportunity to test traditional structures against the nonstructured environment of a hypertext (Joyce, 1988, p. 12). Joyce, author of *Afternoon, a story* (1987), hypertextual fiction, also praises the associative structuring hypertext pro-

---

[4]Blyler (1995) discusses how teachers can encourage social agency in the writing classroom. Blyler's interest is in investing students with analytic skills that not only allow them to operate within the professional discourses they will soon be encountering, but also enable them to critique those discourses and influence them.

vides saying that such structures break down the forced linearity of traditional print-based composing. Hypertext, from the writer's position, thus offers another sort of liberation from print.

Many teachers are taken by these liberating connotations of the hypertext medium and consequently are asking students to compose with these tools. Halio (1996), for instance, has her students write stories by including different media—text, sound, movement, and graphics—into a multimedia narration. In these cases, students are free to add various media elements to their text compositions. Halio notes that such additional elements change students' writing on two levels: First, students become more concrete in their descriptions of various worlds as they add sound, visuals, or movement to illustrate particular ideas. Second, students' narrations evolve into multilayered texts where the new media elements provide new subtexts of meaning (p. 347). Halio also notes that students played more with creating these sorts of narrations and that this element of play seemed crucial in the development of meaningful texts.

Halio's experiments in integrating multimedia into the composition classroom reveal the desire—shared by many in our discipline—to see how traditional composition practices can still be applied to this new medium. Tovey (1998), one such scholar, sees rhetorical notions present in hypermedia and encourages designers of the medium to apply these notions as they develop new projects. Tovey focuses on three elements within hypermedia environments as primary rhetorical devices: metaphors, links, and buttons. Metaphors offer structure by acting on the previous knowledge(s) of users. Links and buttons can be used to create expectations and establish hierarchies. In suggesting the use of these elements, the author emphasizes the importance of "giving flexibility to both author and user" (p. 377) by establishing links that cross over layers and hierarchies. Yet the author simultaneously claims that "all of the possible links in a hypertext, however, should probably not be available—or users might have nothing but labyrinthine webs to be caught in" (p. 376). This fine balancing act that Tovey prescribes reflects a tension between wanting to tie in classic rhetorical notions of "good" composition and a willingness to view the new medium as holding unique qualities on its own, such as the unstructured nature Joyce and Bolter promote.

Similar to Tovey, Odell and Prell (1999) link hypertext to linear ones through such concepts as "given to new" and "creating expectations and fulfilling them" (p. 296). In particular they look at how such hypertext elements as links, buttons, and screen layout form bridges between screens and help to forecast what is to come in future screens. Hunt (1996) also offers insight into the ways rhetoric, specifically that of ethos, can inform the understanding, use, and development of webs. Like Tovey, Hunt notes that the use of metaphors as governing frameworks to organize a site is helpful both for designers and developers. These metaphors, beyond helping to give readers access to the site, also work to create a presence on the Web and thereby create an ethos for that site's organization, group, or individual. This strategy is no different from ones writing teachers try to encourage their students to practice, yet Hunt then shows readers the unique values Web technology offers all its users, namely, the ability to connect to others, forge new relationships, and create community through traditional notions of gift-giving and reciprocity. Such interactive, communal values are not as readily present in print media, explains Hunt, and they are ones that Web developers can appeal to as common values among Web users.

From these three examples, then, one can see an underlining tension between trying to find what is unique about new media technology and discovering how these new media still rely on traditional (i.e., print-based) rhetorical practices. Although Tovey comes closest to balancing traditional rhetoric with the liberating notions of Bolter and Joyce, all three scholars seem particularly interested in striking a middle ground between old and new. Hunt's discussion begins to delve into notions of community that the new technology affords, yet concrete examples of how one works with a physical community and the Web are absent from all three of the discussions. Also absent from the discussion is a sincere examination and admittance of the complexities and problems of the medium. Working with the Web or another hypertext medium is not easy; if nothing else, a learning curve is involved for newcomers. However, when trying to work with communities that do not have ready access to these technologies, these problems are multiplied.

In the following section, I outline my own attempt at bringing together the two worlds of Web design and community service. This case study took place in Troy, New York, and involved a nonprofit organization, three undergraduate students from Rensselaer Polytechnic Institute (RPI), and me. What follows is a brief summary of how this project unfolded.

COMMUNITY CARE AND THE WEB: CASE STUDY OF A COMMUNITY SERVICE WEB DESIGN PROJECT

I was asked to act as liaison between undergraduates who were designing a Web site and their client, a community organization. These undergraduate students were involved in a course called Writing for the Web, which trains students in Hyper Text Mark-up Language (HTML), Web design basics, and appropriate writing for a Web site. As a final project, this course required students to create a Web site for a client in the outside community. One of the three students—we'll call her Janet—had previous contact with a local nonprofit called, for the purposes of this chapter, Community Care. A second did as well; he had gone to this organization with the offer of building them a Web site. Community Care agreed, and I was then asked to oversee the project. Specifically, my role was to ensure that administrators at Community Care were actively involved in the conceptualization and design of the Web site.

To do this, I met several times with them, and we discussed their needs and purposes for their Web site. Later, as versions of the site were completed, we used our meetings to review the students' progress. At the end of the semester, the students turned in a final version of the Web site that met Community Care's expectations. Thus, we all ended the semester feeling good: We felt that the Web site was close to being completed except for minor edits and small technical details.

A few weeks after the end of the semester, however, problems started to arise. I learned that Community Care's editor was intimidated by the number of links in the Web site and by the HTML code. Similarly, the administrator was concerned about how much time Web site maintenance would take away from the editor's other duties. In all, Community Care felt that the Web site was more of a burden than a gift.

Rethinking and Restructuring the Project

I discussed this concern with colleagues and finally decided to make Community Care another offer. I told them that I would help them rebuild their Web site page by page. My intention was to help Community Care not only learn HTML but also give them a stronger feeling of ownership of their Web site. Community Care agreed, and that following summer we scheduled six workshops to take place on campus. Janet, the undergraduate, offered to help in these training sessions, so the two of us began designing workshop sessions together.

Surprises, Surprises, and More (Knowledge) Surprises

Our workshop series started a few weeks later. The first workshop ended in confusion and set the tone for the remaining ones. Several of the participants who had come to the workshop did not have basic computer skills, so we spent much of our time teaching them how to open, close, and save files. This lack of skill was something for which we were not prepared and something to which we had problems adjusting. Our lessons thus ended up being a conglomeration of basic computer skills and HTML. Needless to say, the Community Care employees left the workshop series feeling frustrated.

After the workshop series, later phone calls to the organization revealed to me that Community Care had decided to put the Web site aside for the time being. They seemed a little tired, and, frankly, so was I. As of this writing, it has been more than 1 year since the workshop series, and Community Care has not yet uploaded their Web site to a server.

What Went Wrong?

This story describes a number of dashed hopes. The undergraduates wanted their Web site on the Internet, Community Care had first wanted a Web site and then wanted the skills for maintaining that Web site, and I had wished for a successful union between Web site design and community service. After careful monitoring and adjusting to problems that arose, our shared and separate goals still failed. So what went wrong? One answer is that Web writing, versus more traditional forms of writing, was at the core of this project. Unlike CSL scenarios that focus on creating print-based documents, creating Web sites for nonprofits introduces a whole new set of challenges. First, employees of nonprofits may not have a lot of experience with computers or the Internet. This lack of experience limits how much they can add to the conceptualization and design of a Web site. Further, this lack of experience limits their ability to maintain the Web site once the semester is over. Since Web sites need more consistent updating to remain effective, knowing how to edit a site and upload it to a server, as well as additional technical details, are essential. Finally, in some cases, these obstacles are compounded when the nonprofit does not have computers. Given these issues, then, a Web site offered by a college student in a CSL course might end up more of a problem than a gift. Especially if students disappear after 1 semester.

BEYOND CALENDARS AND CLASSROOMS: BUILDING SUSTAINABILITY

Given this range of concerns, then, and that we teachers are operating under a 14-week semester time frame, how could such challenges best be met? One simple answer would be to mimic Herzberg's model, thus extending the time frame from 1 semester to 2 semesters. This extension would allow students to spend a longer time addressing the knowledge gaps of their clients and still help with the development of a Web site. Yet this approach would not help with access issues. Further, as one Community Care employee told me, "Time moves slowly at a nonprofit." Given, then, the complexity of problems and that community organizations' time schedules are different than ours, 2 semesters still might not be enough.

So how much time is needed? This issue of time becomes the real challenge to those of us accustomed to thinking in terms of semesters and summers. Yet if one looks at other town-and-gown models of integrating technology with community, one sees that long-term commitments are the norm (e.g., Schuler, 1996). These long-term commitments allow scholars to understand better the needs of community clients and come up with well-planned, sustainable solutions to those needs. By sustainable I mean that the solutions last and that the Web sites stay alive and prosper.[5] This is something quite different from the 1-semester CSL course: Community clients are not simply given Web sites at the end of the semester; rather, these deliverables are given and accepted within a larger framework constructed by the joint efforts of teachers and community members. In other words, the ser-

---

[5]The term *sustainability* comes out of the field of economics and generally refers to consideration of future needs in the planed use of present and future resources (e.g., Vazquez, 1999).

vice learning Web design course is embedded within a larger program, one in which faculty discover problems of long-term sustainability and ownership and work to resolve those problems. Students in CSL courses would then work on smaller projects, yet with an insider's view.

Therefore, the question moves away from how to design a better CSL course and focuses more on how to be a true servant to the community through our teaching and our use of technology. With a more sensitive, long-term relationship with our community, we are much less the patronizing experts and much more collaborators in bringing computer literacy and access to our community. We can also ensure that our end projects are sustainable ones that are owned by the community organizations. How then might we do this?

- Think beyond the classroom. Think in terms of a program. Think about funding, resources, the different actors involved, and the needs of those actors. Your students will most likely only help in a small part of this overall program, yet you must still work with the community to help them meet their other needs over a longer period of time.
- Take action. After you have scoped out the potential actors, contact them. Ask for their advice and involve them in creating a larger vision. Is the real goal of the organization urban revitalization? Or is it greater community involvement? How, then, would Web design and students enhance this vision? What other factors would your group need to consider and address?
- Replace the notion of final project with that of ownership or sustainability. With Community Care, the students were encouraged to think in terms of their final project, their Web site, as the determining source for their grade. Yet, as I have described, their final project in many ways ended up being training. If scholars engage in dialogues with community organizations and discover their technological needs and knowledges, then they can figure out the best way their students can help. Should the final project be training? Then the next question should be what kind of training? Training in HTML, Web site design, or a mixture of the two? And who would be there after the semester ends? Would this be necessary? Clearly, a vision aimed more toward sustainability and ownership versus project deliverables changes the goals and the nature of the CSL experience.

CONCLUSION

In conclusion, merging technology with CSL certainly poses some challenges. Not only is one faced with the dilemma of how to help a community gain access to the Internet but also with how to ensure that this access remains and does not end once the semester ends. This last part seems especially challenging because we have grown accustomed to thinking in terms of segmented school calendars.

The movement in composition toward service learning has already shown that scholars are interested in moving out of the ivory tower and into the community. This provides us with a whole new self-image. With the advent of the Internet, the question of how best to integrate the use of technology into our social agendas, specifically those pertaining to community service, presents a number of new challenges. My suggestion is not to strengthen our commitments—they are already strong enough—but to lengthen them. In all, getting away from arbitrary time divisions and ideas that a final project must mean something concrete and visible would allow us to move toward something more durable, more sustainable, and more in keeping with our service intentions.

Response

Bill Friedheim
Borough of Manhattan Community College

A STARTING POINT

The starting and end point for my response to these essays is my own teaching. I begin with the understanding that these highly provocative and instructive essays are not so much about technology as they are pedagogy. New technologies have prompted me to reimagine my teaching, as they have the authors of this volume. So as I read these chapters, I look at them through the lens of my own experience and think about what might apply to my own classroom. The power of a book like this is that it helps create communities of teachers across disciplines dedicated to figuring out how to better promote student agency in learning, sharing, and knowledge building.

In my case, digital media, particularly the Web, have changed the way I approach the history survey because these technologies give easy and quick access to primary sources in multimedia (unimaginable for just a few years ago for community college students); make learning and academic investigation a much more public and transparent activity (through the use of electronic listservs and the Web); and expand the borders of the classroom. Not surprisingly, these same themes run through the essays in this section.

These themes play out in different ways in our classrooms. In mine, I want students to come away with a basic understanding of temporal sequence, which I suspect is a more linear and traditional view of history. History after all does unfold in some kind of chronological sequence. But it is also complex, sometimes unfolding in many different directions and open to competing interpretations. I want students to grapple with this complexity—and here I think hypertext opens new possibilities. But like many of the essayists, I also want to change basic interactions in my classroom—between student and teacher, between student and student, between individual and group inquiry, between received knowledge and inquiry, and between students and the raw material of history (primary sources).

I won't go into any detail about how I have tried to do this: where I have failed, where I have succeeded, and how I tried to sum up those failures and successes. But as my experience increases—and as I read a growing body of literature on how to do this—I feel the need to turn my teaching into research and to figure out how to do such research in discrete, manageable but incremental ways. I need to find out how my students imagine the world; how they make sense of, manage, and synthesize information; and how they grow their own knowledge and experiences.

The essayists in this volume engage some of the same questions about how our students learn and how our teaching best takes advantage of different ways of knowing the world. Their analyses make discussion of questions like these more public, make classroom practice more conscious, and make solutions more collaborative. Even more important, they provide a model for teaching as research.

A BASIC THEME AND SOME TALKING POINTS

Although most of the essays in this section focus on student agency, I want to direct my response to teacher agency, including teaching as research. There is clearly a relationship between teacher and student agency. But the basic argument or theme of my comments is this: Teacher intervention is the key to facilitating active student learning, whether in a traditional classroom or one enhanced by new technologies. The technologies used (particularly the Internet), though powerful, are nonetheless neutral. What is decisive is not technology but how college faculty, administrators, funders, and students use it—and for what purposes.

As I comment on issues of teacher agency that emerge from chapters in this section, I want to examine the larger political, technological, cultural, and administrative context and the particular strategies used by the essayists to promote student-centered, electronic discussion.

TEACHING DRIVES THE TECHNOLOGY,
NOT VICE VERSA

The Internet does not democratize the classroom, decenter learning, redraw the boundaries of knowledge, privilege collaboration over individual scholarship, reorder traditional lines of academic authority, break down the conventional role of writer and reader, or herald a glorious new academic age of community, equality, and inquiry. Nor is the Internet a doomsday machine that will destroy everything we value in the academy. It does not reduce faculty autonomy and labor costs, promote administrative surveillance and control over curriculum, accelerate "an existing trend toward the debasement of education" (Ott, 1998), depersonalize learning, and downplay "the importance of conversation, of careful listening, and of expressing oneself in person with acuity and individuality" (Oppenheimer, 1997).

The Internet does, however, creates possibilities for all these outcomes—both good and bad. The operative word here is possibilities. The Internet is not a power unto itself—a revolutionary or counterrevolutionary force (depending on your view) that will create either a utopian or nightmarish new academic world. Granted, the Internet is a not an ordinary tool but rather a powerful, transformative technology. However, tools and technologies are extensions of human beings. It's human intervention—intentional or not—that ultimately will determine how the Internet changes how we teach and learn.

The essays in this section underscore the importance of teacher intervention in taking advantage of the Internet as a tool for promoting a much more democratic, collaborative, student-centered and inquiry-based classroom. Nancy Knowles and Wendy Hennequin make this very point at the outset of their essay by quoting this cautionary note from Cynthia Selfe (1992):

> [W]hen technology as an artifact of our culture, is employed by teachers who lack a critical understanding of its nature or a conscious plan for its use, and when these teachers must function within an educational system that is itself an artifact of the political, social, economic forces shaping our culture, the natural tendency of instruction is to support the status quo. (p. 30)

In the same article that Knowles and Hennequin quote, Selfe argues:

> Given the embryonic state of our knowledge about what goes on when instruction is carried on in virtual learning spaces . . . , increasing instances of observation and research are essential to directing our efforts over the next decade. Without the information we can gather from such observations, we have little to go on in making decisions about virtual instruction. (p. 33)

Selfe (and Knowles and Hennequin) emphasize conscious reflection and deliberate planning as essential to education in the digital classroom. All of this is labor intensive. Planning means reconsidering classroom relationships; analyzing how our students think and learn; taking risks; summing up successes and failures (for many of us, the latter are particularly instructive); observing, reflecting, and planning; and then observing, reflecting, and planning some more. The process is never-ending, extraordinarily difficult, at times frustrating, and (for me, and I suspect our essayists) exhilarating.

INTENTIONAL MEDIA

I cannot emphasize enough the importance of creating what Randy Bass and Bret Eynon (1998) call intentional media. They write:

> All teachers have intentions when they design and teach a course. In many ways those intentions are a kind of hypothesis, as if to say, "If I teach these particular things, in this particular order, in this particular way, then this kind of learning will probably take place." This mostly unarticulated "course design hypothesis" is loaded with complicated questions and informed by a whole range of knowledge about one's subject matter, one's students, and the learning process. Yet faculty almost never have the opportunity to look at these questions slowly . . . to examine their teaching systematically, and consider their *intentions* in curriculum design for all their assumptions and ramifications. . . . The proliferation of technology in higher education has provided an opening to address our intentions in a new way. (p. 11)

In one sense, digital technologies make student learning much more visible. Fakler and Perisse, among others in this volume, note the possibilities for "surveillance of student work in progress" and of "being able to see what our students are learning." Through the instrument of one of the easier new technologies to master—electronic discussion—teachers can catch a glimpse of how their students perceive the world; how they look at, process, and conceptualize information; and how they turn perceptual knowledge into rational knowledge. The more we observe and understand student-thinking processes, the more we can make media intentional.

Like many faculty, I often think about the special but far too infrequent teaching moments we experience in our classrooms. Sometimes I wonder if this whole period of proliferation of technology is an extended teaching moment. Is it an opportunity to create intentional media and build communities of scholars who reflect on their teaching, write about it, and construct a body of knowledge and research that realizes these visions of academic change and student learning. That's what I think the authors in this volume are about—together, with hundreds of others across the country, they are trying to figure out how to integrate technology into their classrooms in ways that facilitate better teaching and learning.

I think that this is an extended teaching moment because the proliferation of technology, and the speed by which that technology changes, has forced many of us to reconsider our classroom practice. "At no time during my own 40 years in the profession," writes Wake Forest vice president and pioneer in educational technology David Brown (2000), "have I seen so many professors undertaking fundamental remodeling of their teaching approaches. At no time has there been more thoughtful consideration of pedagogy" (p. 3).

At a recent meeting of faculty development coordinators (for integrating technology into the classroom) at the 20-campus City University of New York

(CUNY), one of my CUNY colleagues made an interesting observation. He noted that in his 35 years of teaching there were only two points when a critical mass of CUNY faculty actually talked to one another about their teaching. The first was the writing across the curriculum movement (which first came to CUNY in the late 1970s); the second was the current flurry of activity of teaching with technology. The more that we bring deliberate intent to teaching with technology and the more that we create ever-widening circles of communities of conscious practitioners, the better we can take advantage of this teaching moment.

There's a problem, however. The academy is not necessarily an environment that encourages collaboration, let alone community building. I do not want to overstate the case. There are plenty of examples of collaborative research and teaching at colleges and universities. But they are anomalies (particularly outside the sciences), exceptions to an academic culture and political and economic reward system that promotes (literally) solitary scholarship and teaching.

THE LARGER CONTEXT—ACADEMIC CULTURE

Jo B. Paoletti, Mary Corbin Sies, and Virginia Jenkins see possibilities for changing this academic culture. At the outset of their essay, they announce that they "will argue that one of the most important potential effects of the Internet will be its transformation of humanities scholarship from a mainly solitary pursuit to one that will depend increasingly on collaboration." Given their view of the possible, they foresee such "collaboration . . . gradually break[ing] down the traditional hierarchical paradigm and blur[ing] the division between scholars, students, and interested public."

What strikes me is that this is not a utopian argument about the magical powers of the Internet to transform a stodgy, hierarchical academy into a bastion of equalitarian collaboration, rather, it's a vision of human agency, of communities of students and faculty changing the culture of teaching and learning.

The Internet and other new technologies alter the material context of higher education. But it is faculty, students, administrators, and a lot of external players (politicians, entrepreneurs, corporations, etc.)—and their interactions with one another, with the technology, and with larger political, economic, cultural, and institutional forces—that will create change, for good or bad.

THE LARGER CONTEXT—THE POLITICS
OF ACADEMIC TECHNOLOGY

I think that as educators, if we want to harness the new technology in the interest of progressive pedagogical reform, we need to become more deliberate and reflective in our planning, more organized in our collaboration, and much more political in our everyday academic work. There are many players with many agendas who have a big stake in technology in the classroom. The motives of those who want to use technology to reshape the academy are varied and conflicting—profit, cost cutting, managerial efficiency, administrative control, and pedagogical reform. There are many possible outcomes to this story. At one end of the spectrum, you have Phoenix University, a virtual institution where a part-time contingent labor force teaches 95% of the courses, and administration in partnership with corporate contractors control curriculum. At the other end, you have a growing movement of faculty at the cutting edge of progressive pedagogy and experimentation, such as those in this volume. In the center, you have higher education bureaucracies and institutional structures that will use technology—sometimes intentionally and sometimes not—to reproduce old cultures and power relationships in new forms.

As teachers who see the new technology as an opportunity to reimagine the classroom and reexamine relationships between teaching and learning, student and teacher, and student and student, our tasks are twofold: educational and political. In this case, educational goals must drive the politics. But if we are to build on the best of our traditional classroom practice and use technology to make education more interactive, student centered, and inquiry based, we need to be just as intentional about our politics as about our pedagogy.

The politics of educational technology plays out in many arenas—the department, the college, larger political and funding contexts, and, on some campuses, union-management negotiations. My intent here is not to draw a political map but to suggest that we cannot do what we do best—teach (in this instance with technology)—unless we create a critical mass of colleagues who see the intersection between educational and political goals. It's difficult to succeed in the digital classroom without departmental, administrative, and technical support. A little money wouldn't hurt either. But to garner such support, we need an organized presence that is much more than simply individual faculty members, independent of one another, pleading their cases before the powers that be. Critical masses of faculty need to organize to accomplish the following:

- Persuade departments, administrations, and external funders that we need technical, administrative, and financial support
- Make the case that teaching and learning in our digital classrooms is deeper, more enduring, and more transferable than in most traditional classrooms
- Create alliances—sometimes temporary, sometimes longer lasting—with those who have parallel interests
- Build connections with communities and institutions outside the academy.

The scholarship of teaching is an example of this junction between educational and political goals. On the one hand, such scholarship enables us to assemble a body of evidence documenting what we do and enhancing our own understanding of teaching and learning in our classrooms. On the other, it's something tangible that we can show to departments, administrations, and funders when we seek support.

I think the same is true when we partner with, let's say, museums (Paoletti, Sies, and Jenkins) or community organizations (Prell), or build connections across college lines (Fakler and Perisse, Corrigan and Gers, Knowles and Hennequin). Several essays make the case for the educational benefits of cross-fertilization between college classrooms or between universities and communities, and, in one case, universities and a local museum. But I would argue that it is these kinds of alliances that also strengthen our political hand with funders and academic decision makers when we seek needed technical, administrative, or financial support.

THE LARGER CONTEXT—INFRASTRUCTURE AND SUPPORT

The essay by Nancy Knowles and Wendy Hennequin speaks to the issue of infrastructure—that is, the system of technical, departmental, and administrative support so crucial to sustaining pedagogical reform in the digital classroom. What struck me is that initially there were no support systems in place for their collaboration; they write the following:

> When we began to collaborate on our course, we had no training except for a voluntary workshop on using the Internet provided by a fellow graduate student. . . . We had no departmental or institutional encouragement or rewards for doing what we did, no computer labs for our students to use, no financial remuneration, and no teaching assistants to help us. Any assistance we received, we sought and cultivated ourselves. We simply envisioned the course and worked toward bringing that vision to life.

At first glance, the passage quoted seems to counter my argument about the importance of infrastructure. Even though they were graduate students at the time, Knowles and Hennequin seemed to succeed despite a lack of support. They brought to their project a conquering spirit, a willingness to take risks, and, maybe most important, a vision based on possibility. But there was technical infrastructure. They ended up teaching in mediated classrooms with state-of-the-art computer, video, and recording technology on one of the most wired campuses in the United States, the University of Connecticut at Storrs. But there is no doubt that sheer willpower accounts for much of their accomplishments.

Risk taking, vision, persistence, and stubborn willpower are familiar qualities that characterize pioneers in technology and teaching. But there are only so many pioneers. How do we encourage the vast majority of our colleagues to take the leap? We need to create an infrastructure that does the following:

- Trains faculty, not simply in the use hardware and software but also in ways that encourage them to stretch the technology to extend the boundaries of our classrooms and make them more interactive and inquiry based
- Creates a reward system—money, release-time, promotion, or all three—that nourishes experimentation with technology in the classroom and promotes good pedagogy
- Provides enriched technology for classrooms, students, and faculty, and provides support to maintain that technology
- Builds communities of teachers who are conscious practitioners, who evaluate and learn from what they do in the classroom, share lessons from successes and failures, and organize for more support for innovative use of technology in the classroom.

PARTICULAR STRATEGIES—PEER REVIEW

There are infrastructures—and there are smaller, equally important structures that teachers create in the classroom to facilitate inquiry. I think that many realize that student-centered learning doesn't mean less work for teachers, but more. Much of that work is up front, creating activities, access to resources, and frameworks that enable the student-centered classroom.

For example, Fakler and Perisse created peer evaluation groups where students communicated with complete strangers, first posting draft essays, then critiquing the drafts of their peers, and, ultimately, based on those critiques, revising and finalizing their essays. I'm sure that none of this happened seamlessly. Fakler and Perisse spare us the gory details, but I would assume that there were numerous hours of work to establish and coordinate small groups of students, each group representing three campuses.

Fakler and Perisse provide us with an instructive example of teaching as research. From hypothesis to conclusion, any research involves extensive planning, evaluation, and revision. Fakler and Perisse began with a research question (and implied hypothesis): What would happen if peers were complete strangers to one another? The assumption here is that the partnering of strangers for peer evaluation would yield positive results. The authors investigated the literature on peer mentoring, created mixed-campus groups, and, in their article, report results and develop conclusions. Based on an exploration of relevant scholarship and their own three-campus experiment, the authors amply document the advantages of peer review.

But there were additional advantages as a result of anonymity. The evidence is anecdotal and qualitative—much of it student testimony. One student voice that spoke powerfully to the subject observed:

> I had no idea how much better it is to critique a peer when you don't have to deal with them in class, whether its good or bad critique. I fully support this method . . . I

have always disliked having to look over a classmate's work, especially if I had a relationship with them. I find that I can be so much more honest and evaluate better when I know that I don't have to see that person. Not only that, but it is more accessible when we work on the computer and not with the author. This gives the critiquer a chance to do the critique at a time when he or she can concentrate fully on the piece before him or her.

But the formula is not a simple anonymous peer review = good student writing. The art of teaching is in the following details:

• Creating and managing groups
• Knowing when to mediate and when to let students resolve problems themselves
• Figuring out the creative tensions between group and individual, writer and critiquer, author and audience, student and teacher, and graded and ungraded work
• Managing the relationship between different stages and layers of writing (and creating the space for different kinds of writing)
• Recognizing when and how to model good writing

It's these daily teaching interventions (or sometimes knowing when not to intervene) that often spell the difference between a successful and an unsuccessful learning experience for the student. I was struck by one example in particular in Fakler and Perisse's essay. Initially, they did not grade student critiques of one another's work. Concerned that the quality of student critiques was quite uneven, Fakler and Perisse made an adjustment. Once they decided to grade the critiques, the quality and seriousness of student work in this area improved.

In this connection, it is interesting to note an observation made by Alice Trupe in her chapter and ethnographic study, "Reentry Women Students' Online Collaboration Patterns." One of the students she observed, Peaches, failed to "connect the writing she did in Interchange [electronic conferencing] with the writing she did for her final portfolio because she rightly deduced that what counted (i.e., what her grade would be based on was the essay, regardless of her classroom participation in computer)."

In a sense, I am comparing apples and oranges because the group Trupe studied was not doing the kind of peer evaluation of early drafts that Fakler and Perisse's students did. Nonetheless, there is a common question: What kind of incentive do we give students to interact with their peers online? Many of us, myself certainly included, have wrestled with this question—not always with satisfactory solutions. How do we create structures that encourage serious, probing online interaction between students? To what extent do we model and showcase good interaction? What criteria do we establish for grading such participation? Should we grade it at all?

There is no one-size-fits-all answer. But I'm impressed by how Fakler and Perisse struggled with the issue, summed up that graded critiques might motivate stronger student response, developed criteria for grading, and provided models of good critiques. Teacher intervention and constant fine-tuning helped realize the potential of online interaction between students (and strangers) on three separate campuses.

PARTICULAR STRATEGIES—ETHNOGRAPHIC STUDIES AND COLLECTING DATA ON STUDENT LEARNING

What's particularly important in all of this is discovering how students see the classroom, perceive their own learning, and locate themselves and their particular life experiences in relation to school, technology, teachers, other students, and bodies of academic knowledge. Class, race, gender, and age, as well as life experiences specific to an individual, shape students' perceptions of themselves as learners. At community colleges, like the one where Alice Trupe based her study or the Borough of Manhattan Community College where I teach, there is often a big cultural, racial, class, generational, and sometimes gender divides between students and faculty. At any school, it's important that faculty probe and try to understand who their students are, but particularly so at public 2-year institutions with very diverse, nontraditional student populations.

We need more ethnographic studies like Trupe's. I don't think we can extrapolate universal truths from a few student profiles, and Trupe of course makes no such claim for her small cohort. But the value of Trupe's work is in combination with what I hope will be a growing number of such studies. The more students we profile, the better we can understand what's unique to their particular experiences and what is general to larger patterns of student perception and cognition. As Trupe's study and several others in this volume show, the public nature of student discourse online makes student thinking and learning much more observable.

PARTICULAR STRATEGIES—COLLABORATIONS THAT EXPAND CLASSROOM BOUNDARIES

In separate articles, Christina Prell and Jo B. Paoletti, Mary Corbin Sies, and Virginia Jenkins use electronic media to stretch the boundaries of such discourse and interaction beyond the classroom and into world outside the academy.

Prell, focusing on linking composition and service, describes a developing CSL movement that has grown in variety and strength in recent years. But she pushes this initiative a step further, exploring the "relationship between service learning and hypertext and what such a union might afford the composition teacher."

Paoletti, Sies, and Jenkins describe a collaborative undertaking that integrates teaching, research, and service, enlisting students and faculty at the University of Maryland and curators at a local museum into a university-community partnership. Like Prell's effort, the enterprise has a lot to tell us about the relationship between learning and hypertext.

Both essays are rich in insight but differ in one very fundamental way: Prell's project, by her own admission, went wrong. The Maryland project, although confronted with some sensitive problems about intellectual ownership and working relationships between teachers, students, teaching assistants, and museum curators, nonetheless was a success. What's instructive is that we often can learn just as much from our missteps as our successes. (My own experience with technology has certainly improved because of insights gained from summed-up failures.)

Prell documents the tension in her CSL project between the aims of enriching students' writing (in this instance, hypertext) and community service, between the needs of the classroom and those of the community. Part of the appeal of CSL is that it adds an important dimension to student writing, grounding it in a practical world of everyday needs and problem solving. Students' "final projects" undoubtedly broaden skills, making their writing more practical and connected. But Prell asks that although such projects may serve the needs of the classroom, do they necessarily serve those of the community?

Prell concludes that "a vision aimed more towards sustainability and ownership [by the community] versus project deliverables [to a teacher for a final grade] changes the goals and nature of the CSL experience." The implication is that such reframing yields benefits both to the community and the classroom. Prell observes:

> By sustainable I mean that the solutions last and that the Web sites [the deliverable produced in this instance for a community organization] stay alive and prosper. This is something quite different than the 1-semester CSL course: Community clients are not simply given Web sites at the end of the semester: rather, these deliverables are accepted in a larger framework constructed by the joint efforts of teachers and community members. In other words, the service learning Web design course is embedded within a larger program, one in which faculty discover problems of long-term sustainability and ownership and work to resolve those problems. Students in CSL courses would then work on smaller projects, yet with an insider's view.

Prell suggests that sharp contradictions arose between classroom and community goals in part because "Web writing, versus more traditional forms of writing, was at the core of this project." I'm not sure that hypertext necessarily makes the problem more difficult than traditional writing. In the history of town-and-gown relationships, patronizing academic attitudes are certainly not unique to Web-based collaborations. Whether it's the traditional written word, hypertext, or other forms of educational expertise, there are plenty of examples of condescending academic saviors turning off patronized community-based organizations. Faculty and students need to work hard, no matter what the deliverable, at creating relationships with communities based on mutual benefits and respect.

Nonetheless, it is a tall order to expect students to learn skills in Web construction, become proficient in communicating in a different medium, and then train others in developing and sustaining Web sites. This is particularly so when the others are members of community organizations and lack the skills let alone the basic technology to create and maintain Web sites. We come from an academic culture that too often encourages intellectuals to inflate their importance by mystifying their expertise. There is a built-in temptation for intellectual condescension when there is a monopoly of technological skills, hardware, and software on the academic side of a partnership and a big technological deficit on the community side. That is why we need to learn the lesson Prell's experience teaches us: think in terms of sustainability and long-term partnerships based on mutual benefit and exchange.

Prell also reminds us of the potential of hypertext and multimedia to make student composition more tangible in description and multilayered in analysis. The Virtual Greenbelt Museum and some of the student work showcased in University of Maryland American Studies syllabi very nicely dramatize this potential.

The Paoletti, Sies, and Jenkins essay nicely mirrors Prell's in certain important respects. First, the Virtual Greenbelt (VG) essay demonstrates the wisdom of Prell's call for the creation of more sustainable and mutually beneficial structures of university-community partnership. Whatever problems the VG museum confronted (about intellectual ownership, decision making, control, planning, and leadership), it nonetheless built ongoing structure: a product of its many different constituencies struggling over issues of responsibility. Sustainability explains a large measure of the project's success.

We're told that the VG "owes its existence to Jo Paoletti's frustrations with the temporal and spatial limits of the traditional classroom." I suspect that this is precisely what tantalizes so many of us about the Web: the possibilities of extending traditional academic boundaries. But once we go beyond these limits, we have to negotiate what the authors call a "web of relationships." The possibilities are exhilarating, but the work is daunting.

Proof is in results. The Virtual Greenbelt provides tangible evidence that collective learning "promote[s] greater intrinsic motivation to learn, more frequent use of cognitive processes such as reconceptualization, higher-level reasoning, metacognition, cognitive elaboration and networking, and greater long-term maintenance of the skills learned" (Paoletti et al. quoting Johnson & Johnson).

It is through collaboration that University of Maryland American Studies students learned and practiced visual literacy, peer edited one another's work, completed complex research tasks and cooperatively built course Web sites. For example, Paoletti et al. comment on the super artifact project in AMST 205:

[It] required a heavy time commitment from the instructor, but it has fostered more interaction between teacher and students, and between students and students. It has generated a sense of collegiality that is often lacking in the regular classroom situation. The devise of having to work out technical problems to produce the final Web site engaged students in more intense discussion of content issues as well.

This led to still another level of collaboration between students and curators of the Greenbelt Museum to design a VG exhibit. What I find notable here is that the "heavy time commitment" by instructors Paoletti, Sies, and Jenkins resulted in a structure that nourished collaborative learning.

The virtual exhibit also demonstrated the capability of hypertext "as a means to help students see and develop interdisciplinary connections in their course work" and to explore "material culture using text, movement, image, and sound." In final projects and portfolios, Paoletti, Sies, and Jenkins required that students hyperlink their work with classmates' in ways that deepened that interdisciplinary connection. In some examples of student work, the mandated connections are awkward; in others, they begin to open up many different pathways to the understanding of particular material objects. What we see are new possibilities for student construction of knowledge: less linear, more open ended, more multimedia, and more cross-disciplinary. My argument about all of this is not to diminish the scholarship, creativity, and imagination of good student work, but rather acknowledge the labor and intelligence of teachers who facilitate that good student work.

CONCLUSION

As we make our teaching with technology more intentional, we need to become more rigorous and focused about how we measure the outcomes of student learning and the effectiveness of classroom strategies. To a remarkable degree, the essays in this section have done just that. However the authors of these chapters have done more: They have shared their experiences as part of a larger discourse and deepened our understanding of the uses of electronic discussion.

Lee Shulman (1999) has written:

Learning is least useful when it is private and hidden; it is most powerful when it becomes public and communal. Learning flourishes when we take what we think we know and offer it as community property among fellow learners so that it can be tested, examined, challenged, and improved before we internalize it. (p. 11)

In this volume, learning has become public, powerful, and communal.

III

FACULTY COLLABORATION AND ELECTRONIC MEDIA

12

Writers Anomalous: Wiring Faculty Research

Cheryl Reed
Naval Health Research Center, San Diego, California

Dawn M. Formo
California State University, San Marcos

A few years ago, when job offers spun the members of our informal writing gang to universities on opposite coasts and points in between, we decided to form an electronic writing group. To its gregarious instigators (one of whom counted computers and writing among her research foci), collaborative e-mail seemed an ideal forum for keeping up serious scholarly contact. Electronic exchanges, we reasoned, would have the immediacy of speed-of-light transmission (when servers were up); the intimacy of a cozy chat among friends; a flexible asynchronicity that could accommodate preposterous work schedules; and technical support from staffs at the four universities that provided us with e-mail accounts. Artlessly, we invited new colleagues, a graduate student, and an undergrad to this democratic, bodiless, genderless, narrative-driven writing space. As a wry comment on all this multivocal heterogeneity, we decided to call ourselves, Writers Anomalous.[1]

Our failure was complete.

Here's what didn't happen. We did not become "electronic" (or any other sort of) collaborators (Bonk & King, 1998; Lunsford, 1990; Peck & Mink, 1998; Sharples, 1993; Stillinger, 1991). We did not critique the social or rhetorical implications of technological capabilities (see Barrett, 1989; Birkerts, 1996; Bolter, 1991; Edwards, 1995; Ermann, Williams, & Shauf, 1997; Heim, 1993; Kiesler, 1997; Landow, 1992; Markley, 1996; Pimentel, 1995; Schroeder, 1996, among many others). We didn't form the "electronic community" or the productive computer-mediated interaction imagined in the classic computers and writing theory

---

[1]Thanks to Sam Reed, Cheryl's witty son, for suggesting this name.

assembled by Handa (1990a), Hawisher and LeBlanc (1992), Hawisher and Selfe (1991a), Holdstein and Selfe (1990), and Selfe and Hilligoss (1994). We did not become "wired women" (Cherny & Weise, 1996). Quite simply, we couldn't make the leap from being autonomous, univocal authors to being the kind of engaged collaborators we'd been telling our students *they* should become.

Member of Writers Anomalous who were more comfortable with face-to-face interactions couldn't seem to find a similar level of interpersonal connection electronically. The colleagues we'd invited to join the group communicated, offline, with whatever group member had sponsored them. Student members continued to discuss texts and institutional politics for the most part with their mentors. Rather than utilizing the group e-mail we'd set up, one key group member telephoned or snail-mailed whomever she needed to contact. The individual who'd received this nonelectronic communiqué then e-mailed it to other group members!

After several months watching our terminal patient die, Dawn and Cheryl held a short memorial ceremony and decided, "What the heck? *The two of us* can still collaborate online." And so, we wrote a book (and this article)—across distance, across computer platforms, through asynchronous schedules, around crashed hard disks, and despite erratic levels of tech support. We had a wonderful time. We are still friends.

What made the second collaboration different from the first? We believe, to enlarge on a theme raised by Jay David Bolter (1991) and discussed from a psychological standpoint in Sherry Turkle's work (1995), that our successful collaboration emerged from a happy synergy of technology and personal, rhetorical, writerly acts. Our interactions, both online and off, created a series of "writing spaces" (some electronic, some not), which were flexible enough to accommodate both our intent (to write a practical, rhetorical guide to the academic job search) and our idiosyncratic writing and communicating styles.

In fact, despite the tendency of computers and writing research to argue that the integration of technology in society changes or even degrades interpersonal dynamics, we found our own electronic collaboration moving in the opposite direction. Rather than finding the dynamics of our friendship altered by electronic media, we discovered that the patterns inherent in our friendship structured the dynamics by which the electronic media functioned *for us*. Writers Anomalous had, in effect, committed virtual suicide by trying to make the electronic venue act like the verbal. Our collaboration on *Job Search in Academe* (1999), on the other hand, looked for—and found—ways to make deep professional and personal connections in a different medium. At times, the electronic allowed us to communicate with more immediacy and intimacy than scheduled, face-to-face writing jags ever had.

THEORETICAL SUPPORT FOR WIRED FACULTY

Research into collaborative theory, as well as collaborators who've discussed their own joint writing processes, have long hinted that the act of collaboration is a powerful force that overshadows both the medium and the physical space in which writing is produced. Here's what writers and theorists have been saying:

- Writing together transforms the way you think. In the preface to *Madwoman in the Attic: The Woman Writer and the Nineteenth-Century Literary Imagination* (1979), Sandra Gilbert and Susan Gubar write, "The process of writing this book has been as transformative for us as the process of 'attempting the pen' was for so many of the women we discuss. And much of the exhilaration of writing has come from working together" (p. xiii).

- The collaborators' decision to create a single, collective, or interdisciplinary voice is foundational to collaborative writing, for it can define not only the text you are creating but the relationship between/among collaborators. The five authors of *Feminist Scholarship: Kindling in the Groves of Academe* (1985) explain that one of their significant discussions as collaborators was whether to write together or to assign sections according to field specialty. They chose "to write the book jointly rather than assign each of us to write a discrete section," so they could "reflect [the] interdisciplinary goal" of feminist research (DuBois, Kelly, Kennedy, Korsmeyer, & Robinson, p. vii).

- Writing together encourages a kind of play with ideas and with self that doesn't usually exist for the solitary writer and researcher. At the 1993 convention of the Conference on College Composition and Communication in San Diego, California, Mary Field Belenky, Blythe McVicker Clinchy, Nancy Rule Goldberger, and Jill Mattuck Tarule described their experience as collaborators of *Women's Ways of Knowing* (1986) as "pajama parties." They enjoyed weekends together writing in their pajamas.

- Writing together is a risky feminist act. For Carey Kaplan and Ellen Cronan Rose (authors of *Doris Lessing: The Alchemy of Survival* (1989b), *Approaches to Teaching Lessing's "The Golden Notebook"* (1989a), and *The Canon and the Common Reader* (1988)), the decision to begin writing together 20 years ago was decidedly a feminist act in an academy that not only represented a male establishment but also seriously questioned the validity of coauthored texts by humanities scholars. As risky as writing together was professionally (would they get tenure?), writing together helped them navigate the academic pathways.

- Writing together is how most writers outside of the humanities compose. To write collaboratively as a humanities scholar is risky scholarship. In the preface to *Singular Texts/Plural Authors* (1990), Lisa Ede and Andrea Lunsford explain:

> We began collaborating in spite of concerned warnings of friends and colleagues, including those of Edward P. J. Corbett, the person in whose honor we first wrote collaboratively. We knew that our collaboration represented a challenge to traditional research conventions in the humanities. Andrea's colleagues (at the University of British Columbia) said so when they declined to consider any of her coauthored or co-edited works as part of a review for promotion. Lisa's colleagues (at

Oregon State University) said so when, as part of her tenure review, they supportively but exhaustively discussed how best to approach the problem of her co-authored works. (pp. xi–xii)

While the collaborative act has been well explored, writings on *electronic* collaboration appear less often. Mary Peterson's "Life on the Internet: Portrait of a Collaboration" (1993) briefly describes the friendships that she developed with retired Professor Phillips, someone she met through the CLINTON list, and Dr. Robert Werman, a medical doctor in Israel often posting to the HUMANIST during the Gulf War. Peterson claims in a matter-of-fact tone, "Using the Internet, faculty at institutions thousands of miles apart can send correspondence, work-in-progress, or research data to their colleagues" (p. 10).

Many faculty rather routinely use the Internet in pursuing collegial contacts and scholarly research. In fact, a January, 1999, thread on the Alliance for Computers and Writing electronic discussion list had several list members describing their own use of e-mail in writing and research projects. Yet, to date, the body of research on electronic collaboration appears to be overwhelmingly aimed at *student* collaboration, perhaps because the old push for the isolated scholar is still alive in the humanities. While Michel Foucault's 1969 essay "What Is an Author?" persuaded us that the autonomous author is a recent invention, Martha Woodmansee's research (1994) into the cultural construction of authorship reveals that the notion of the isolated scholar still strongly controls our thinking. Promotion and tenure policies, copyright laws, and a generalized cultural mythology that advocates "pulling one's own load" still represent collaborative scholarship as a rather suspect practice (pp. 24–25). Yet, Lisa Ede and Andrea Lunsford's (1990) research into writing across the professions (collected in *Singular Texts/Plural Authors*) reveals that outside of the arts and humanities, writing together is a significant part of what it means to be a professional.

The direction of these very different areas of research tells us that in writing *Job Search* and possibly in writing this article we have done and are doing three taboo things:

- Collaborating
- Using electronic media to coauthor *scholarly* research
- Talking about things many people would like ghosted (e.g., that the job interview involves rhetorical skills that can be improved, or that electronic media lends itself to *scholarly* production)

Discovering again these taboos as we collide repeatedly with them, we break them again. We know that as writers we produce better texts by writing together. And, though for the most part our writing was driven by concern for all those jobless grad students, standing on street corners, waiting for a school bus that might never come, we had endless conversations about the strangely bifurcated scholarly conversation regarding collaboration, as well. Our need to share our "insiders' " tips with grad students entering a dismal market outweighed our concerns about the professional sagaciousness of collaborating. The synergy of our experience as collaborators (and specifically as electronic collaborators) will keep us doing it—that humanities triple taboo: electronic collaboration about the unspoken realities in the academy.

GETTING WIRED

Our exploration of the rhetoric of the job search in a postmodern academy required constructing a series of fragmented-but-connected writing scenes, a textual body expanded beyond the usual appendages by our access to, and ease with, electronic media. Because the composing task continued to morph as the text developed, at different stages of the writing process, we used electronic media differently. Our electronic collaboration included the following:

- Personal e-mails that batted around ideas as they developed
- E-mail calls for contributions to electronic discussion lists
- E-mail conversations with key contributors and, later, savvy peer readers of emerging drafts, which led to at least one enduring professional friendship
- Chat room discussions with hopeful job candidates
- A Web page previewing key items (such as a list of tricky interview questions) in our book, which was not yet on the market
- Word processing activities through physically exchanging laptops, e-mailing attachments to each other, and literally having four hands on one keyboard at times
- Revisions, editions, and negotiations with our publisher and printer via e-mail across 3,000 miles

Let's look at a few of these writing spaces to see how two tenure-track professors got wired.

Scene 1: A beach in Southern California. We walk barefoot in the sand, talking. The air is balmy and moist, and we gratefully breathe out the city smog, relinquishing the hyper pace of the city for the rhythmic rush of cool water around our ankles. Tourists in gaudy giftshop beach outfits yell delightedly as sea gulls steal their Doritos. A dolphin surfaces—way out, past the surfers, the swimmers, the land-bound. Joggers puff past pale flabby sunbathers as kids wash new-found treasures in the gentle waves. Nobody really sees us; like the others, we're simply part of the scene. Bodies, identities, histories, and desires are very much secondary to a sort of anonymous, autonomous connection. We begin writing our book in this other space.

Job Search began as an article, an article that we were writing collaboratively when we both lived in San Diego. Our collaboration was not exclusively electronic when we wrote the book (it has been almost exclusively electronic in writing this article, though, save a few phone calls). When we began the article that grew into the book, we found walks along the Torrey Pines Beach quite useful (and beautiful). While we often packed paper and pencil (sometimes even a laptop), by the time our toes were in the sand, we had abandoned tools, including the rudimentary pencil. Our ideas about the first job search article were interspersed with energized conversations about our personal and professional lives. If someone had been following our path through the sand, we are almost certain they would have had difficulty following the paths of our conversations. We looped our personal and professional lives and ideas about the article in and out of our fast-paced discussions, discussions in which we both developed the habit of finishing each other's sentences. These conversations, usually somewhat frenetic, sent us back to our separate computers so we could collect what we had just organized (and reorganized again) as we walked along the beach.

Physically writing that original article was a semester-long, weekend writers' experience. The space in which we wrote seemed to determine the platform we used. When we met at Cheryl's, we collaborated by sitting very comfortably in beach chairs on the living room floor talking, thinking, and writing—often simultaneously. To get the ideas down, one of us (usually Cheryl) typed on her Mac laptop, while the other talked and jotted ideas down on paper. Cheryl's family wandered in and out of our writing space, commenting on our content, our shared job search histories, and our writing process. When we met at Dawn's, we couldn't get comfortable on the floor to write (no beach chairs!). Instead we pulled two chairs alongside her desktop IBM; Dawn typed as she and Cheryl talked. Those "I've-got-an-idea-moments" usually cued the person not typing to be quiet for a moment as the other communed solely with the text and the screen. At times we'd switch roles and a new pair of hands would be moving across the keyboard.

By the time we agreed to write a full-length book for a particular publisher, each of us had secured tenure-track jobs on opposite sides of the country. Certainly we wouldn't be able to write our book together on companionable weekends face to face. We now needed to engage computer technology fully, something that Cheryl had been kindly nudging Dawn toward for months. "We'll use e-mail," Cheryl commented nonchalantly and frequently. While Cheryl quickly imagined creating our text together online—e-mail chats, MOO sessions, asynchronous writing jags, and online research interviews, Dawn didn't quite get it. She thought about ways to use the computer of course, but she also assumed the following:

- Their Fed Ex bills in the year ahead would be quite high.
- Their phone bills would likely be even worse.
- Collecting data from new faculty candidates, important research for the book, would be costly and time-intensive.

All of Dawn's assumptions were wrong.

The means by which we collaborated were in the middle of a paradigm shift, as were our lifestyles. We wanted to write together, and we knew that we couldn't wait, as had Lunsford and Ede, for snail-mail to deliver our ideas to each other. Our publishing deadline alone wouldn't allow for 5-day mail delivery pauses. We also knew that as collaborators, we enjoy writing *together*. We looked forward to creating our collective voice. In that sense, we are very much non-hierarchical (dialogic) collaborators, personalizing in our own writing process the antidote to painful disconnects we saw harming so many hopefuls in the postmodern job search (Ede & Lunsford, 1990, p. 134). The natural link to maintain this kind of writing relationship and to meet publishing deadlines was technology. Being wired with word processing, e-mail, and the Internet meant we could write *together* (perhaps even more productively than when we lived in the same city).

Yet, we realized that, having been trained in only traditional forms of scholarly production, we would have to carve out our own writing spaces as we wrote.

Scene 2: A chat room in cyberspace. Dawn logs in from California, Cheryl from Pennsylvania, and the other participants from all points in between. As the invited guests in this session, we're expecting to engage in mock interviews with grad students who want preinterview practice answering tough questions, but we find participants reluctant to expose themselves even in this relatively anonymous space. Faced with only a computer screen, able to log in with pseudonyms, physically absent, potential interviewees still feel the intense intimacy of job interview rhetoric. So, as with most electronic chats, comments overlap each other, questions are answered out of synch with the thoughts currently appearing on screen, and pairs or small groups of chat participants step "outside" for a private exchange from time to time. The overall feel is one of friendly interest and collegiality. We never interview anyone.

Along with face-to-face workshops conducted with graduates entering the academic job market, this MOO session convinced us that our book needed to be written. Even in a supposedly nameless, genderless, status-free environment, potential faculty candidates revealed that they were too intimidated to participate in online mock interviews. Their reluctance to practice and their specific questions about what to anticipate in real interviews told us exactly what we'd long suspected: Graduate students need coaching in how to use their rhetorical training to help themselves secure employment. Responses to the Web page describing *Job Search* (a link to Cheryl's faculty home page) reiterated this need in a backhanded way. Although we'd included several pages of information and activities to get job seekers thinking critically about their own interview processes (along with, of course, *Job Search* publishing information), a few Web surfers chastised us for not putting *more* information on the links.

Meanwhile, the diversity-embracing other-space we'd entered at the beach continued to evolve in cyberspace. Our collaboration opened up into virtual infinity when we posted a call for contributions to *Job Search* on two large discussion lists. The prolific, immediate response (via e-mail, of course) was both gratifying and daunting. The day after the call was posted, Cheryl had already received some forty responses, and contributions continued to stream in over the next few days. Overnight, we had a pool of savvy contributors who took our book beyond the lived and theorized experience of two employed PhDs. Now we had musings, speculations, success stories, horror stories, advice from Search Committee members, lists of questions candidates might ask or be asked, and even sample letters sent to various powers-that-be in the search process. Some contributors sent full-length, polished narratives, which, sadly, we could only excerpt. Others offered to review drafts of chapters and provide quick and vital feedback.

The informal, chatty, semi-anonymous character of such e-mail chatter provoked just the tone we wanted to create in *Job Search*. We laughed out loud (and winced) when one witty contributor wrote, "My story is the one graduate students scare each other with around the campfire at night." Other responses generated longer exchanges of commiseration, mutual support, spin-off projects, and, ultimately, new friendships. Although we didn't realize it at the time, these exchanges were beginning to form the style of presentation we would later adopt for the entire book.

The generic characteristics of e-mail would also play a part in our ability to realize our multi-vocal, empowering vision for *Job Search*. Later in the writing process, when we had a draft ready to send to the publisher, we used electronic media to ensure contributor anonymity—not just in e-mail exchanges, but also in the printed version of our text. Because the serial, disembodied e-mail exchanges Cheryl had initiated with electronic personas unveiled some intimate, and often politically charged, real-life situations, Cheryl decided that each contributor should see how his or her words were being adapted to the book *before* the manuscript went to the publisher. She used standard word processing capabilities to search the entire, 250-page manuscript for items from each of the some 60 contributors. She then e-mailed each contributor a cut-and-pasted document that showed *where* quotes or narratives were being used (context) and *how* they were used *throughout the text* (scope). Poignantly, many contributors who'd written freely of their experiences in the relative anonymity of an e-mail exchange (or even the friendly comfort of a face-to-face interview) asked that their names be omitted from the *written, published* work. The "virtually verbal" interchange over e-mail was fluid, malleable, and non-threatening. Once those virtual exchanges were envisioned as part of a static, "official" print text, however, their writers applied less forgiving criteria to their own words—and began worrying about the real-world repercussions of what they'd written.

Incidentally, the cut-paste-and-e-mail maneuver, while designed to inform contributors how their words were being used, also showed us that we'd drawn too heavily from one or two contributors. While it was almost impossible to see how much any one contributor had been quoted *throughout the book,* cutting and pasting all items from an individual writer into one document let us visually compare length. When we saw that a particular contributor had been quoted much more frequently than the others, we went back to our collection of e-mail responses: Was there a similar thought hiding among underquoted or even omitted contributors? The unexpected boon of Cheryl's e-mail blitz, then, was textual balance. Ironically, given early computers and writing speculation that *electronic* spaces were more democratic, we used electronic media's ease of global revision and immediacy of communication to make our *print text* more inclusive, multi-vocal, and, of course, *safe* in which to express ideas.

> *Scene 3: A funky coffee shop in San Diego. We are again face to face, so to speak. A mother and her small son play pool across the cavernous, cozy room; nearer, a writer bends gently over his work as 60s music feeds our souls (Cheryl)/distracts us (Dawn). The proprietor's tie-dyed shirt is the real thing. Ideas are flowing, and we talk almost as much as we write. Although we both have brought our laptops, we opt to use Dawn's IBM, taking turns capturing the spell of each other's words. This session is very much about tone; in fact, we write ourselves a note: "Our book should read like tips from a couple of inside secrets over latte." The rhythmic, sometimes frenetic clicking of the keys is punctuated by spontaneous, largely ineffectual gestures at the screen and exclamations like, "Press the arrow! Hit return!" Because we are working cross-platform, there are also dark times when Cheryl moans, "Ohmagod what did I just do? On my machine that button would have—" The verbal weaves in and out of the electronic: Magically, the conversation that's been developing over several years and countless philosophical crises takes on an ephemeral substance in the little glowing box we balance in our laps.*

Simultaneously with, and juxtaposed by,

> *Scene 4: A small apartment in San Diego. Dawn's on the couch reworking the introduction (again) on her IBM laptop; Cheryl sits in a beach chair on the floor drafting chapter 3 on her Mac. The concentration is intense. We are simultaneously deep within the images dancing across our screens and acutely aware of each other's thinking processes. From time to time, one of us will talk out of the page we're writing, eyes not lifting from her own screen, as if both of us are reading the same words: "I really think we need to—". Sometimes we talk eye to eye, leaning back from our laptops as if to disconnect from the machines temporarily. Suddenly, it's 3:00 a.m. This scenario plays again and again as the book takes shape.*

When we tried to compose *Job Search* the traditional way—birthing a text as a series of multi-vocal, but rather linear, unified, writing acts—we bogged down.

While we, for instance, started out trying to work together on one chapter at a time, moving through sections only as we agreed on each section's content and tone, we quickly abandoned this for a more workable, and at times desperate, dissolution of our writing acts into fragmented bits of process: "You have a lot to say about negotiating contracts; why not work on that? And I want to pull the bit about telephone interviews out of its existential void."

Of course, both composition and collaboration theories have long acknowledged that writers and collaborators don't compose in a linear path. However, electronic media allowed our emerging text to be so malleable that we could, in a sense, always be working on the whole text all the time. Even the smallest bit of text or process (e.g., revising a paragraph, inserting a quote, deleting a reference) was always already part of the imaginary whole. With a flick of the keyboard, portions of text could be moved from one section or chapter to another (and back again, as we hashed through points), inserted seamlessly into the emerging text(s) of nascent chapters. Rather than merely dumping extraneous text into a running file of outtakes or jumbled notes-to-self about how to approach a later section *when we got to it* (in essence starting the writing process over and over again) we "grew" nascent chapters even as we ostensibly concentrated on other sections. Much like the overlapping, rather non-linear threads of a rapid-paced chat room exchange, we wrote and revised points, lines, and whole chapter sections as they came up in our collaborative, multi-textual, and face-to-face conversations.

We dipped in and out of contributor e-mails; responses from peer readers on earlier drafts (zapped off as e-mail attachments, still dripping and bloody from our composing session); the taped voices of interviewees; our own face-to-face interactions; and even a video series of Leonard Bernstein's *Young People's Concerts* suggested to us by Cheryl's son Sam. When both laptops were engaged, we even borrowed the desktop in the next room to surf a library database or check for e-mail responses to queries sent out earlier. With dual composers and keyboards, conversations not directly related to the main thread of a particular discussion could continue on their own as viable asides, with Dawn or Cheryl diverting quickly from one chapter to develop bits of another, then coming back again to the main project when that thread of discussion rounded out for the moment. Like a text-based electronic chat, our text became a running transcript of developing themes. Importantly for us, no emerging ideas got lost.

The ease with which we could move, replace, insert, and otherwise order text as it emerged, in fact, produced text we hadn't originally envisioned. Rather late in our writing process, we rather masochistically decided that we "really ought to" compose a chapter of practice interview scenarios and found with some relief that the bulk of that chapter had already emerged via overlarge example sections we'd cut-and-pasted into two other chapters. A discussion of interview ethics—originally intended as a sort of parting editorial comment for job hunters-turned-Search Committee members—became a substantial, contributor-driven discussion of professional ethics. A quick note about negotiating job offers evolved into a full chapter that included some very detailed advice about faculty contracts. Each of these had grown out of verbal conversations quickly recorded as text as we developed other chapters.

Using the computer, then, allowed us to write and to converse simultaneously. Sometimes the conversations were spoken aloud, but quite often the conversations were written into the text in the form of multiply-bracketed, all-caps comments (easy to search for electronically, and not likely to get retranslated as hieroglyphics during e-mail transit). Some comments were requests for specific help: [[[I CAN'T GET THIS QUOTE TO FIT. HELP!]]] Many comments were mutual encouragement, such as, [[[I LOVE THIS LINE!]]], which we, quite ingenuously, occasionally forgot to delete before sending our drafts out to peer readers. Some comments took the form of questions about the structure of a paragraph or chapter: [[[DO WE NEED THIS HERE? SHOULD IT GO IN THE ETHICS SECTION?]]]

Engaging in writerly production and editorial comment simultaneously allowed us to keep our fingers moving across the keyboard, collecting most of the questions, concerns, and tangents along the way. Our tangents often were our humorous interjections to ourselves or to each other. They expressed our immediate, intuitive reactions to the professional issues we were analyzing, our act(s) of living and working in particular institutional contexts, and our own characteristic styles of negotiating these constructions. These very private notes to our collective authorial self would never appear in the finished text, and weren't intended as such. Yet, these meta-notes about our process and our relationship in many ways drove the writing process.

While we often wrote together, we also worked individually on first drafts of several chapters with the intent of swapping computers a day or two later. When we swapped computers, the other's text became our own. Our comments didn't usually take the form of tentative questions; instead, we simply changed each other's texts. As we became more aware of our collective voice, we felt less and less that such "take-overs" were usurping each other's words. Rather, we were writing from one mind with two nodes (one in Dawn's head, one in Cheryl's)— connected, it seemed, on the computer screen.

Scene 5: Somewhere over the Atlantic. Cheryl's travel grant has run out. We are again 3,000 miles apart. Our publisher sends us e-mails subject-lined "30,000 feet over the Atlantic," requesting additions, consults, and clarifications. Manuscript revisions fly coast-to-coast, often overlapping each other. We date them now, instead of giving them names. We are a very long way from the beach.

Added to the impact of technology on our own collaborative writing and research experiences was its influence on other key contributors to our writing careers: the publishers. While we created the first draft of our book manuscript in the same city, the day Cheryl left San Diego, the entire manuscript was sent via e-mail attachments to both Cheryl in Pennsylvania and our publisher in Virginia. For the six months following, we sent many versions of the completed manuscript back and forth via e-mail attachments. The publisher also solicited peer review readers for our manuscript via the Internet.

In the months ahead, with the exception of the book galleys, nearly all of the revision suggestions between Cheryl and Dawn and among Cheryl, Dawn, the publisher, and the copyeditor came via the computer. The computer and its technological capabilities kept all of the contributors of our publishing project connected throughout the process. More importantly, that connection impacted our workload in unexpected ways:

- **Technology allowed our work to be continuous.** A happy distribution of authors along bi-coastal time differences allowed us to burn the candle at both ends. Cheryl could write a substantial part of the day on the East Coast and still get Dawn a revision by the latter part of her workday—or send the 2:00 a.m. version of the manuscript to Dawn so she breakfasted over it while Cheryl slept. (Incidentally, it wouldn't have worked quite as well the other way around: A night owl on the West Coast and an early riser on the East would've clashed terribly. And Dawn will tell you that phone calls from the East Coast can come awfully early.)

- **Technology both decreased and increased our workload.** Electronic media created a level of efficiency, thoroughness, and speed that we could not have accomplished otherwise. Ironically, the speed with which we were able to revise and conduct online research resulted in more work. Three months before the manuscript was to be published, our editor asked for additions that in a pre-electronic era would have likely been saved for a sequel.

- **Working cross-platform enabled/mandated us to distribute key tasks.** Since our copyeditor and editor requested an IBM-formatted text, Dawn assumed responsibility for organizing and distributing the manuscript to these folks throughout the revision process (a huge and often thankless task). At very different points of the writing process, Cheryl used her Mac to initiate calls for contributions, assemble contributor submissions, or follow up on release forms, contributor acknowledgement preferences, and address changes. This made our workload appear unbalanced at different points of the process—one of us always appeared to be busier than the other.

Yet we found, throughout this process, that we needed each other professionally to conavigate the fragmented postmodern researcher/teacher's reality. And we found that we slipped quite easily into using technology to achieve this personal and professional connection. Getting wired together helped us mutually utilize and process the "vast networks of information" one of Dawn's savvy students, Leiana Naholowaa, writes about:

Jean-Francois Lyotard's *The Postmodern Condition: A Report on Knowledge* encapsulates the present-day reality of post-industrial societies where data banks "transcend the capacity of each of their users" and so become " 'nature' for postmodern man" (51). The interconnectedness enhanced by computers and sophisticated technology requires an honest assessment of the processing, creating, arguing, and employing of vast networks of information.[2]

PUTTING THE PIECES TOGETHER

For us, electronic, collaborative research and composing represented the most appropriate medium for producing *Job Search,* a project that is informed in theory and practice by the postmodern thought that reality and truth are complex, fragmented, and multiply-authored. In writing *Job Search* and this article, we in many ways enacted a postmodern construction of meaning. We were always dealing in pieces of text:

- Putting pieces together across platforms, with contributors from across the country, with an editor who was often writing to us from 30,000 feet above the Atlantic, and with a copyeditor in New York
- Taking pieces apart in order to discard them in some cases and, at other times, to attach them to other pieces
- (Re)constructing these fragmented truths in such a way that they make meaning for graduate students questing after academic employment

In a very refreshing way, producing *Job Search* together also caused us to create the real writing spaces we ask our students to inhabit. Most of us have now designed many of our classes as rhetorical spaces filled with shared learning and writing. Research by Ede and Lunsford (1990), Stanley Aronowitz (1992), and Carol L. Winkelmann (1995) convince us that the non-academic professional world and the democracy in which we intend for students to participate fully is all about what Winkelmann labels the "corporate" or collaborative text (p. 437). The continued research of Kenneth Bruffee (1993), Anne Ruggles Gere (1987), Patricia Sullivan (1996), and Karen B. LeFevre (1987) persuades us to develop our collaborative learning pedagogies. Technology, we would add, encourages new ways of thinking and composing *together,* which in turn results in texts that could not have occurred otherwise.

Serendipitous emergence and revision of text is, of course, part of the writing process, itself. However, electronic collaboration enhanced our ability to grow a multi-vocal text by keeping us in touch with many voices, texts, and conversations simultaneously as part of our writing process. Like the interactivity be-

---

[2]Quoted from Naholowaa's teaching philosophy, written in response to a seminar assignment.

tween Web links Alice Trupe (2002) notes in her forthcoming Web text on basic writers and electronic media, "This interactivity changes the ways in which connection between chunks of text is established, through associative linking rather than through verbal cues." Our mix of media/multiple texts funneled right into our growing text onscreen rather than remaining separate. Electronic capabilities, in a real sense then, allowed for the development of a third collaborator: our computers.

13
What's in a Name?
Defining Electronic Community

Donna N. Sewell
Valdosta State University

The impetus for this chapter began in a physical community, the English Department at Valdosta State University. As colleagues left for graduate school or other jobs, we used e-mail to stay in touch, allowing my local community to expand to reach now-distant colleagues. Thus, electronic media, for me, have always been connected to notions of community. Using e-mail led me to other electronic media: discussion lists and MOOs. Each forum felt like a community, each with a distinct culture. Because of this connection between technology and community in my experience, claims by scholars such as Stephen Doheny-Farina (1996) that community does not exist online startled me. Believing myself to be a member of several electronic communities, I wondered why someone would deny their existence. This chapter, then, derives from my exploration of such denials. Although incorporating scholarly discussions about communities, this chapter springs from the physical and electronic communities that I inhabit.

Community traditionally refers to a physical place where people gather. Electronic communities work much the same way, except that the gatherings occur via data streams and electric current rather than in a physical place. Although many people perceive such electronic forums as wonderful resources, as founts of knowledge to access when needed, some of these forums go beyond mere resources to become communities. This chapter explores definitions of community, searching for ways to understand better the implications of these definitions as well as their appropriateness for electronic media. To make this analysis more concrete, I examine a community in each of two electronic media, discussion lists and MOO environments.

ELECTRONIC MEDIA

First, I review the distinctions between discussion lists and MOO environments. An electronic discussion list is a low-tech environment that uses listserv technology to distribute electronic mail to all subscribers; subscribers may then post messages to the group and receive e-mail in return, with the system creating a one-to-many exchange.[1] Peter Kollock and Marc A. Smith (1999) describe discussion lists as "push media" because once people subscribe "messages are sent to people without them necessarily doing anything" (p. 6). The particular discussion list analyzed in this chapter is WCENTER, a discussion list for individuals interested in writing center issues.[2]

MOO environments are also low-tech, requiring only telnet access to connect to the environment, though most users rely on a client (software designed to ease communication). MOO environments allow synchronous communication in addition to the asynchronous communication of discussion lists. MOOs facilitate more than mere chat, however, differing from Internet Relay Chat and other real-time chat programs in that they allow users to extend and interact with the environment.[3] This chapter analyzes Tuesday Café,[4] a weekly

---

[1]Discussion lists are more generally known as electronic lists, or e-lists. Readers may have heard terms such as listserv, named in this chapter; listproc; and majordomo associated with e-lists, but these are actual technologies that enable e-list management, not the e-lists themselves, an important distinction for readers to keep in mind.

[2]An archive of WCENTER discussions is available online at http://www.ttu.edu/wcenter, and subscribers may post to the e-list by sending an e-mail to wcenter@ttacs.ttu.edu. Readers interested in subscribing may do so by sending an e-mail to listproc@listserv.ttu.edu, with no subject line and with the message text reading only "subscribe wcenter firstname lastname," where obviously the subscriber's real first and last names are substituted in the command line.

[3]MOO is an acronym for MUD Object-Oriented, meaning that a MOO is a specialized type of MUD. MUD is an acronym for Multi-User Domain or Multi-User Dimension. MUDs derive from fantasy role-playing games, such as *Dungeons and Dragons*. MUD players, meaning anyone who's logged in and online, can create fantasy worlds. For more information about educational MUDs, readers should see Cynthia Haynes and Jan Rune Holmevik's *High Wired: On the Design, Use, and Theory of Educational MOOs* (1998). For more general scholarship about MUDs, readers should reference the Lost Library of MOO, located at the following URL: http://www.hayseed.net/MOO. Should the Web address change, readers can locate the Lost Library by using a search engine, with "Lost Library of MOO" as the search term. Other authors in this volume also discuss MOOs; see chapters by Sherry Turkle and Karen McComas, in particular.

[4]The Tuesday Café is sponsored by the Netoric Project, a computers and writing community founded in 1993 by Tari Fanderclai and Greg Siering. In 1995, Netoric won the Contributions to the Community Award in the Teachers and Technology Awards contest sponsored by the Alliance for Computers and Writing, Sixth-Floor Media, and McDougall-Littell. In 1998, James A. Inman became a cocoordinator, followed in 1999 by Cindy Wambeam. Readers can learn more about Netoric by visiting its Web site at the following URL: http://bsuvc.bsu.edu/~00gjsiering/netoric/netoric.html. Should the Web address change, readers can locate the site by accessing a search engine and using "Netoric project" as a search term.

gathering of computers and writing specialists on Connections, an educational MOO.[5]

WCENTER

WCENTER, a discussion list focusing on writing center issues, functions as an environment in which newcomers can ask questions, directors can brainstorm about problems facing their centers, and all participants can do everything from formally planning conference presentations to informally sharing birthday greetings. The boundaries between personal and professional issues are permeable, allowing for many crossovers. The list averages about 20 postings a day, though 140 messages were posted on March 3, 1999. The 140 messages were prompted in part by a request to know what reading was on everyone's bedside table. Simone Gers (1999), a WCENTER subscriber, posts about her perception of WCENTER:

> I have learned so much about issues that are relevant to my situation and have shared several wcenter discussions with colleagues. I enjoy the friendly banter and camaraderie. These aspects of the list suggest to this neophyte that the group is friendly and open to new voices. The side threads are not hard to follow. What keeps me reading is the level of discussion—I appreciate the thoughtful posts.

This description of WCENTER captures why people continue to subscribe. WCENTER allows writing center staff to focus on relevant issues: tutor training, job announcements, certification programs, and electronic tutorials. Yet, as Gers points out, WCENTER is also a friendly place, one encouraging to newcomers and open to personal musings. WCENTER is one of the most welcoming electronic or physical communities I know.

Tuesday Café

The Netoric Project's Tuesday Café allows computers and writing professionals to gather once a week for an hour to discuss relevant issues. During September of 1999, topics included handling inappropriate behavior on electronic discussion lists, politicizing the computer writing classroom, and indexing comics at the University of Florida. Because attending the Tuesday Café requires real-time communication, fewer people participate, but the participants are often more intensely involved in the discussion. Cafés may average approximately 10 participants at any meeting, but half of those are often involved in the public discussion. Because MOOs allow for private conversations as well, one cannot be completely sure how many conversations occur simultaneously.

---

[5]Readers may connect to ConnectionsMOO via telnet at connections.moo.mud.org:3333, if comfortable with using telnet. Readers who want to learn more about Connections should consult information and documentation provided by Tari Fanderclai at the following URL: http://web.nwe.ufl.edu/~tari/connections

Tari Fanderclai (1996), one of the founders of the Tuesday Café, aptly describes the social nature of MOOs:

> As with asynchronous forums, I am connected to people who share my interests, but MUDs provide something more. For example, the combination of real-time interaction and the permanent rooms, characters, and objects contributes to a sense of being in a shared space with friends and colleagues. The custom of using one's first name or a fantasy name for one's MUD persona puts the inhabitants of a MUD on a more equal footing than generally exists in a forum where names are accompanied by titles and affiliations. The novelty and playfulness inherent in the environment blur the distinctions between work and play, encouraging a freedom that is often more productive and more enjoyable than the more formal exchange of other forums. . . . MUDs provide a sense of belonging to a community and encourage collaboration among participants, closing geographical distances among potential colleagues and collaborators who might otherwise have never met. (pp. 228–229)

As Fanderclai notes, meeting in real time encourages a greater sense of community than is achieved simply by asynchronous communication. Misunderstandings can be corrected without too much of a time lapse, and the interactivity of the environment encourages playful behavior, all of which help build community.

CHARACTERISTICS OF COMMUNITY

Definitions of physical and electronic community abound. John Unsworth (1996), one of the cofounders of *Postmodern Culture*,[6] offers a typical definition: "Community is generally a function of shared location, shared interests, and sometimes shared government and shared property; in order to deserve the name, a community needs more than one, though not necessarily all, of these attributes" (p. 138). I use these criteria of traditional communities to examine electronic communities before examining what is missing from current definitions of them: interaction, communication, obligation, and emotional connection.

Shared Location

If shared location generally contributes to community, do electronic communities count? Do they exist in a place? Stephen Doheny-Farina (1996) argues that the notion of place precludes electronic forums from being communities: "A community is bound by place. . . . It isn't a place you can easily join. You can't subscribe to a community as you subscribe to a discussion group on the net. It must be lived. It is entwined, contradictory, and involves all our senses" (p. 37). Doheny-Farina

---

[6]Readers may learn more about *Postmodern Culture* by visiting its Web site: http://jefferson. village.virginia.edu/pmc

misses the experience of electronic communities by suggesting that they are not "lived . . . entwined, contradictory" (p. 37). Participants in active electronic communities know how interwoven our online lives can be with our physical lives. Clear lines of difference do not exist between the two. Susan C. Herring (1996a) writes:

> From my observations, academic list subscribers do not view the activity of posting as targeted at disembodied strangers. Their addresses are people with whom they either have a professional relationship, or could potentially develop such a relationship in this future. This is likely to increase (rather than decrease) inhibition, since one's professional reputation is at stake. (p. 487)

For many electronic community members, our online and physical lives complement each other as we e-mail people across the hall and across the country. We may never meet all the other community members, but we do meet some of them. Members of WCENTER gather for a breakfast at the Conference on College Composition and Communication (CCCC) each year in addition to seeing each other at conferences, and we often order pins in advance so that we can easily recognize other subscribers at CCCC. Many Tuesday Café participants attend the annual Computers and Writing Conference.[7]

Although these electronic places cannot be reached by car or boat, they are accessed by using e-mail or telnet addresses. Because typing in the wrong address prevents one from finding that community, many electronic communities, whether they are discussion groups or MOO environments, do inhabit a common space on a network—or at least, the means of communication occurs via a shared space on a network. The space may not be a physical one in which we can stand, but it exists in a more diffuse way, the location stretching to reach its members, in the same way that my local sense of community stretches to include colleagues attending other schools.

MOOs go even further than discussion lists in creating an electronic place, as Michael Day, Eric Crump, and Rebecca Rickly (1996) note:

> Further, the text-based virtual reality in the objects and rooms of the MOO creates a metaphoric place, a constructed world, a fictional work/playspace . . . teachers and researchers use their imaginations and the MOO environment to create a sense of *place* in which productive and creative interaction can occur. (p. 294; my emphasis)

---

[7]The annual Computers and Writing conference began in the early 1980s and is now a prominent event for scholars interested in the impact of information technologies on writing studies. The conference is currently overseen by the Conference on College Composition and Communication's Committee on Computers in Composition and Communication, more informally known as 7C's.

Visitors to Connections and Diversity University,[8] educational MOOs, enter varied "rooms" that suggest certain uses, allowing some behaviors and discouraging others. Upon entering Tuesday Café, one sees the following text:

> A comfortable little cafe with rough wooden floors and walls. A long counter with a row of tall wooden stools in front of it lines about half of the north wall. Rhet stands behind the counter, ready to serve. A large menu is posted on the wall behind him. Near the counter is a booth with a table between two long benches. A small whiteboard listing upcoming discussion topics rests on a stand in a corner. At the west end of the room is a projector. Next to the projector is a recorder. In the center of the room is a polished wooden table surrounded by matching chairs. The hallway lies to the south. (January 1, 2001)

This textual description allows visitors to imagine this gathering place as a café, complete with Rhet, the server, and a menu posted on the wall. Rhet is more than mere description; because of Rhet's programming, visitors may order virtual food and drink by typing in the correct commands. This quick glance at Tuesday Café should help readers get a sense for the type of place that can be created electronically through text. In addition, computers and writing professionals gather every Tuesday night in real time to meet and discuss relevant issues, using shared time to heighten the sense of shared space.

Before moving away from a discussion of place, I must emphasize that not everyone agrees that shared location is required for community. Steven G. Jones (1995b), in "Understanding Community in the Information Age" writes:

> Definitions of community have largely centered around the unproblematized notion of place, a "where" that social scientists can observe, visit, stay, and go. Their observations had largely been formed by examination of events, artifacts, and social relations within distinct geographic boundaries. (p. 19)

Jones dismisses the notion of shared location as a criterion, preferring instead the work of Thomas Bender, who defines community "not as places but as social networks, a definition useful for the study of community in cyberspace for two reasons. First, it focuses on the interactions that create communities. Second, it focuses away from place" (p. 24). Although place may be problematic as an attribute of electronic communities, if we insist on using it, many groups meet that criterion. If, as Jones and Bender suggest, interaction is a better criterion, electronic communities certainly foster interaction, at least from a core of the group. Later in this chapter, I analyze interaction more fully.

---

[8]To learn more about Diversity University, readers should access the project's Web site at the following URL: http://www.du.org Additionally, readers should see Karen McComas's chapter in this volume.

Shared Interest

Shared interest, Unsworth's second criterion, is the least controversial element of electronic communities. Even scholars who refuse to accept such collectives as communities agree that shared interests identify these groups. However, shared interests are the only attribute of communities that some people grant to electronic forums. Such forums attract interested groups of soap opera fans, computers and writing participants, writing center staff, and fiction writers. These environments are accessible via e-mail or telnet, and their logged discussions, whether asynchronous or synchronous, are posted on the Web at certain locations.[9]

Shared Government

Shared government is harder to see online, but means of handling grievances generally evolve in electronic communities. These communities often create lists of frequently asked questions (FAQs) or Web pages that suggest appropriate and inappropriate behavior. FAQs may be posted regularly on discussion lists, and newcomers may be required to read community policies before requesting a character on a MOO. The Netoric Project's Web page, for example, provides information on the purpose of its Tuesday Café meetings and establishes policies for researchers. With discussion lists, inappropriate e-mail postings are dealt with in varied ways according to the character of the community. WCENTER, for example, tends to be very gentle with newcomers who behave inappropriately, a response which fits the tone of most writing center conversation.

Online communities, like physical ones, often contain individuals who are more involved than others. Marc A. Smith (1999) writes:

> Some newsgroups are populated by a core of dedicated participants who contribute much of the value found in the group. A core group of posters can act as a means of socialization, ensuring that group experiences and lessons are conveyed to the next generation of participants. The presence or absence of a core group who produce a significant and disproportionate amount of the participation may explain why some newsgroups are more ordered and productive than others. (p. 210)[10]

To examine WCENTER carefully, I downloaded the WCENTER archives for January through May 1999. Defining dedicated participants as those people who posted at least once a month for any 3 of those 5 months, I narrowed the 366 posters down to 138, 38% of the poster population. This 38% posted 2,091 messages, 79% of all the messages posted to WCENTER during this time. These numbers support Smith's claim that a core group of participants set the tone of an elec-

---

[9]As of this writing, Tuesday Café logs reside at http://bsuvc.bsu.edu/~00gjsiering/netoric/logs and WCENTER archives at http://www.ttu.edu/wcenter

[10]A newsgroup is similar to a discussion list: They are both asynchronous and low-technology.

tronic community. Thus, shared government occurs, in part, via shared cultural norms.

In the final analysis, however, the technology of discussion lists and MOOs is hierarchical. In electronic environments, all users usually have a voice, but not all users have a vote. Final authority resides with the list owner[11] of a discussion list and the highest ranking wizard[12] of a MOO. The technology allows these authorities to delete subscribers, moderate the list, and control the groups in other ways. WCENTER and Connections have generous leaders who do not exercise much control over participants, but the potential exists.

Shared Property

Shared property is probably Unsworth's criterion of traditional community least relevant to electronic community. On e-mail discussion lists, property exists only in the written texts that evolve from the discussions, and each post belongs to an individual. The text, though linked to an individual, enters the consciousness of community members and becomes part of the community's history as a result. The question, then, focuses on how shared that property is. All that really seems to be shared in discussion lists is the culture of the list. No tangible property exists.

Because characters (or players, as they are known) on some MOOs may contribute to the environment by building rooms and worlds online, MOO builders do have a sense of shared property. An example of this shared property can be seen by using telnet to log in to Connections and walking to Americana, a realm created by junior-level American literature students from Valdosta State University.[13] This world contains rooms modeled after "The Cask of Amontillado," "Rappaccini's Daughter," and "The Yellow Wallpaper." The opening description for Rappaccini's Garden reads as follows:

> A sunny garden with high stone walls to keep out curious eyes. A crumbling marble fountain at the center of the garden gushes sparkling water. A marble basin in the fountain's pool holds a shrub with purple flowers. Flowers cover almost every surface of the garden, climbing the walls and draping broken statuary. Pale yellow

[11]The responsibilities of listowners vary from e-list to e-list. On e-lists where the discussion is unmoderated, meaning that all messages immediately go out to subscribers without needing to be approved first, the listowner does not have to be very active. However, on e-lists where the discussion is moderated, the listowner must approve each message, which can be time-consuming.

[12]Readers no doubt notice the title's reference to fantasy and role-playing games. Wizards in MOOs are essentially administrators, but in that role, they face many challenges, from creating new players and allocating account space to managing social conflicts that occasionally arise. For a particularly telling example of social conflicts in MOOs and a critical discussion of its implications, readers should see Julian Dibbell's *My Tiny Life: Crime and Passion in a Virtual World* (1999).

[13]Telnet software is available on most computers. Instead of entering a Web address, such as http://www.valdosta.edu users would enter a telnet address, such as the following for ConnectionsMOO: telnet://connections.moo.mud.org:3333

flowers bloom on a vine entwined around a statue of Vertumnus, god of seasons. An antique sculptured portal covered in poison ivy leads east to another wing of the garden. A carriage waits at the entrance of the Garden to take you to Florence. A purple flower is lying on the ground. You see a scroll here. (January 1, 2001)

These worlds, however, are electronic ones, so seeing this world as an example of shared property means accepting the notion that property does not have to be tangible.

Most of the attributes Unsworth lists for traditional communities (shared location, interests, government, and property) fit electronic communities at least loosely. However, even if they do not fit, Unsworth states that meeting all these criteria is not necessary. Thus, even using traditional standards, electronic communities deserve to be identified as communities. Earlier, though, Stephen Jones (1995b) disagreed with the notion of place as relevant to electronic community, arguing instead for community to be defined by interaction. I now return to the notion of interaction as a means of identifying community membership before examining communication, obligation, and emotional connection.

Interaction

Electronic communities are readily available for people with access to computers and modems, but community requires more than mere physical access. Such access is a necessary but insufficient indication of community, as Derek Foster (1997) notes: "Virtual communities require much more than the mere act of connection itself" (p. 29). When computer users visit an electronic community merely to discover information, they are accessing resources, not joining a community. Using a community's resources is akin to lurking—reading the community's conversations but not participating. Although some may want to equate reading with participating, such participation benefits only the reader and not the community. Lurkers are present in the environment, in that they access the materials, but they are not part of the community. They observe the community interaction without participating in it. Lurkers do so for varied reasons. In a post to WCENTER, Lady Falls Brown (1998) reports on Muriel Harris's study of WCENTER lurkers who lurked because they were shy, did not have sufficient time to interact, or just wanted information from the group. Jeanne Simpson (1998) responds by noting that some lurkers indicated that they do not post on WCENTER because they feel intimidated to do so.

Richard C. MacKinnon (1995) helpfully distinguishes between users (the people who sit down to use the computer) and personae (online characters) and uses these distinctions to argue that lurkers are not community members:

A user can read and contemplate the words of another user, but unless there is a visible (i.e., written) response via his persona, the action of reading and contemplating goes unnoticed. If a user is unnoticed, then he or she is not interacting with other users. Because personae are created as a result of interaction, reading and contemplating alone are insufficient to generate or maintain the existence of a persona. (p. 119)

For MacKinnon, readers (or computer users) cannot be members of the community because they do not exist within the community. MacKinnon continues, "The passive user remains outside the boundary of Usenet existence and his or her actions are unnoticed to 'life' within" (p. 120). To deny someone community membership is not to say the person does not exist. MacKinnon's distinction allows us to say that the user/lurker exists, but a persona does not. Other list members (lurkers and personae) do not know that other lurkers exist until they identify themselves, at which point they are no longer lurkers. Why does such a distinction matter? Because interaction matters to a community. Without interaction, a community will die. Conversely, with too much interaction, a community will drive off members because people cannot read all the messages. I do not want to suggest that all WCENTER subscribers should post or unsubscribe. Perhaps what is missing is more precise terminology to account for the hundreds of subscribers to electronic discussion lists, for the thousands of avid but silent readers of Usenet groups, for the people who attend MOO meetings but do not speak.

In "The Internet and Its Social Landscape," Steven G. Jones (1998) states that lurking is not a social behavior and compares it to other reading activities: "There is a remarkable parallel between lurking and reading. . . . Reading, and print culture generally, have been criticized for the ways they isolate individuals . . . , promoting a sense of the imagined, the 'read about,' rather than engagement with the world" (pp. 13–14). Reading gives one a sense of the community without making one a part of it. People often listen in on electronic forums for awhile before interacting with members. Lurking allows one to get a feel for the community before participating in it. I visited schools before I attended them or joined the faculty, and I visit community organizations before deciding to participate, so the word *lurk* connects to physical life. Lurking, though a practical and useful behavior for many reasons, is insufficient for community membership.

Communication

Community membership requires communication, and the line between communication and community is blurry. Derek Foster (1997) writes, "But though communication serves as the basis of community, it must not be equated with it. One can communicate with another individual without considering that person to be a member of one's own community" (p. 24). Like access, communication is required for community but insufficient alone. Michael Day, Eric Crump, and Rebecca Rickly (1996) note the importance of play in online environments, such as Tuesday Café:

There is always a specific and timely topic of professional interest, but the conversation is frequently littered with personal asides, jokes, gossip—all the social "glue" that holds communities together and helps people enjoy with others what *could* be just another dreadfully dull working meeting. (p. 303)

This description gets at some of the qualities that make the Café meetings a community rather than simply a gathering of academics. Community members interact on varied levels and in varied ways—they talk, they joke, they argue. They often do more than converse.

Obligation

Another aspect of community is obligation. Dave Healy (1997) writes, "In the face of increasing geographic and occupational mobility, we need ways to maintain connections and connectedness, and the Net provides almost limitless possibilities. But the Net is also an apt medium for contemporary Americans because it doesn't *demand* anything" (pp. 64–65). Healy is right that electronic forums do not demand participation; that decision is the individual's. However, obligation distinguishes community members from lurkers. Community members feel obliged to the community; they feel commitment. This trait is not something that can be identified by counting posts. Instead, this criterion is more internalized, one that only a person can determine rather than an outside observer. This type of characteristic underscores the need for qualitative research in electronic communities.

Community membership is determined, in part, by the individual and that individual's interactions with others. Considering how e-mail and e-lists function should enable me to read or delete messages freely, but a sense of community changes this for me, compelling me to read messages and participate in communities that are important to me. If I delete messages without reading them, I lose part of the discussion, treating it more like a resource than a conversation. I choose which forums to join as communities and which ones to treat as resources. Although I agree with Foster that conversation doesn't equal community, I do think the two are closely related, agreeing with Eric Crump (1998), who writes, "Conversation is the stuff from which communities grow" (p. 190).

Emotional Connection

Foster (1997) argues for a movement away from objective criteria in analyzing community: "That which holds a virtual community intact is the subjective criterion of togetherness, a feeling of connectedness that confers a sense of belonging" (p. 29). Foster's point gets at the heart of community. If community is created in part by a feeling of togetherness, that feeling cannot exist independently of people. If community means a group of people with common interests who interact (professional, personally, or both), then WCENTER and Tuesday Café both function as communities for some of the people who participate. Electronic communities involve emotions as well as reason.

Community is not a static object or place, but people with common goals. As with face-to-face communities, members are welcome as long as they pitch in, help out, express interest, participate. Joining a community requires commitment and active participation, though people are also welcome to lurk. Analyzing criteria for community membership is not meant to exclude anyone but to make the terminology meaningful.

IMPLICATIONS

This chapter accepts Unsworth's criteria (1996) for communities, which call for some of the following: shared location, shared interests, shared government, and shared property. However, in agreement with MacKinnon (1995), Foster (1997), and Healy (1997), I add interaction, communication, obligation, and emotional connection to Unsworth's list. These criteria, derived from reading in the field and applying those definitions to my own online practices, need to be measured by others. Humanities researchers need to study online communities in more depth, spending more time online as participant-observers, moving research beyond the broad outlines sketched in this chapter. Time spent in the field may turn out to be time spent online for researchers interested in ethnographic research.

Researchers need to use qualitative research methodologies (interviews, participant-observers, textual analysis) to investigate electronic communities, examining at what point a forum moves from resource to community for users. Case studies are needed that follow people online, learning how people create and join communities, when they move from lurking to participating (if they ever do), and why such changes do or do not happen. To echo Lynn Cherny's statement (1999), more such studies are sorely needed:

> There is a pressing need for consensus on—or at least discussion about—what the research questions are, what study methods are appropriate, what case studies exist already, what the research coverage is for different kinds of computer-mediated communication, and where the good comparative studies are that build on existing work. (p. 2)

Some research into electronic environments is already happening, in fact. *Electronic Discourse: Linguistic Individuals in Virtual Space* by Boyd H. Davis and Jeutonne P. Brewer (1997) exemplifies the close textual analysis needed of these online forums. This book investigates students' computer conferences over 3 consecutive semesters, analyzing their linguistic features. Cherny's *Conversation and Community: Chat in a Virtual World* documents a year-long ethnographic study of a MUD community.

I hope this chapter engages readers in the definitional issues surrounding electronic communities and piques readers' interest in further research. As scholars spend more time online, we need to pay attention to how this time is being spent and what kinds of electronic environments we create for our colleagues and students.

14

Cow Tale: A Story of Transformation in Two MOO Communities

Karen McComas
Marshall University

> *Every now and then something in our deeper selves enables us to realize*
> *that what truly counts in life is not a matter of what is in you or what is in*
> *me but of what occurs between us. That divine spark of relationship may*
> *be the most fundamental life force of all.*
> —Albert Murphy (1981, p. 473)

In 1986 I left my clinical practice as a Speech Pathologist in the public school system to accept a faculty position in the Department of Communication Disorders at Marshall University in Huntington, West Virginia. Through teaching academic courses and supervising clinical practice, I uncovered a passion for teaching and learning I did not know existed within me. As the semesters rolled by, I felt an urgency to talk with my university colleagues about teaching and learning and a need to share what I was reading, thinking, and observing in my classroom. Although I had daily contact with others on the faculty, our conversations most often focused on the business of teaching, not the craft of teaching; problems with the students, not the development of students; and lack of learning, not the process of learning. Without nourishment, my passion for teaching and learning was in danger of extinction.

The spring of 1993 proved to be a pivotal point in my teaching career. Two events that spring, a writing across the curriculum (WAC) training retreat and the acquisition of a modem for my computer, spawned my professional renewal and transformed my understanding of what it means to say, "I am a teacher." Back then, at the end of my 7th year of teaching, I thought these events were unrelated.

The WAC retreat revealed to me how little I knew about teaching, learning, and writing and prompted me to begin thinking deliberately about these processes. During this 2-day retreat, workshop facilitator Barbara Walvoord challenged me to question my assumptions about teaching, learning, and writing; to revise my course objectives and outline; and to craft new assignments. My new modem, acquired that same spring, gave me access to the Internet, inviting me to explore cyberspace. Together, the retreat and the modem led me to two online MOO communities that were instrumental in orchestrating my professional renewal and transformation.

How these two MOO communities inspired, generated, and supported my transformation as a teacher and a learner is the focus of this chapter. I begin with yet another attempt to explain, in words, the concept of an online environment called a MOO, even though a MOO can only truly be understood through experience. Then, I briefly introduce the two MOO communities primarily responsible for my transformation. Next, I describe what occurred in these communities. Finally, I examine the development of these communities and the individuals within them in the context of transformational learning.

WHAT IS A MOO?

A MOO is a text-based virtual reality environment that resides as a database on a computer, usually called the MOO server, connected to the Internet. MOO (multiuser, object-oriented) systems are direct descendants of a game system called a Multi-User Domain (MUD). The first MUD, developed in Great Britain in 1979, provided gamers with interactivity in their virtual worlds (Anderson, Benjamin, & Paredes-Holt, 1998, p. 144). After a decade, in 1989, MUD evolved into a system called TinyMUD that allowed players to modify and extend the virtual world. Within a year the first MOO, LambdaMOO, appeared, providing a gathering place for socialization on the Internet. Two years later, MOOs evolved into environments developed to support specific professional groups and purposes. The first of these, MediaMOO, led the way for innumerable professionally minded MOOs. Some of these MOOs catered to narrowly defined professional communities, such as BioMOO (a MOO for biologists); others catered to more broadly defined professional communities, such as Diversity University MOO (a MOO for educators from all disciplines; Haynes & Holmevik, 1998, pp. 2–3). MOOs typically have a basic theme or mission around which they are developed and built. Diversity University MOO has an educational theme and mission and was built around the theme of a college campus. LambdaMOO, developed around the theme of a fraternity house, exists as a social organization, instead of a professional one.

It is important to note that despite the game-based history of MOOs, they have received considerable scholarly attention in recent years. MOOs have been used as "interactive research spaces" (Dobler & Bloomberg, 1998, p. 68) and for collaborative instruction (Kemp, 1998, p. 145). LinguaMOO, serving the University of Texas at Dallas Rhetoric and Writing program, hosted an online dissertation defense in 1995 (Grigar & Barber, 1998, p. 192). Numerous university writing centers use a MOO as an online host where writing tutors meet with student writers (Kemp, 1998, p. 146). Haynes and Holmevik (1998, pp. 1–2) suggest that MOOs are a bargain to develop and maintain for schools and universities, providing an Internet technology that works across platforms and systems.

Users connect to a MOO using a telnet program or other MOO programs that are typically referred to as clients. Multiple users can connect to the MOO server at the same time and interact with the MOO server by issuing commands. The MOO database contains virtual objects that respond to user commands with output that corresponds with the respective input command. These virtual objects are simply clusters of MOO commands to provide users with virtual experiences reminiscent of real experiences. For example, a virtual slide projector is a cluster of MOO commands that textualize ways in which a person might interact with a nonvirtual slide projector. To show a virtual slide on the MOO, the user would type: "show 3 on myproject" (where "3" represents a slide number).

People who connect to MOOs are called characters or players. When connected to a MOO, characters can talk with each other in real time and build virtual places or objects (e.g., a virtual tape recorder or a virtual office), thus extending and personalizing the MOO environment. Characters who belong to a MOO community have names but not necessarily their given names. Visitors to a MOO connect using special guest accounts with names like "Blue-Guest" or "Red-Guest." In a MOO, characters communicate by typing and listen by reading. If a character talks, the textual rendering of that action looks like this:

```
Karen says, "I am talking."
```

If a character communicates nonverbally, the textual rendering appears as:

```
Karen waves.
```

And, if a character's thoughts are communicated, the textual rendering uses thought bubbles and looks like this:

```
Karen . o 0 (I wonder why that happened?)
```

Finally, conversations in a MOO follow different conventions than oral conversations. MOO conversations occur in real time, meaning that the transmission of one character's command to the MOO can result in an almost immediate effect on the screens of all other characters within the same MOO room. MOO dialogues are characterized by multiple threads (more than one topic serving as the focus of the discussion) and multiple voices (more than one person talking at the same time). Transcripts of MOO conversations reveal and preserve the rich nature of these interactions, in terms of both breadth and depth. Reviewing transcripts of MOO discussions, my journals, and electronic mail messages reminded me of the critical role these multiple threads and multiple voices played in my transformation. Where possible, I use excerpts from these artifacts to reinforce my belief that "what truly counts in life is not a matter of what is in you or what is in me but of what occurs between us" (Murphy, 1981, p. 473).

THE COMMUNITIES

The two MOO communities that serve as the focus of this chapter occurred some 4 years apart (1993 and 1997) and at two different MOOs (MediaMOO and Diversity University MOO). Undoubtedly my first experience at MediaMOO was the catalyst for my ultimate transformation as a teacher and a learner. I am equally convinced that without success at MediaMOO, the second community at Diversity University MOO (DU MOO) would not have been conceived.

```
Well, I discovered Media[MOO] . . . I am in a writing across the
curriculum (WAC) experiment on my campus and I was looking for a
mailing list about writing and I found MBU[-L] and I subscribed
to that for awhile . . . that  was  at  the  time  when  the
[MediaMOO] address got posted . . . I telnetted to MediaMOO and
visited a couple of times . . . I'm easily addicted to things
. . . all of a sudden I was spending twenty  hours  a  day  on
MediaMOO.  (K. McComas, personal communication, Sept. 27, 1993)
```

Amy Bruckman, then with the Epistemology and Learning group at the Massachusetts Institute of Technology (MIT), developed MediaMOO as a host to media researchers, including teachers experimenting with the use of computer technologies in the teaching and learning process. Although I first visited MediaMOO to attend discussions about computers and writing, hosted by the Netoric Project, I soon became part of a smaller community using MediaMOO as its base.

```
the second time I connected, Pokey paged me . . . I was still a
guest at that point . . . and asked me to join him and try out
the ping pong table. (K. McComas, personal communication, Sept. 27, 1993)
```

Although I had no idea as to Pokey's identity, his friendly invitation and my already established fascination with the MOO compelled me to join him. Pokey became my first mentor at MediaMOO. In fact, his virtual ping-pong table was the direct catalyst for a virtual audiology lab I later developed at MediaMOO. In addition to providing me with models of various ways to use MOO objects, Pokey invited me to join a MOO communication channel called FC (for Future Culture).

Communication channels on a MOO enable individuals to communicate with others on the same channel, even though they may be in different rooms on the MOO. Using the channel allowed us to work in our own project spaces yet gave us ready access to others on the same channel and working in their own MOO spaces.

For the next year, the FC group guided me through the conception, design, and programming of a virtual audiology lab, a virtual space I hoped would enhance my students' understanding and skill in audiometric testing and diagnosis. By the end of the year I was ready to introduce the virtual audiology lab to my students. Because MediaMOO was developed to host researchers and not classes, I had to secure a permanent home for the virtual audiology lab, a place designed specifically for teachers and their students. My search for such a home led me to Diversity University (DU) MOO, the home of the second online community discussed in this chapter.

`Greetings, fellow mootchers`. (P. Ventura, personal communication, Sept. 17, 1997)

Four years after I joined the FC community at MediaMOO, PaulV-mt's message of introduction christened the class titled Teaching in MOOs: Technological and Pedagogical Perspectives as the MOOTCH class and class participants as MOOTCHers. The purpose of the class was to bring educators together on a MOO (DU MOO, in this case) so they could learn more about MOO technology and the pedagogical implications of such technology.

`LucyS-mt says, "Hi, all; my name is Lucy Schultz; I teach in the English Department at the University of Cincinnati; I'm in rhetoric and comp, and my particular research interest is the history of writing instruction in 19`[th] `century schools."`
`JanineC-mt says, "I'm Janine, from Missouri and I teach middle school."` (MOO Transcript, 1997)

Roll call revealed 14 teachers of middle school, high school, and higher education present and accounted for. Class members represented the disciplines of engineering, international studies, composition, computers, journalism, religious studies, business, and communication disorders. We hailed from two countries (Argentina and the United States) and represented nine different states. We committed to meeting 1 evening every week for 16 weeks.

Originally designed with two major units of study (the technology and the pedagogy), the course evolved into an experience with four distinct phases of development, instead of units of study. Instead of adhering to an organization system based on content, the class organized itself based on process. In retrospect, these phases of development that evolved in the MOOTCH community were the same stages of development I experienced in the FC community at MediaMOO. Specifically, I familiarized myself with the MOO by playing with the environment.

Then I moved into a period of building my personal space within the MOO. Next, I reflected my new thinking about teaching and learning by revising my MOO space. Finally, I shared my work by publishing for the larger MOO community.

Playing

There is much research into the role of play in the learning of preschool children, including the shift away from playing to learn once children begin their formal educational experiences and the value of play in transformational learning (Corbell, 1999; Resnick, 1996, 1998). Most newcomers, in spite of their purposes for visiting a MOO, initially engage in a period of play and experimentation. While playing, users test the limits of the MOO, locate models of MOO objects and spaces, and gain new insight about ways to use those MOO objects and spaces in the teaching and learning process. Through play, users engage in low-stakes activities, thus allowing an element of trust to develop among community members. For these reasons, a period of play is critical to the ultimate success of a community and the ultimate transformation as teachers and learners among community members. Although play is critical to community development, how individuals and communities play varies. Important variables influencing community play are the purposes or missions of the MOO, the learning community, and the learners themselves. In both of the MOO communities previously described in this chapter, learners entered the communities through play. How each community played, however, differed significantly.

As a host to media researchers, the mission of MediaMOO was to provide a mechanism for researchers to experiment and collaborate to develop new understandings of how technology facilitates or changes how people learn. The FC community had a particular interest in experimenting with new ways of presenting reality. Individuals within the FC community came from a wide variety of backgrounds (including a professor of history, a high school student, a college student, and others working in various segments of the computer industry) with particular purposes and diverse interests. These factors, purposes, and personal interests influenced how the FC community played. The primary focus of this play was on exploring the boundaries and possibilities of the MOO. Working individually and collaboratively, FCers developed and played with a wide variety of MOO objects. I modified an object, creating a red convertible to transport me from one location to another on the MOO. Another character modified the same object to create a charging ram. Still another character modified the same object and created a ferris wheel later used in a virtual amusement park.

Given MediaMOO's mission to host media researchers, this mode of play was understandable and expected. DU MOO's mission, on the other hand, was to host teachers and their students. The MOOTCH community formed for the specific purpose of learning how to use MOO technology to facilitate the teaching and learning process. MOOTCHers, although representing diverse disciplines and grade levels, were all teachers. Individuals within the MOOTCH community possessed different strengths and weaknesses as teachers and different instructional challenges in their current teaching assignments. Although the FC community played with creating new worlds, MOOTCHers expressed a desire to integrate the MOO into their existing teaching environments. These were the factors influencing how the MOOTCH community played. Reflective of the strong influence of language in their backgrounds, MOOTCHers first played with words and language:

```
Paul-mt . o 0 (I am. . .)
Van Faussien . o 0 (therefore I think!)
SallyF-mt . o 0 (I'm pink therefore I'm spam?)
Rebecca-ETS [to David-ETS]: You've been soooo quiet.
David-ETS . o 0 (been thinking) (MOO Transcript, 1997)
```

The fascination with words and language, established in the first meeting of this community, persisted as a theme throughout the entire course. Indeed, after our final class meeting, Bernardo wrote that he did not expect to experience withdrawal from the course. Instead, he introduced us to a new word saying he planned to experience:

'saudade,' as our brazilian neighbours say. It is like having a sweet slight pain in the chest, every time you recall and long for all the nice things you went through. (B. Banega, personal communication, Feb. 23, 1998)

In addition to playing with words and language, MOOTCHers also engaged in imaginative play referred to as harmonic convergence. Harmonic convergence occurs when the interactions and events occurring at a physical location merge with the interactions and events occurring on the MOO. Although still regarded as playing, because of its dependence on the ability to suspend disbelief, harmonic convergence represents a more advanced form of play indicative of the developing relationships among community members. Including individuals in the physical location into a MOO conversation is one example of harmonic convergence. Another form of harmonic convergence is illustrated in the following excerpt, initiated by Bernardo as he ate pizza in his physical location during one of our MOO meetings:

```
Bernardo-mt holds up a BIG sign: help yourself with the pizza,
    now!
JanineC-mt mmmm good pizza
MichelleR-mt says, "Yum-sausage and pepperoni!"
rrosell says, "crust is great"
Alan-ETS [to Bernardo-mt]: I make pizza. How'd you get such a
    good crust?
```

```
Bernardo-mt [to Alan-ETS]: just a special trick from Argentie
  makers!
```
(MOO Transcript, 1997)

Riding virtual ferris wheels, thwarting the advance of a charging ram, engaging in word play, and experiencing harmonic convergence were all activities reflecting the developing confidence of the learners in these two communities and of the growing level of trust among the community members. The simple act of play prepared these communities for the next challenge in the MOO, that of building a personal space within the MOO.

Building

As a virtual environment, a MOO is extended beyond its basic architecture when users create new MOO spaces and objects. As a text-based environment, a MOO obtains physicalness from specific and deliberate use of language to describe these MOO spaces and objects. This process of extending the MOO is referred to as building. Although building within the MOO requires the execution of relatively simple MOO commands, creating a physicalness to those new spaces requires creative manipulation of language. By building, individuals make a commitment to the larger MOO community, articulate their conceptions of the potential of the MOO, and reveal something of their personalities. As individuals progress to customizing their MOO spaces, they find it necessary to collaborate with others and to clearly identify the purpose of their newly created space and for whom the space is intended.

Most individuals are eager to build a room as their first construction act on the MOO. In doing this they make a commitment to the MOO in two ways. First, the creation of new spaces makes the MOO a different environment, a bigger environment. Second, a room on the MOO provides their MOO character with a home. As their MOO home, their new room becomes the place they find themselves when they log in or the place they return to whenever they issue the home command within the MOO. It is in the naming and describing of these MOO spaces that individuals give clues to their personalities. Often, as I did with my first building efforts at MediaMOO, these spaces are depicted as offices. Through my room description (the words visitors see when they visit my room), I conveyed my love of books and my quest for a quiet and peaceful work environment. Some members of the MOOTCH community expressed their personalities with more creative names for their virtual homes. DJ's Pad and Pen Palace was home to a composition teacher with a fondness for pens; Judge Wanda's Courtroom was home to a criminal justice teacher; and the Siouxland Airport was the home to a creative writer interested in how stories might develop from role playing in a virtual airport populated by strangers.

Effectively building MOO objects or rooms naturally elicits collaboration among members of a community. As builders create descriptions of their MOO spaces, they often ask for feedback to ensure that others perceive the room in the way it was intended. In addition, although building a room is relatively simple, customizing that space can prove to be more challenging. MOO rooms come in a variety of types, each with different capabilities, such as the capability of adding seats to the room. Executing specific commands accesses room capabilities. Collaborating with others during the process of customizing a room allows builders to work at their own pace and to capitalize on the expertise of others, as shown in the following excerpt:

```
PaulV-mt [to WandaH-mt]: "pretty neat, Wanda. Seems like a just
    place
WandaH-mt [to PaulV-mt]: Kind of stupid, I know, but I'm learn-
    ing a lot.
PaulV-mt [to WandaH-mt]: "It's great. How did you do the fur-
    nishings?
WandaH-mt [to PaulV-mt]: I have a lot of furniture in the room,
    but you have to type look.
WandaH-mt [to PaulV-mt]: You type the command @addfurn
PaulV-mt [to WandaH-mt]: "Do you then get a list of available
    items?
WandaH-mt says, "Yes."
PaulV-mt [to WandaH-mt]: "As soon as my room is ready I want a
    window with a view
JanineC-mt says, "Wanda I like your room"
WandaH-mt [to PaulV-mt]: A window with a view?
WandaH-mt [to JanineC-mt]: Thanks. . .it is a work in progress.
PaulV-mt [to WandaH-mt]: Yes, overlooking a pond with Canada
    Geese
WandaH-mt [to PaulV-mt]: Canada Geese?
PaulV-mt [to WandaH-mt]: "yep, the big black and gray ones that
    fly south every winter (MOO Transcript, 1997)
```

In addition to naming and describing these MOO spaces, builders must decide the type of room they want. Some rooms provide the ability (through the available commands) to regulate the speaking abilities of the room occupants. Other rooms are portable in that they are a room within a room and can move from one location to another. Still other rooms can be customized to emit messages at specified intervals. Confronting these choices requires builders to consider the intended audience and the purpose of the MOO space. In other words, builders must reconceptualize, or revise, their understanding of the MOO as a learning space.

Revising

> To exist humanly is to name the world, to change it. Once named, the world in its turn appears to the namers as a problem and requires of them a new meaning. (Freire as cited in Hobson, 1998)

As discussed in the previous section, using a MOO places us in a new world. This world is revised as we build rooms and create objects of our own, thereby extending the boundaries of the usual teaching and learning environments. Whether these new spaces serve to replace or merely supplement old spaces, they render the old spaces obsolete. Consequently, old ways of thinking about teaching and learning are also rendered obsolete. Confronted with a new environment we are forced to revise our conceptions of the processes of teaching and learning because the spaces where we teach and learn are different.

```
but I kept thinking, "What am I doing here?" and "Why am I here?"
and "What could I do here?" . . . and I thought, "if someone can
make a ping pong table so that I, as a user, think it's really a
ping pong table, why can't I make an audiometer so people really
think it's an audiometer" and it's just sort of evolved from
there. (K. McComas, personal communication, Sept. 27, 1993)
```

At MediaMOO the idea to create a virtual audiometer was the point at which my notions of teaching and learning were challenged. Simply creating an object and naming it "audiometer" would not serve my purposes because the object did not function as an audiometer. I needed to develop a new object and program that object to function as an audiometer. This solution placed me in the role of the learner as I had no previous programming experience. Fighting my initial urge to abandon the project, I turned to the FC community for help.

After expressing their faith in my ability to learn how to write programs, the FCers directed me to tutorials (in the form of virtual videotapes at MediaMOO) and a copy of a MOO programmer's manual. Their facilitative style suited my own learning style: "I like to try it myself . . . I would hack on something for a long time . . . eventually I broke down and asked for help" (K. McComas, personal communication, Sept. 27, 1993). Learning in this way forced me to understand, not just know, the MOO programming language and conventions. With each bit of understanding came increased independence and confidence in my ability to create meaning out of new information. In this way, my understanding of how people learn was radically altered. In addition, my understanding of how to facilitate this kind of learning was reconstructed:

```
that really had me stumped . . . and so he [Pokey] helped me
with that . . . it was a "for" statement . . . actually what he
helped me with didn't work but it set off a light bulb in my head
and I managed to get it [to work] and that was pretty much the
pivotal point for me . . . I felt empowered by that . . . I had
gotten something to work. (K. McComas, personal communication, Sept. 27,
1993)
```

Although MOOTCHers did not need to learn how to write functional programs, they also found revision necessary as their next step. As they sought to extend their instructional environments to the MOO, their previous notions about teaching and learning spaces and processes were no longer functional. Thus, MOOTCHers faced the task of reconceptualizing MOO space to achieve particular teaching and learning objectives. In other words, newly built rooms could only be adequately customized after careful consideration of specific instructional contexts.

One MOOTCHer, rrosell, demonstrated this reconceptualization, or revision, process when he informed us he should "start in a simple way and develop" as his confidence developed. Furthermore, he expressed his desire to avoid "getting in over [his] head" and his concern about getting caught up in the "technology of the MOO." Finally, he suggested using the MOO as a forum for open class discussions because the MOO "opens the possibility for a real Platonic dialogue" (R. Rosell, personal communication, Dec. 3, 1997). That is, he wanted to use the MOO to create learning experiences that did not routinely occur in his traditional teaching environment. This revision process reflected rrosell's coming to understand his purpose for using the MOO, along with an increase in his confidence to do so. In fact, as an example of his increased confidence, rrosell suggested that the class continue beyond its original end date to provide everyone an opportunity to teach their first MOO class with a trusted audience, the rest of the MOOTCHers. To everyone's delight, he volunteered to present first.

Publishing

As these two communities developed through the phases previously described, individuals and the communities themselves underwent significant changes. Of primary importance were the increased levels of confidence with MOO technology; a dependence on one another for guidance, support, and feedback; a different understanding of the teaching and learning processes; and a high level of trust and respect for one another. These changes paved the way for individuals within each community to publish their work for others. In other words, they went public with their work.

At MediaMOO, the FC community cheered loudly and set off virtual firecrackers when my first virtual patient responded to a hearing test by "raising" her hand. In addition, they insisted I present my work to the larger MOO community with an Open House (a common method of introducing new projects). Although I could not imagine many would be interested in visiting a virtual audiology lab, approximately 35 people attended the Open House. Quite frankly, at the time I did not care if anyone attended. I had designed and programmed a sophisticated project. When I started, 4 months before the Open House, I had never written a line of programming code in my life. My success empowered and transformed me. Four years later, when I started the MOOTCH class, I wanted those learners to gain power through their own transformations.

As rrosell of the MOOTCH community suggested, we extended the MOOTCH class to allow for presentations. These presentations represented first attempts to publish, or make public, people's work. Each presenter would have the opportunity to set up a class of MOO student characters, develop learning objectives, design a learning activity, and customize their MOO space appropriately.

```
Dear Van,
I am approaching Monday night feeling a lot more confident than
I did two weeks ago when we all decided to try this experience.
(R. Rosell, personal communication, Dec. 3, 1997)
```

The plan was simple. We agreed to use a system of peer coaching to provide assistance and support for each presenter. I coached the first presenter. The first presenter then coached the second presenter, who, in turn, coached the third presenter. This system was effective on two levels. First, each presenter had a substantial support system in a peer who had already successfully completed a presentation. Second, peer coaches, in the very act of coaching, deepened their understanding about and facility with teaching in a MOO. Through these experiences, each member of the MOOTCH community was empowered and transformed. Additionally, they were ready to take their work to a broader public, their own students:

```
I appreciate your indulgence for allowing me to demonstrate my
mastery of moo-technique; and whatever the outcome, I can tell
you that the experience of putting the material together was
very instructive and that without it I would still be back in
the Moodle ages. I am consoled by the fact this experience will
make it possible for me to bring my classes to the moo. Without
it, I know it never would have happened. (R. Rosell, personal communi-
cation, Dec. 9, 1997)
```

TRANSFORMATIONS

Douglas Robertson (1996, p. 42) distinguishes between types of learning as either simple or transformative. Whereas simple learning is additive—that is, the learner adds new knowledge to an existing paradigm—transformative learning "causes the learner's paradigm to become so fundamentally different in its structure as to become a new one." This kind of learning becomes necessary when conditions are such that the old paradigm is no longer functional (Mezirow, 1998).

As demonstrated in the two MOO communities described, transformational learning can be the unintended outcome of an experience. When I first logged into MediaMOO, curiosity, not a desire for transformation, led me there. Likewise, MOOTCHers came to the MOO to learn more about teaching on MOOs, not to be transformed. Taylor (1998) identifies three critical conditions for fostering transformative learning. Examining these two MOO communities within the framework of these conditions provides an explanation for the transformations that occurred and insight into how transformation might be fostered in other learning communities and environments.

The first condition necessary to facilitate transformative learning suggests that the learning conditions must "promote a sense of safety, openness, and trust" (Taylor, 1998, p. 48). The play experienced in both MOO communities provided the opportunity to create a safe, open, and trusting climate for learning to occur. Earlier in this chapter, I described how play contributes to the development of trust among the members of a learning community. Of equal or greater importance in creating a safe, open, and trusting climate during community play is the tone set by the facilitator. Through careful observation of learners, facilitators can design and pace learning opportunities and model behaviors conducive to the development of an appropriate learning climate. One member of the MOOTCH community described this type of facilitation in this way:

> You paddled a steady course, taking us into depths of a new sea, but you also kept us within sight of land. (P. Ventura, personal communication, Feb. 22, 1998)

A second condition for fostering transformative learning suggests the use of "instructional methods that support a learner-centered approach; promote student autonomy, participation and collaboration" (Taylor, 1998, p. 48). In the development of these two learning communities, the building and revising phases provided these conditions. First, and perhaps most important, the learners in each of these communities determined the particular project they wished to work on. At MediaMOO, I decided to create a virtual audiology lab to address a particular teaching problem I faced. In the MOOTCH community, WandaH-mt created a project (Judge Wanda's Courtroom) to provide her high school students an opportunity to participate in a trial. As each person decided on a project, they moved forward at their own pace, thus working autonomously. During these times of building and revising, learners actively participated by building and customizing their own spaces.

Finally, a third condition is the use of learning activities that "encourage the exploration of alternative personal perspectives, problem posing, and critical reflection" (Taylor, 1998, p. 49). As we customized our MOO spaces, we posed problems and generated solutions to resolve those problems. After implementing these solutions, we critically examined the results and began the process over again until we achieved the desired results. Even the experience of publishing our work with our peers resulted in reflection and further revision. Fifteen minutes into rrosell's presentation, a MOO broadcast message announced an emergency shutdown of the MOO. While waiting to reconnect, rrosell sent e-mail to the class:

```
Dear Fellow MOOTCHer's,

A Higher power must have intervened to save us all this evening.
I have no regrets that things ended as they did, as I began to
realize that my questions needed more explanation and background
information. (R. Rosell, personal communication, Dec. 9, 1997)
```

Clearly, the phases of development followed by the individuals in these two MOO communities resulted in more than simple learning. Although some simple learning did occur (such as learning the commands to create objects), the cumulative effect of these experiences resulted in transformative learning. Our transformations, though individual, were fostered by the growth and development of the communities within which we worked. Bernardo-mt describes the connection between transformation and community in this way:

```
Those "strangers in September" have got closer class after
class, helping each other, coaching each other, making fun [of]
each other, constructing community and friendship all together.
Monday nights are different for us now . . . as Forrest Gump
would say, "we got to MOO". What a good example of collaborative
learning . . . How we profited of each other['s] expertise, of
each other['s] judgement . . . and what [an] unruly bunch of ea-
ger MOOers we have become. (B. Banega, personal communication,
Feb. 23, 1998)
```

The real measure of transformation lies in how the newly developed paradigm presents itself in our old teaching spaces. Reviewing the history of my teaching practices reveals two changes that occurred as result of my own transformation. First, an increased emphasis on facilitating student learning (instead of directing it) appeared. I began to require students to read new material before class and to use class time for discussion instead of lecture. Second, an increase in my use of small group work reflects the value I placed on peer collaboration as a learning strategy.

What these two MOO communities revealed to me in writing this chapter, however, gives me a new understanding of transformation and poses new questions or problems to consider. Specifically, how can I create conditions in my classroom that will facilitate transformation within each of my students? How can I encourage them to begin with play? What would that play look like? How can I facilitate the building, or explication, of students' existing paradigms about learning and what they are learning? How can I structure learning activities that will render their existing paradigms dysfunctional and how can I foster revision of these dysfunctional paradigms? Finally, in what ways can students reflect their new paradigms and how can this work be made public? The challenge is daunting; the reward is to once again transform myself into a new teacher and learner.

15

The Collaboration That Created the Kolb-Proust Archive: Humanities Scholarship, Computing, and the Library

Caroline Szylowicz
University of Illinois at Urbana-Champaign

Jo Kibbee
University of Illinois at Urbana-Champaign

This chapter presents a case study of the successful collaboration that produced the Kolb-Proust Archive for Research at the University of Illinois at Urbana-Champaign (UIUC). Centered on the prolific and influential French writer Marcel Proust, the Kolb-Proust Archive exists as both a physical and virtual repository of the primary research materials of Proust scholar Philip Kolb and serves as a unique research tool for scholars worldwide. The creation of the archive is the result of extensive collaboration among humanities scholars, the university library, the Department of French, and agencies on the UIUC campus and abroad. Beginning with a brief discussion of humanities computing and libraries, the paper then describes the content, evolution, and significance of the archive, with a focus on the human and electronic collaborative efforts that have been indispensable to its existence.

COMPUTING, THE HUMANITIES, AND LIBRARIES

Humanists have long enjoyed a symbiotic, collaborative relationship with libraries. The print collection of books, journals, manuscripts, and archival materials is vital to their scholarship, providing the raw material from which new ideas are generated and developed. The products of this scholarship, books and journal articles, then become part of the library's collection for future scholars to draw on. It comes as no surprise, therefore, that humanities scholars have traditionally held greater affinity for the library than for the computer lab. A recent investigation

255

(Masey-Burzio, 1999) of the reaction of humanists toward computing technology reveals a healthy skepticism about value and ease of access to online information. Humanities faculty participating in a focus group at Johns Hopkins University Library voiced concern about technical barriers to access (e.g., low end machines) and the difficulty of finding useful material. They reported frustration with the level of discussions on listservs, the poor quality of digital images, and the difficulty of viewing lengthy texts on a computer screen.

Advances in information technology, however, have had a significant impact on both libraries and humanities scholarship. Card catalogs gave way to online catalogs, which in turn enabled the development of union catalogs, such as Online Computer Library Center (OCLC's) WorldCat, now indispensable for identifying and locating materials held in libraries throughout the world. Converting print indexes, such as the Modern Language Association International Bibliography, to online databases has provided more flexible search capabilities. Networking technology offers library users the possibility of remotely (i.e., from home or office) consulting library catalogs and indexes, full-text materials (e.g., literary texts, journal articles, reference works), images, and multimedia. The Internet has opened new possibilities for scholarly communication and collaboration through e-mail, chat spaces, and file transfer. Using a computer, humanities scholars can communicate with colleagues and access primary material for research—texts, images, videos, musical scores, and recordings. Although the promise of technology may still exceed the reality (poor image quality, copyright limitations, accessibility problems), computing technologies nonetheless hold the potential to expand humanities scholarship by bringing digitized resources to the scholar's desktop.

Computer-assisted humanities research has led to increasingly sophisticated research. In the area of literary and textual studies, for example, computers enable the preparation of concordances and provide the basis for stylistic analyses, vocabulary studies, authorship verification, and the collation of variants for critical editions. Professional organizations, such as the Association for Literary and Linguistic Computing[1] and the Association for Computers and the Humanities,[2] support and promote this work. The creation of digital text archives, such as the Thesaurus Linguae Graecae[3] and the Trésor de la Langue Française,[4] offer scholars unprecedented access to a wealth of resources that might otherwise remain unavailable or difficult to use. The advent of the Web, coupled with advances in image digitization, javascript, and multimedia, has changed the picture (literally), and computing in the humanities has expanded beyond the digitization and analysis of texts alone.

---

[1]See the Association for Literary and Linguistic Computing Web site at http://www.kcl.ac.uk/humanities/cch/allc/
[2]See the Association for Computers and the Humanities Web site at http://ach.org/
[3]See the Thesaurus Linguae Graecae Web site at http://ptolemy.tlg.uci.edu/~tlg
[4]See the ARTFL Web site at http://humanities.uchicago.edu/orgs/ARTFL/

These digital projects provide fertile ground for innovative collaborations. Indeed, the production of online resources often mandates collaboration because they fall at the intersection of scholarship and technology. Libraries are increasingly becoming key players in these collaborative initiatives. The Center for Electronic Texts in the Humanities[5] at Rutgers University and the University of Virginia's Electronic Text Center[6] play an important role in the research, development, and dissemination of electronic texts. A number of libraries now host etext centers, which range from basic collections of commercial full-text CD-ROMs to fully equipped centers involved with the creation, training, and use of digital texts. One of the foremost projects is the American Memory Project,[7] a major component of the National Digital Library Program at the Library of Congress, which provides access to more than 50 multimedia collections of digitized documents, photographs, recorded sound, moving pictures, and text from the library's Americana collections.

The Kolb-Proust Archive for Research, described in the next section, takes the idea of networking and digitization in yet another direction. Instead of limiting itself to mounting only historical and literary texts, images, and multimedia for scholars to use, the archive encourages intellectual interaction by making available the research notes that humanities scholar Philip Kolb painstakingly developed in studying Marcel Proust and in editing Proust's correspondence. In doing so, the archive has become both a physical and digital resource center for the study of Proust and the turn-of-the-century European culture in which he flourished.

OVERVIEW OF THE ARCHIVE

The Kolb-Proust Archive for Research was established in 1993 at the library of the University of Illinois at Urbana-Champaign and offers researchers an extensive collection of research materials relating to Marcel Proust and his historical and cultural milieu. The physical archive consists of a collection of books (Proust's works, memoirs, biographies), newspapers and journals, biographical documents (birth, death, marriage certificates), and photocopies and transcriptions of letters by and to Proust, sorted by correspondent. At the heart of the archive lies Kolb's life's work: more than 40,000 handwritten cards on which he recorded information he found in the letters of Proust and in other sources (e.g., newspapers), which he used to annotate Proust's correspondence (see Fig. 15.1). Because Proust did not date his letters, Kolb did a prodigious amount of detective work in ordering the correspondence, working from details of Proust's extensive

---

[5] See the Center for Electronic Texts in the Humanities Web site at http://www.scc.rutgers.edu/ceth/

[6] See the Electronic Text Center Web site at http://etext.lib.virginia.edu

[7] See the American Memory Project Web site at http://memory.loc.gov

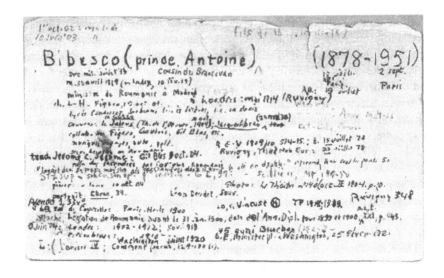

FIG. 15.1. Example of Philip Kolb's handwritten card.

descriptions of his daily activities and from various records of historical events. The cards, organized by category, include a chronology of Proust's life and contemporary events, a bio-bibliography of real and fictitious people mentioned in Proust's correspondence, and a file of literary citations found in the correspondence, as well as several others. The goal of the archive is to digitize the information on these cards and to create an extensive and interactive online research tool.

The nature and content of the archive reflect the rich and enigmatic nature of its subject, Marcel Proust. Born in Paris in 1871, Proust is a major literary figure of the 20th century. His major opus, *A la Recherche du Temps Perdu (Remembrance of Things Past)*, is notable for its influence on the development of the 20th century novel. Proust was also a prolific essayist and correspondent. His voluminous letters offer reflections on people and events, detail his daily activities, and provide depth of insight into the author rarely encountered in the correspondence of other writers. Moreover, Proust had a large and varied circle of correspondents, and therefore his letters constitute a significant source of information about Parisian life and French culture at the turn of the century. Among the numerous people, events, and topics represented in the correspondence are the Dreyfus Affair, the trial of Oscar Wilde, the separation of church and state in France, World War I, Richard Wagner, Russian ballet, and the theory of relativity. Unraveling, organizing, explicating, annotating, and indexing this correspondence was the focus of Kolb's scholarship and laid the foundation for the Kolb-Proust Archive for Research (see Fig. 15.2).

The archive itself retains both a physical and a virtual presence as its collection is digitized to take full advantage of the electronic medium. As a physical location, the archive began in Kolb's former study in the university library, and was

FIG. 15.2.   The Kolb-Proust Archive Web site.

subsequently moved to a location adjacent to the Modern Languages and Linguistics Library. Kolb's original collection of research materials has been preserved and continues to grow with the addition of newly published works relevant to the activity of the archive. The room accommodates a small staff, a print collection, card files, computer equipment, and workspace for visiting researchers. The virtual archive[8] consists of the first series of documents from the collection that have been digitized, tagged in SGML (Standard Generalized Markup Language) and made searchable. Currently available are a bibliography of texts by and about Proust (1884–1991) and a chronological file of events (1633–1909). Text-retrieval software originally acquired for the library's Digital Library Initiative (see Fig. 15.3) offers powerful search capabilities. One innovative feature of the Web site combines SGML tagging with the use of authority lists to increase the precision of personal name searches, which are particularly important to historical research.

COLLABORATION IN THE ESTABLISHMENT
OF THE KOLB-PROUST ARCHIVE

Kolb's family and assistants recognized the value of the collection for Proust scholars and other researchers interested in the period and urged that the docu-

---

[8]See the Kolb-Proust Archive Web site at http://www.library.uiuc.edu/kolbp/

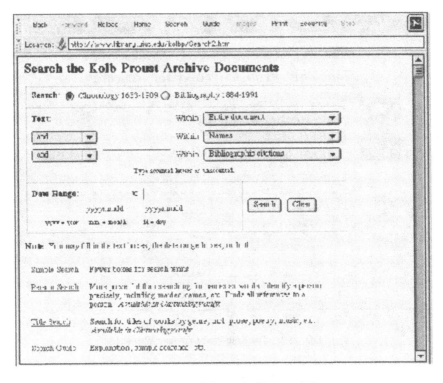

FIG. 15.3. The Kolb-Proust Archive search form.

ments be maintained and made accessible to the public. From this initial idea, the archive developed as a joint effort among several campus units who saw potential in the Kolb collection and who collaborated to bring the project to fruition.

The environment at the University of Illinois was ripe for this initiative. The Advanced Information Technologies Group (AITG)[9] had been established as a joint venture of the National Center for Supercomputing Applications (NCSA), the College of Liberal Arts and Sciences, and the university library, to stimulate and support the use of information technologies for instruction and research in the humanities and social sciences. The AITG provided seed money, consulting, and technical support to those whose research, instructional projects, or both, involved the use of innovative technologies, with the goal of using the projects as a test bed for developing applications that could be shared. The multidisciplinary nature of the collection (French intellectual life at the turn of the century), the well organized structure of the card files, and the wealth of data they contain appealed

---

[9]See the Advanced Information Technologies Group Web site at http://www.library.uiuc.edu/aitg/

to the AITG. The project provided an opportunity to develop a model of organization for humanistic information, to investigate the means of representing its complexity, and to use technologies such as text encoding, hyperlinking, and networking to expand search and retrieval capabilities. These technologies also offered potential solutions to problems of preservation and access because some materials are on acid paper.

The initiative moved forward with the cooperation of multiple campus units. The university archives, a unit of the university library, contributed to the preservation of the original materials, sharing expert advice and supplies. The availability of an electronic version of the files will ultimately help to preserve the original documents by limiting their use. In the meantime, the university archivist has provided a list of guidelines for the proper storage and handling of documents until the original materials can be retired and permanently transferred to the university archives repository. The staff took immediate steps toward preservation by moving documents away from sources of heat (not always an easy arrangement in cramped quarters) and removing paperclips and other metal attachments to prevent both damage to brittle paper and stains caused by rust. They placed the contents of cardboard folders and boxes into acid-free folders and containers to slow down the acidification process. Within the folders all documents were laid flat, and brittle newspaper clippings were photocopied onto archival paper. The university archives provided all the supplies necessary to extend the life of the collection. In addition, the university archives transferred and processed several drawers filled with Kolb's administrative and teaching records, which are traditionally collected for campus records but which were of no direct use to the Kolb-Proust Archive.

As one of the major beneficiaries of the Kolb-Proust Archive, the Department of French has been a major partner in the project. Along with the AITG and the library, the French Department has participated in the crucial task of securing funds to maintain the archive by writing grant applications and obtaining support for the project from the Campus Research Board's Critical Research Initiatives fund. The French Department has provided support and resources, including graduate students to help with the digitization process and the interpretation of Kolb's abbreviated notation system. The archive, in turn, offers research opportunities for departmental faculty and students and attracts visiting scholars and graduate students with a potential interest in Proust studies.

An increased interest in Proust studies is reflected in seminars offered by faculty and visiting scholars, most notably in an international Proust 2000 Symposium, which was held at the University of Illinois in the spring of 2000. The symposium, organized jointly by the French Department and the university library, with the collaboration of many campus units, provided Proust scholars with a venue to share their latest research and publicized the important resources of the university in the area of Proust scholarship. It featured plenary sessions with international speakers as well as informal roundtables and music and art events. The latter included a performance of Claude Debussy's opera *Pelléas et Mélisande* by the University of Illinois Opera, a recital of chamber music relating to Proust's work by the School of Music faculty, and an exhibit of original manuscripts, documents, and decorative arts prepared by the university's Krannert Art Museum in collaboration with the Rare Book and Special Collections Library and the Kolb-Proust Archive.

The university library plays a critical role in the success of the archive, offering space, personnel, and technical expertise. The establishment of the Kolb-Proust Archive allowed the library to forge a new direction in the creation and dissemination of unique materials in the humanities. The library currently supports two large-scale digital projects: the Digital Imaging Initiative (DII)[10] explores ways to use electronic means to preserve and promote the use of digital images throughout the campus and scholarly community; the Digital Library Initiative (DLI)[11] provides access to full-text articles from selected journals in the sciences and engineering and seeks to use text-retrieval technology to effectively search these technical documents. The Kolb-Proust Archive has benefited from the DLI's research and technical expertise because the DLI team has shared its knowledge of indexing and searching SGML documents during the implementation of the Web interface to the Kolb-Proust collection.

Collaboration has also involved individuals and institutions outside the university. Scholars contributing to the project reside in several states of the United States as well as in France. These collaborators are on site from time to time, as their schedules allow, but the bulk of the work is done through fax and electronic mail. The archive's director, Dr. Katherine Kolb, is a professor of French language and literature at Southeastern Louisiana University who visits the campus periodically to attend meetings. French scholars likewise play a role in the archive's development. With funds provided by the library, the campus, and the French government, the archive hired Dr. Françoise Leriche, a Proust scholar from the University of Grenoble, who spent two summers in Illinois participating in the early stages of the project. Her knowledge of Kolb's files and work methods contributed to the design of the templates that were used for entering the contents of the card files. She undertook the publication of several of Proust's letters that Kolb had not been able to include in his edition and selected other letters for an anthology of the correspondence. Her research to annotate and date the remaining unpublished letters and prepare the selection of letters for the anthology makes full use of the resources of the Kolb-Proust Archive but is conducted almost entirely via remote access.

---

[10]See the Digital Imaging Initiative Web site at http://images.grainger.uiuc.edu/
[11]See the Digital Library Initiative Web site at http://dli.grainger.uiuc.edu

FUTURE COLLABORATION

In the near future the Kolb-Proust Archive will enter another stage in the creation of the digital collections it seeks to offer to the humanities community. Because the correspondence of Marcel Proust is the cornerstone of the project, the next logical step is to add a full-text electronic version of Proust's correspondence to the files already online or currently being digitized. However some non-technological issues need to be addressed, namely cost and copyright. Initially, the Kolb-Proust Archive tested scanning coupled with optical character recognition (OCR) software as a way to process the 21 published volumes of correspondence, but this proved ineffective for the needs of the project. The process was too slow to apply to the more than 10,000 pages of the correspondence, and the OCR software was unable to handle imperfections in the paper, uneven contrasts, broken characters in the typeface (*d* interpreted as *a*, *i* interpreted as *l*) and diacritics in the very small font used for annotations. The percentage of errors was very high, and comparing the digitized and printed texts (to locate and correct errors) took as much time as keying in the text manually. The 21 volumes of correspondence were typeset such that the publisher could not share an electronic copy with the archive, and consequently the text will have to be entered manually by a commercial data entry company.

A potential partnership with the ARTFL project[12] (American and French Research on the Treasury of the French Language) at the University of Chicago could help to lower the cost of digitization. By commissioning a common electronic version of the correspondence including basic structural SGML codes such as headers, paragraphs, and notes, the archive and ARTFL can both achieve their goals. The common version will be customized later to meet each project's specifications. For example, the archive will proceed with the more elaborate SGML encoding, which is applied to its other collections. The ARTFL project, which is particularly interested in the content of the letters, could use the mark-up to adapt the digitized text to its own search tools by eliminating secondary text (such as notes) delineated by specific codes. Moreover, because the ARTFL project has already secured the rights to many published French texts, collaborating with them will aid in negotiations with the publisher of the correspondence. In recent years, the publisher has expressed some legitimate concerns about the potential loss of revenue if the correspondence is freely available on the Web and about the possibility of piracy. One way to address these concerns would be to include the publisher as a partner in the project by developing a fee-based service, at least for the full-text portion of the Web site. A system of registration to access the site and the use of electronic watermarking (codes encapsulated within the electronic files to verify their authenticity) would certainly deter piracy and provide legal recourse,

---

[12]See the ARTFL Web site at http://humanities.uchicago.edu/orgs/ARTFL/

should such an incident happen. At present, negotiations are still underway to reach an agreement that will satisfy both parties.

In addition to the text of Proust's correspondence, the digitization of original Proust letters housed in the University of Illinois Library's Rare Book Room has also been considered to be linked to the electronic texts of the letters and card files. Creating and managing large numbers of digital images presents a completely different set of challenges. For that task, the Kolb-Proust Archive would collaborate with the DII and rely on their standards and methods. One of the DII goals is to establish best practices for digitizing various classes of visual materials. Over the past 5 years it has developed and tested guidelines for digitizing more than 10 special collections of the university library. The DII guidelines cover all steps of the digitizing process, including digital capture, storage, description, rights, and delivery of digital resources. For the Kolb-Proust Archive, the DII could formulate recommendations on the type of equipment needed for the capture (e.g., scanner, digital camera) based on the extent, format, and condition of the original letters; the optimal resolution, format, and size for the digital images; storage and network access; and description standards agreeable to both the Rare Book and Special Collections Library and the Kolb-Proust Archive. The archive could provide funding to implement the process, storage resources for the digitized images, and all the data necessary to identify and describe each letter precisely.

SERVICE TO THE SCHOLARLY COMMUNITY

Since its inception, two overarching goals of the Kolb-Proust Archive have been to provide universal access to otherwise unique or hard-to-find materials and to share its skills in text digitization and encoding. Access to the archive is achieved in two distinct ways. The physical archive opens its doors to visitors wishing to use its printed collections and has attracted scholars who use its collection in conjunction with the Rare Book and Special Collections Library and the Proustiana section of the Modern Languages and Linguistics Library for their research. The digitized portion of the collection is available on the archive's Web site.

In the first 5 years of its existence, the Kolb-Proust Archive contributed significantly to several humanities projects. Its collections have been used in the publication of additional letters of Marcel Proust as well as for the revision of a volume of translated letters. In 1993 a team of Japanese Proust scholars undertook the compilation of a detailed index of Proust's correspondence to complement the volumes published by Kolb. When the index was nearing completion, the Kolb-Proust Archive received a set of proofs and supplied additional biographical information that had never found its way into the published correspondence, thereby completing the tool and rendering it more useful to Proust scholars.

Availability of the Web site also generated an unanticipated service: reference questions regarding Marcel Proust began to arrive shortly after the home page was mounted in 1995. As use of the Web became more popular and the archive gained visibility through the expansion of its site, requests increased steadily. Most of these questions are sent by scholars throughout the world who post their questions directly to the Web site, but a notable portion of questions are referred to the archive by other libraries. From basic queries about the titles of Proust's works to copyright concerns or complex bibliographic queries, amateurs of Proust, scholars, and students are shaping a professional service component in the archive. What started as an occasional service to individuals is slowly turning the archive into a clearinghouse for all things Proustian. A list of links to other Web resources relevant to Proust studies has been added, as well as a page of calls for papers and conference announcements. Visitors to the virtual archive are encouraged to send any comments they may have about the Web site. Based on these comments, as well as on suggestions from several scholars, the staff of the archive hopes to expand this aspect of service by adding new resources such as a Proustian who's who, which would list scholars and students, their interests, and their recent works. In addition the archive plans to open a space for discussion about Proust and related subjects in the form of a mailing list, bulletin board, or both. The archive would then become a space for people from various geographic locations and professional backgrounds to discuss and exchange ideas and information.

The archive also hopes to share its technical expertise by serving as a model for similar humanities projects. Recently, the archive was contacted by a scholar at the University of Toronto who is developing a project concerning the thousands of letters received by the French writer Emile Zola (1840–1902) that were donated to that institution. Such an endeavor could benefit from the expertise gained in the establishment of the Kolb-Proust Archive for tasks such as organizing and processing large numbers of letters, identifying necessary access points to the documents, managing indices, transcribing and dating the letters, and so on. Besides supplying the technological and methodological advice, the archive would also provide a large corpus of reference materials because Proust and Zola were born only a generation apart. If this collaboration materializes, it will be the archive's first attempt to share its technology and resources to aid in the establishment of another humanities research project. Ultimately, cooperation with other institutions could lead to shared networked resources on various aspects of society and culture at the turn of the 20th century.

SUMMARY AND CONCLUSION

Despite humanists' legitimate concerns about the relationship between scholarship and information technologies such as the Internet, it is clear that technology has the potential to take humanities scholarship in new and productive directions. Technologically enabled projects such as the Kolb-Proust Archive move beyond simply mounting and distributing historical and literary texts and provide a sophisticated search engine for accessing research notes and primary source materials. Such a project involves a team effort with multiple players—scholars, academic departments, computing centers, and other units on campus and abroad. The university library, the traditional locus of humanities scholarship, plays a key role in coordinating and facilitating the necessary collaboration of these players. As this project demonstrates, however, electronic collaboration is also critical to such an undertaking. Without the ability to communicate and share information electronically, the complex collaboration would have been virtually impossible to accomplish. As a result of this successful application of digital and networking technologies, the Kolb-Proust Archive not only makes research materials available but also serves as a catalyst for future scholarly collaborations.

Response

T. Lloyd Benson
Furman University

One of the sure signs that a technology has transformed society is that people begin to treat it as part of the landscape. Carol Sheriff's (1997) recent reappraisal of the impact of the Erie Canal on antebellum America, for example, shows that within a decade of its completion users stopped marveling at the canal's low cost and breathtaking 5 m.p.h. speed and turned to debating how to cope with its impoverished workforce and its disruptions of traditional property rights. What once had overcome nature soon became part of nature. Similarly, Tom Standage (1998) shows in his book *The Victorian Internet* how the world's first electronically mediated communications tool (the telegraph) was created in controversy but became ubiquitous. Promoted in utopian terms by its backers and dismissed as cumbersome or useless by its critics, over time it proved essential to news and commerce. The ultimate sign of its success was that it did the right things but did not do them well enough. Within two generations inventors had outmoded it with the telephone and radio, not as a rejection or abandonment of the concept but rather as a testimony to its powerful collaborative potential.

The maturation of computing technology brings the academic community to a similar stage. Even the most vocal Luddites on college faculties have come to rely on word processing, e-mail, and fax machines. Library card catalogs have become museum pieces, replaced by rows of sleek terminals capable of serving up not only call numbers and locations but also the complete texts of books and articles. Computing technology's omnipresence is illustrated by the fact that the loss of network services can shut down a campus. As previous generations discovered, our reliance on the technology heightens our awareness of its shortcomings.

For humanistic faculty this problem is especially visible. Collaborations that a decade ago seemed impossible now seem ordinary. Few people a generation ago could have imagined the Humanist discussion list, for example, which in formal terms is daily interactive interdisciplinary humanities computing seminar whose participants include scholars on every continent.[1] It is a remarkable joint effort. Yet what in the abstract seems so amazing can become a daily exercise in frustration. Every morning I confront a dozen or so messages from Humanist. Most of these I delete after reading their titles. As luck would have it, more than once a deleted message has proven during later conversation to be pivotal, leaving me to scramble through the Humanist archives to restore it. The ones that do filter through range in quality from irrelevant to sublime. If it is a long message, I often print it out, knowing that I will read it more carefully on paper than on the screen. For me this pervasive and essential technology has its problems.

FACULTY COLLABORATION ISSUES

Close reading of long texts is one of the defining acts of humanities scholarship. Our endeavor is still driven by a quasi-monastic model in which the intense solitude of study is a characteristic, if philosophically questionable, ideal. That we do this core task of reading by ourselves helps to explain why humanities faculty have been the last among the disciplines to adopt formally collaborative strategies. Historically, humanities faculty have been strongly individualistic in their attitudes. Both by attracting people with a disposition against strong authority and by socializing them to work on projects alone, the humanities disciplines have favored the solitary researcher (Ladd & Lipset, 1975). Ironically, as Ronald Schleifer (1997) has pointed out, this seemingly private act takes place in the midst of a pervasively powerful community. It requires an extensive advance collaboration in the form of literary and historical apprenticeship, and it is done in the framework of multi-referential and dynamic semantic contexts.[2] That computing technology only partly lends itself to sustained reading of longer texts adds to the difficulty of such work. Despite such challenges, these four essays show that electronic collaboration has transformative potential for teaching, research, and writing.

The issues they raise can be broadly grouped into three categories. The first set includes those technical and structural conditions that shape the collaborative environment. Current technology is not always humanities friendly, but Cheryl Reed and Dawn Formo show how powerful generic word processing and e-mail can be in the hands of capable users. Karen McComas and Donna Sewell show

---

[1]To learn more about HUMANIST, visit http://www.princeton.edu/~mccarty/humanist/humanist. html

[2]For working faculty's limited understanding of these disciplinary differences, see Judith A. Langer, "Speaking of Knowing: Conceptions of Understanding in Academic Disciplines," in Anne Herrington and Charles Moran's *Writing, Teaching, and Learning in the Disciplines* (MLA, 1992).

similar results with MOOs and discussion lists, neither of which requires equipment more complex than text-based server programs, telnet, and terminal software. The Proust archive developed by Jo Kibbee and Caroline Szylowicz required a more formidable array of scanning, storage, and indexing tools to create but can be accessed using any generic Web browser. That such straightforward means can generate such productive results should confirm the technology's value.

Another of the advantages of the computer's simplicity, and especially its demand for literal interpretations and regular structures, is that it forces us in the humanities to be more conscious of how we interpret and teach the specifics of our texts. It is said that you never really understand something until you explain it to someone else, and computers demand that information be spelled out in excruciatingly regular detail. Yet it is also easy to imagine how a more powerful collaborative software environment could enhance the processes each of these authors went through and the results they achieved. These works underscore the need for evolution in the tools and standards for scholarly writing, as well as in the apparatus of text description, commentary, and classification.

The second set of questions concerns the intersection of these electronic environments with the personal and professional dimensions of the humanities endeavor. In any collaboration the psychology and backgrounds of the participants are crucial to its success. As Reed and Formo discovered in organizing "Writers Anomalous," not everyone has the time or the temperament to engage in electronically mediated collaborative scholarship, nor does any particular collaborative structure meet the needs of all participants. Although a diversity of views and needs is one of the most important strengths of any collaborative effort, it multiplies the potential conflicts. Faculty also have to consider how collaborative projects will be perceived by departmental colleagues, tenure and promotion committees, and administrators. Professional organizations have begun to develop standards for reviewing electronic projects (most notably the *M. L. A. Guidelines for Evaluating Work with Digital Media in the Modern Languages*), but implementation varies widely from campus to campus.[3] Such instability can make faculty hesitant to participate in collaborative efforts, especially when projects cross campus boundaries. Moreover, unlike in the sciences, social sciences, and professional programs, there is little incentive in the form of large grants to attract and sustain collaborative proposals in the humanities. In disciplines where the monograph is the staple commodity, shared authorship is deflationary. Some of the major scholarly organizations have recognized this problem and called for a decoupling of the linkage between promotion and publication, but such reforms seem far off (Butler, 1995; D'Arms, 2000; Teute, 2001). One might also mention the lurking intellectual property issues that shadow every networked digital project.

---

[3]See the MLA guidelines at this URL: http://www.mla.org/www_mla_org/reports/reports_main.asp?mode=subpage&area=&sub=rep1180040418

In MOO-space, for example, sponsoring institutions, programmers, and participants might all claim common law copyright on the objects and communications that exist there. One can easily imagine the unfortunate scenario where litigation becomes the only way to resolve these claims.[4]

The third set of issues, those of community and governance, arises from the other two. Although some scholars doubt that electronic gatherings qualify as authentic communities, e-groups certainly face the same problems encountered by real-world groups and require many of the same kinds of solutions. Above all, each of the successful collaborations described in these essays required the trust of the participants. Even the much-maligned lurkers remain on lists only as long as they are confident in the authority of the information being received and the quality of the discourse. Creating trust and value in faculty collaboration is an art rather than a science, but some systematic principles can be discerned from these essays. The work of pedagogist Parker Palmer (1998) offers a useful framework for interpreting them. Learning spaces, he argues, embrace a series of paradoxes. They should be bounded by guidelines that provide focus and motivation but should be open to a diversity of discovery paths. They must be safe and reassuring, but "charged" enough to demand experimentation and exploration of the unknown. They should "invite the voice of the individual and the voice of the group" (p. 74–77) so that participants can express their minds freely in a forum where the group can give corrective feedback and call for refinements. They should promote quiet individual reflection as well as encourage forthright speech. Although Palmer associates these ideas with the teacher–student relationship, they make just as much sense for enabling faculty collaboration.[5]

To apply such a list of principles requires considerable skill from organizers and conveners. All four essays show the importance of mentoring and group cooperation in the early stages of a project. Because electronic faculty collaboration in the humanities is a recent phenomenon, individuals with enough expertise to serve as mentors and core participants have been rare, but the situation is improving. Judicious intervention and role modeling by such people can prevent collaborations from becoming either chaotic or authoritarian. As linguists Boyd Davis and Jeutonne Brewer (1997) discovered in their rhetorical analysis of student discussion groups, postings by the earliest and most eloquent members were closely emulated by later submitters. Setting the right tone for the collaboration is a diffi-

---

[4]For the implications of intellectual property on scholarly work and the widespread confusion about fair use and its implications, see National Science Council, Committee on Intellectual Property Rights and the Emerging Information Infrastructure (2001), "The Digital Dilemma: Intellectual Property in the Digital Age" http://books.nap.edu/html/digital_dilemma and David Green, editor, "Copyright and Fair Use: Town Meetings 2000," National Initiative for a Networked Cultural Heritage http://www.ninch.org/copyright/townmeetings/report2000.pdf

[5]For an interesting systematic study of personal factors relating to trust in group activities, see Robert Wuthnow, "The Role of Trust in Civil Renewal," in Robert K. Fullinwider, editor, *Civil Society, Democracy, and Civic Renewal* (1999).

cult skill. All four essays provide insights into how such good practices might be established.

Writers Anomalous

If digital communication is indeed an electronic frontier (with humanities faculty among its most recent settlers), then one might expect to find cyberspace equivalents to the community formation process that occurred on the historical frontier. Historians of American settlement have identified a number of aspects to this process, including migratory streams where new settlers abandon some of their old customs and simplify and reemphasize others; develop new customs in response to the new environment; form community networks, institutions, and rules of etiquette; struggle to take advantage of local resources; form communications and market exchange links with established communities in older areas; promote efforts by community boosters to attract new settlers; and adapt struggles to gain legitimate recognition for the new communities by existing political authorities and to convince these authorities to provide the infrastructure needed to sustain the new communities.[6] The story of Reed and Formo's collaboration has interesting parallels with several of these factors.

Their case study of faculty writing collaboration underscores how much the success of joint projects depends on what participants bring with them to the group. Recent research by Nonaka and others into successful collaborations in the corporate environment shows that informal communities of practice worked best when members shared a significant body of tacit knowledge (implicit ideas that they exchanged through war stories, rather than formal instruction) and where the participants' roles were interdependent rather than duplicative. The combination of shared implicit knowledge and diversity of skills allowed successful teams to overcome the uncertainties of project work in a competitive environment. Indeed, researchers confirmed the commonsense insight that groups that never encountered ambiguities or disorientations became complacent and intellectually stagnant (Nonaka, 2000). They also found that groups with too little tacit knowledge or shared experience and with sharp imbalances in knowledge or economic power proved far less stable and far more likely to have hierarchical management controls imposed from above than did equivalent but balanced groups (Leonard & Sensiper, 2000).

Reed and Formo's discussion of how they solicited and incorporated contributions from academics is instructive. The rapidity and abundance of responses to their call for contributions to *Job Search* provide direct evidence of how natural

---

[6]See, for example, Christopher Morris, *Becoming Southern: The Evolution of a Way of Life, Vicksburg and Warren County, Mississippi, 1770–1860* (New York: Oxford University Press, 1995), Don Harrison Doyle, *The Social Order of a Frontier Community, Jacksonville, Illinois, 1825–1870* (Bloomington: University of Illinois Press, 1983); Darrel E. Bigham, *Towns and Villages of the Lower Ohio* (Lexington: University Press of Kentucky, 1998).

electronic technology has become for ground-level faculty collaboration. One can imagine how this process would have played out 20 years ago. The authors might have telephoned department chairs, contacted friends, given panels and posted announcements at professional conferences, and published notices in the *New York Review of Books* and professional newsletters. This would have been a time-consuming process. Moreover, because all of these contacts would have been initiated by the authors, the result would have excluded anyone not in their academic circle and specialization. There would have been fewer equivalents to the cross-posting phenomenon so common on discussion lists, which allows queries to reach far across traditional disciplinary and occupational boundaries. Reaching their focus group of people just entering the profession would have been especially challenging. Thus the technical obstacles to such a project might ultimately have made it impractical. The electronic forums used by Reed and Formo not only speeded the data collection process but also increased the project's serendipitous inclusiveness.

The authors provide interesting examples of software adaptation. They discovered that environments such as MOOs may be too forthright, too public, and too bereft of body-language clues for such an intimate activity as a mock interview in front of strangers. The contrast with the spontaneous outpouring of stories about the job search process that they received after posting to discussion groups is indicative of each media's unique strengths. The call for narratives placed the writing process much more under the control of those who submitted. They could compose at their own pace and use their own organizational schemes. A mock interview in MOO-space with no time for reflection and with an agenda controlled by the interviewers can only have been intimidating. The experience of Karen McComas illustrates that such high-stakes exchanges are possible only after a community of trust has been established. It is also interesting that the core of Reed and Formo's collaboration was conducted using the oldest and most widely understood forms of software, especially word processing and e-mail.

These tools have consequences. For humanists especially there are limitations to the current technology. As Edward Ayers and Will Thomas (2001) indicate in their experimental electronic article for the *American Historical Review*, computers and HTML have been best suited for those arguments and forms of text that can be broken into screen-sized chunks, hierarchical content trees, and unidirectional hyperlinks with only a single destination. Such constraints place severe limits on argument structure. Most humanities scholarship requires greater levels of referential complexity than existing software (much of it developed for the commercial market) can provide. Their efforts to tell the story of two small communities over a narrow 15-year time span required integrating HTML, scanning and digital imaging technology, databases, statistical analysis software, geographic information and mapping systems, indexing tools, and word processing. No existing program proved capable of interconnecting such a varied pool of resources into the cohesive, multipathed, and richly textured narrative that Ayers and his associates sought to produce. They were fortunate to develop a staff of program-mers and data consultants who could provide the linkages and develop the tools they needed for such a demanding task. As Reed and Formo's collaboration dem-onstrates, integration in just one software environment can be difficult enough. The strange scenario of the two of them passing a laptop back and forth to com-plete their writing task is a sure sign of both the technology's value and the tech-nology's limitations. A recent survey by noted software expert Jon Udell (1999, 2001) of collaborative technologies in the sciences points to similar limitations and opportunities.

Defining Electronic Community

Cultural commentators, suggest sociologists Barry Wellman and Milena Gulia (1999), typically overstate the extent and intensity of relationships in actual com-munities. For at least the last century, they note, individual relationships have been "geographically dispersed, sparsely knit, connected heavily by telecommu-nications (phone and fax), and specialized in content" (p. 187).[7] The ideal of the rural neighborhood may have been lost in practice, but as Susan Sessions Rugh has shown, late 19th century migrants to towns and cities held tightly to romanti-cized memories of the rural folk community, despite their urban environment. These settlers created a myth that still exerts a powerful hold on contemporary ar-chitecture, community planning, and real estate promotional efforts (Jackson, 1987; Rugh, 2001; Vickery, 1983). Given this history it is no accident that discus-sions of community have been transported to the digital world. Community, Ayers (2000) notes, "is one of the most appealing metaphors of the Web" (p. 8). Donna Sewell's explorations underscore the power of this ideal and demonstrate the validity of its application to scholarly humanities gatherings, such as WCENTER and Tuesday Café.

Sewell presents a useful series of conditions for community, including shared location, shared interests, shared government and property, interaction and com-munication, shared obligation and commitment, and emotional connection. One might add to this list a common language and a shared sense of history, factors that in real communities have been important sources of unity and identity. It is revealing that users of MOOs and other cyber-chat environments have developed their own codes, syntax, and slang, much of which is uninterpretable by outsiders. Even in the foreshortened time span of cyberspace, expertise in this language serves to distinguish experienced users from newbies, highlighting the impor-tance of a shared chronology of experience in defining community identity. That

---

[7]See also Carl E. Schorske, *Thinking With History: Explorations in the Passage to Modernism* (Princeton, NJ: Princeton University Press, 1998), especially pp. 37–54, for even earlier examples of intellectuals lamenting the supposed loss of community in the face of urban, industrial, and technolog-ical change.

both language and history are also crucial components of nationalism should underscore their importance to group identity. These conditions contribute to shared interests and common obligation and are the essential foundation of an emotional connection among community members.[8]

Application of this framework to these essays reveals how greatly digital technology has already affected community formation in the humanities and how much promise it has for future group efforts. Taken alone, Reed and Formo's *Job Search* project would probably not meet Sewell's criteria. Participants contributed by telling stories to the authors but had few interactions with each other. Publication of *Job Search* completed their formal obligation. It might have closed the partnership between Reed and Formo as well, except that e-mail and the shared history of writing provided a potential means for their continued relationship. With digital groupware it is easy to see how the transient experience of any single project of fixed duration could be transformed into an electronic community. Given a Web site, a discussion list, and a MOO, the contributors could readily serve as counselors to each other, could add new insights to the book, and could even mentor new participants in the job search process. As Internet designer Amy Jo Kim (2000) shows in *Community Building on the Web*, information about organizational strategies and the groupware to enable such collaborations is now available and well understood. Groups that use Web sites, discussion boards, and chat rooms, she argues, are able to sustain community interest quite effectively.

The criteria for community match well with the expression of McComas and Sewell in the MOO environment. It takes only modest imagination to see a MOO session as occurring in a real place. One might compare such a session to actual workplace communities, where individuals spend most of their time in cubicles or on separate floors and where cross-partition banter to unseen neighbors is only occasionally punctuated by face-to-face encounters at the water cooler. Indeed, one might feel more identification with a regular MOO acquaintance than with someone in another academic specialty who has an office down the hall. Both authors provide ample evidence of the commitment, reciprocity, and emotional connections characteristic of noncyberspace communities. Although neither author explicitly discusses whether digital environments pose unique community formation problems for humanities faculty, their experiences do confirm that such communities are possible and valuable.

The requirement of interaction and communication would seem to rule out the Kolb-Proust Archive as a true digital community. As an information resource rather than a communication site the archive does not meet one of the crucial tests. Yet the archive promotes indirectly the collaboration of hundreds of Proust scholars around the world. If not itself a community, it contributes to community by being a common landmark, in the way that Central Park in New York City is a

---

[8]On shared identification and history see Nathan Gardels, "Two Concepts of Nationalism: An Interview with Isaiah Berlin," *New York Review of Books*, 21 November 1991.

landmark gathering place for the city's residents. Although most users will visit the archive as lurkers, neither depositing information nor conversing with each other, their presence will have important community-building consequences. By providing a shared experience that most Proust scholars will henceforth be expected to know about, the archive will reinforce collaboration both intellectually and affectively. Further evidence of this collaboration into community process can be seen in the fact that the site has already become a catalyst for face-to-face meetings.

There is the danger, of course, that all of the communities described in these essays could become overly narrow, self-selected, and autodidactic. The Internet already contains more scholarly resources than any one group of individuals could ever master. We turn out of necessity to portal pages and search engines that depend on the key words we already know to take us to places we find similar or appealing. Site-customization tools allow us to filter out information we find uninteresting or distasteful. It will not be long before the automated "we think you might like the following products . . ." tools now commonplace on commercial sites such as Amazon.com show up in scholarly contexts. The philosophical heuristic for these systems is a cluster-analysis test that finds patterns of similarity and difference, but not of creative non sequitur. The tragedy of these algorithms is that as they improve linguistically and statistically they get better at filtering out inconvenient but potentially crucial juxtapositions. It is true that such filtering mechanisms also exist in real communities. They are far less efficient and far more disrupted by the multidimensionality of natural language, however, than are comparable Internet tools. Just as the Internet can open us to new communities, it can also lead us into an ever-narrowing spiral of overspecialization. In the process, it could reinforce an unfortunate tendency that the humanities disciplines were already prone to in the predigital era.

Cow Tale: A Story of Transformation in Two MOO Communities

One can find encouraging signs about current digital technology in Karen McComas's experiences with MOOs. Like Sewell, she found MOOs to be richly interactive, intellectually stimulating, and emotionally satisfying. Her appealing chronicle of growth as a scholar and teacher in MOO-space underscores how far software environments have come in enabling collaboration. It also illustrates the importance of the human dimension to making such collaborations a success. It is worth noting that both connectivity tools and people were crucial to the success of her projects. Like Reed and Formo, she participated in community formation processes that depended on the creation of trust among the participants. This, in turn, required the skills of experienced core participants and role models. Her encounter might have been very different without the friendly invitation and mentoring provided by the participant called Pokey. McComas also praises the FC participants for bringing her gently into the world of programming. In her words, "their facilitative style suited my own learning style." The software's capabilities provided the attraction; the people using it caused her to improve her skills and to reexamine her teaching.

That the MOO environment promoted intellectual growth through experimental collaboration is not surprising. McComas's developmental experiences of playing, building, revising, and publishing nicely dovetail with the learning dualities articulated by Palmer. In a MOO environment, a sense of structure and a diversity of discovery paths are literal concepts as well as metaphors. Likewise, rooms can contain both individual and group voices, with plenty of encouragement and corrective feedback. Her descriptions of the transformations provoked by a collaborative digital environment also parallel the discoveries of Nonaka and other organizational theorists. As teachers and early adopters of technology, McComas's MOO associates all shared a large, tacit base of common information, experience, motivation, and temperament. This was reinforced by the common history of learning about the MOO and each other through a joint task and shared adventures. There was also enough diversity of skills and background to make interaction valuable to everyone, but not so much to generate a backlash against power inequalities. We, along with McComas, can use her story to reflect intelligently about our teaching and collaboration styles.

As McComas's example so effectively shows, MOO technology's multiple communication methods and programmability are powerful collaboration enablers. Of all the technologies discussed in these articles, however, MOOs demand the most from their users. At the very least, one must be proficient at the keyboard. In crowded rooms, the pace of exchanges can leave slow typists and deliberate thinkers behind. Fortunately, MOOs can overcome this through their technological flexibility. Thinkers with a contemplative style can find quiet areas and posting techniques more suited to their own preferences. A more difficult requirement is the need for a spatial and visual imagination. Text-based MOOs work best for people who can translate text descriptions into mental images. For those users for whom spatial and visualization skills are not a strong point, the MOO environment can be frustrating. McComas found MOOs to be a good match for her learning style. Those with other learning styles are likely to have different experiences.[9] Likewise, although McComas found customization through room construction and programming to be a useful exercise for examining her teaching objectives, "digging" in MOOs is not for the faint of heart. In my own experiences, while teaching programming to architects and engineers as well as during

---

[9]There is extensive research into learning styles. One influential model that categorizes student aptitude into visual, reading, verbal, and kinesthetic learning style categories can be found in Richard M. Felder and Linda K. Silverman, "Learning and Teaching Styles in Engineering Education," *Engineering Education*, LXXVIII (April, 1988), pp. 674–681. Also relevant is Anthony F. Grasha's *Teaching with Style: A Practical Guide to Enhancing Learning by Understanding Teaching and Learning Styles* (San Bernardino, CA: Alliance Publishers, 1996).

conversations with colleagues in computer science, I have learned that many individuals have no cognitive disposition or patience for even simple programming tasks. Content analysis specialists who select people for work in text-coding situations report a similar diversity of aptitudes (Shapiro & Markoff, 1998). As a unique genre with its own characteristic modes and assumptions, the MOO environment may be well suited for some kinds of faculty collaboration in the humanities but not for others.

Humanities Scholarship, Computing, and the Library: The Collaboration That Created the Kolb-Proust Archive

For electronic document collections to be worth the considerable expense and time they involve (not to mention the steep learning curve they demand of any traditionally trained humanities scholar), they must transcend the limitations of the old paper and microfilm technologies. A useful list of criteria is offered by documentary editing authority Mary-Jo Kline (1998), who argues that digital collections must either supply multiple witnesses and forms of transcription; offer creative new ways of selecting and arranging the documents; provide richer forms of annotation, commentary, and discussion; or extend the collection into audio and visual media. In light of the Sewell essay and the success of the Kolb-Proust Archive, one might add to Kline's criteria the test of whether a digital project can foster new ways of community and faculty collaboration.

The rapidity with which archives and libraries have adopted digital strategies reflects the scholarly benefits and pragmatic difficulties of managing the information explosion. Issues of improving access, overcoming limited physical space, preventing physical deterioration, and improving indexing efficiency make digitization very attractive. Two projects, JSTOR and Making of America originated in concerns over shelf space but have become mainstays for research. Far more users access the image files of these two projects than ever used the bound volumes. But there are cautionary tales. Nicholson Baker's (2001) controversial *Double Fold: Libraries and the Assault on Paper* describes an almost Baudrillardian nightmare of simulacra, in which the rush to scan and discard the originals has resulted in the destruction of thousands of pages of irreplaceable newspapers and books. Others have warned of how limited the life span of electronic materials can be. As digital preservation expert Howard Besser points out, people who still have WordStar files on 8-in. floppy disks or data deck stored on computer punch cards can appreciate how poorly digital resources migrate across time (Baker, 2001; Besser, 2000; Derian, 1994; JSTOR, 2001).

The Kolb-Proust Archive is one of many interesting prototype projects in recent years that provide a model for good practice and illustrate the subtle revolutionary potential of electronically enabled collaborative research in the humanities. Although it has been common for archives to acquire the research notes of major scholars, it has been rare for these collections to be published. Researchers wanting to study these materials had to travel to the archives. The resulting difficulty in determining how scholars came to their conclusions and what things they chose not to incorporate made their research process only as clear as the footnotes they provided. In the new digital project, in contrast, one can see not only the finished product but also the blueprints and prototypes. Those research references that the scholar chose not to include may be the most valuable resources of all. This digitally enhanced exposure of the epistemological dimensions of a research project has the potential to be as important to scholarly methods as was the invention of the footnote. Increasingly, scholars are going to expect to see the data behind an author's final conclusions.

Here is another case where digital serendipity is important. As Kibbee and Szylowicz point out, the materials in their project may be as useful to scholars with a general interest in "Parisian life and French culture at the turn of the century" as it is to the specific community of Proust scholars. The unexpected flurry of reference questions generated by the archive is testimony to how smothered intellectual cross-fertilization was by earlier technologies. I have seen the same thing with my own effort, the Secession Era Editorials Project.[10] Each week I receive a dozen or so e-mails generated by the project. These range from requests for advice from fourth graders who are writing about John Brown and inquiries from academics wanting to know more about the concordancing tools available on the site to queries from genealogists wanting to know about a relative whose name appeared in one of the editorials. David Seaman of the University of Virginia's Electronic Text Center tells me that they receive so much correspondence that they have added extensive FAQ pages to keep their mailboxes from overflowing with requests. Gregory Crane (1998) of the Perseus Project says of this phenomenon:

> We can see by the patterns of use and the mail that we receive the stirrings of a vast audience, hungry for ideas and for that practice of thought to which we, professional academics, have been privileged to dedicate our lives. Ten year olds read about the ancient Olympics; military officers at foreign posts read Thucydides; bankers examine Greek vases during lunch time pauses in their work, and adult learners in the kitchens of rural homes look up words in our electronic Greek lexicon as they work their way through Plato.[11]

These exchanges require data standards. One of the reasons scientists were able to use computers for collaborative work far earlier than were humanities scholars is that computers are much better at handling the controlled concept of

---

[10]See http://history.furman.edu/~benson/docs/index.htm

[11]The Perseus Project is located at http://www.perseus.tufts.edu It may be relevant that I have made frequent use myself of the Perseus Project to track down classical quotations used in mid-19th century American political speeches.

numbers than the chaotic richness of words. At the machine level, the numeric representation of things as basic as alphabets are still being worked out. It remains a challenge to include English, Greek, and Hebrew characters in the same electronic document, not to mention Chinese, Arabic, and Cherokee. HTML, MIME (for mail), and PDF have become the prevailing interchange standards for networked exchange, but even these are woefully inadequate to the humanities scholar's task (Ayers & Thomas). For projects such as the Kolb-Proust Archive, the two crucial standards are the Text Encoding Initiative (TEI) and the Encoded Archival Description standard (EAD).[12] These provide a rich set of rules for structuring and categorizing textual data. Kibbee and Szylowicz's discussion of the markup of name references in documents to have them conform with a standard indexed authority list is a good example of how TEI coding systematizes document language and enables scholarly access.

Standardization might not be without its costs. In the large context, these markup schemes might be viewed as manifestations of the useful but insidious homogenization, bureaucratization, and rationalization of society that theorists such as Max Weber (1963) and Joseph Schumpeter (1950) warned against a half century ago. In such a world it is the technicians and managers who legitimize what is common, marginalizing those needs and innovations that they cannot themselves anticipate or imagine. For example, because humanists tend to think in terms of authors and voices, schemes such as the TEI tend to be rich in tools for labeling people. Many humanities projects make extensive use of them. Although markup languages have provisions for standardizing dates and places, however, these are less well articulated and far less commonly employed in digital document collections than are naming schemes. Of all the leading humanities-related projects on the Web that I am aware of, only the Perseus Project and the Valley of the Shadow Project have incorporated an interactive facility for analyzing the spatial dimensions of their collections.[13] Having a map of Proust's Parisian geography might be an interesting complement to an understanding of his web of personal relationships. Finally, the TEI and similar markup structures have an implied system of knowledge hierarchy (the tree structure) that has as much to do with how computer scientists structure data for rapid access as it does with how humanists organize their views of the world.[14]

One means of meeting this challenge is through site interactivity and user feedback. Kibbee and Szylowicz report some promising initiatives at the archive for

---

[12]See the TEI Consortium home page http://www.tei-c.org and the EAD home page at http://www.loc.gov/ead for more information.

[13]For perceptive comments about the importance of geography, see Gregory Crane, "Designing Documents to Enhance the Performance of Digital Libraries," *D-Lib Magazine* (July, 2000) at http://www.dlib.org/dlib/july00/crane/07crane.html.

[14]On the philosophical and interpretive challenges of markup languages, see David Barnard, Lou Burnard et al., "Hierarchical Encoding of Text: Technical Problems and SGML Solutions," *Computers and the Humanities* XXIX (1995), pp. 211–231.

fostering this kind of dialogue. These components will create community, while allowing outsiders to redefine who belongs in the community. At a more technical level, they can provide commentary about what improvements are needed. Another solution is to make the electronic portions of the project open source. This allows users not only to work with the interactive query tools provided by the archive but also to reconfigure the markup to meet their own needs. There are serious governance issues to be resolved under such a model, but the experiences of the open source software community suggest that cooperative collective management of sources is not only pragmatically possible but also promotes broader creativity and cross-fertilization. At an intermediate level, tools are beginning to emerge for networked commentary and annotation. The World Wide Web Consortium's new Resource Description Framework (RDF), Xpointer hyperlink language, and Amaya/Annotea annotating-browser-editor prototype are the first indication that digital tools for this core set of humanities activities are at last on the horizon.[15] When fully realized these will allow the equivalent of marginal notes to be scribbled in the corner of any document in a digital archive and for such marginalia to be made available to any other scholar who might be interested. These tools may also make possible a semantic web, where documents are linked according to some recognizably human framework of meaning rather than raw lexical matching (Berners-Lee, Hendler, & Lassila, 2001). It will soon become essential that the assumptions behind these tools and algorithms be made available on Web sites and that consideration of their methods be included as part of the apprenticeship process in humanities disciplines.

IMPLICATIONS

As Reed and Formo note in their essay, faculty collaboration in the humanities disciplines is still a risky activity. Although some forms of collaborative technology (especially word processing, e-mail, and discussion lists) have become an almost invisible foundation for academic teamwork, other digital tools are either less respected or less developed than they deserve to be. Because digital collaborations in the humanities force scholars to take on roles not traditionally associated with their specific disciplines, it is difficult for their traditional colleagues to understand what they are doing. Because digital projects require the expertise of librarians, programmers, statisticians, graphic designers, standards organizations, and philosophers, as well as humanists, it may not always be clear which discipline owns a project. Such interdisciplinary cooperation is mostly a good thing.

---

[15]Information on the RDF protocol is available at http://www.w3.org/RDF/ See also Eric Miller, "An Introduction to the Resource Description Framework," *D-Lib Magazine* (May, 1998) at http://www.dlib.org/dlib/may98/miller/05miller.html For Xpointers, see http://www.w3.org/XML/Linking. For Amaya and Annotea, see http://www.w3.org/2001/Annotea/

For those charged with evaluating these endeavors and their participants, however, the old rules apply only partially.

There are lessons to be learned from the scientists' experience in collaboration. Because digitized humanities information is suddenly so valuable and so applicable to so many kinds of circumstances, it is beginning to be commodified. As the most important documents collections get absorbed by Chadwyck-Healey, Gale Group, and Dow Jones, humanities faculty will soon encounter the same problems of contending with the corporate world that the grant-driven sciences have been struggling with for years. It is difficult to predict how reward structures in the humanities will evolve, but it seems likely that as in the sciences, those entities with the deepest pockets will have a profound effect on the research agendas of those who would collaborate in textual and critical scholarship. Even those who do not seek grants will be guided by the sheer magnetism of these dominant resources.

Evidence from these four essays suggests that the benefits of collaboration outweigh the risks. Reed and Formo's diverse insights and discovery of a combinatorial writing voice show us how we might reconceptualize the authorship process in digital environments. Sewell and McComas offer the theory and practice of community formation and provide specific insights into how MOOs and other groupware might change how we teach and learn. Kibbee and Szylowicz demonstrate that even an ostensibly passive information resource like the Kolb-Proust Archive might energize new collaborations and bring together individuals from different backgrounds to interact around a common focus. The history of earlier technologies suggests that the unanticipated consequences of innovation may be as fruitful as those their inventors intended.

IV

ELECTRONIC COLLABORATION
AND THE FUTURE

16

Imagining Future(s): Toward a Critical Pedagogy for Emerging Technologies

Timothy Allen Jackson
Ryerson University

> *New technologies = new perceptions. Reality is a man-made process.*
> —John Brockman (1996)

> *I am convinced that at present humanity is going through a bifurcation process due to information technology.*
> —Ilya Prigogine (1999)

> *But I am a prisoner of hope.*
> —Cornel West (1995)

As academics, we are in the consciousness business. This is the end game of our efforts, if we are serious about the impact of ideas onto the body public as opposed to purely instrumentalist motivations driven by market forces that ultimately produce supply-side pedagogies. This is the essence of what it means to profess. The moral and ethical responsibility of this mission is and has often been at odds with the needs and demands of industry, yet this responsibility is what separates the academic project from industrial training proper. Historically, education has been the servant of industrial production with a decidedly less-than-critical pedagogy.[1] Henry Giroux (1983) suggests a "radical pedagogy needs to be

---

[1]Critical pedagogy takes as a central concern the issue of power in the teaching and learning context. It focuses on how and in whose interests knowledge is produced and passed on and views the ideal aims of education as emancipatory. A more expansive definition is available at http://www.csd.uwa.edu.au/altmodes/to_delivery/critical_pedagogy.html

informed by a passionate faith in the necessity of struggling to create a better world" (p. 242).

As academics in democratic societies, we hold a moral and ethical responsibility to remain engaged in these ideological struggles for the present as well as future benefit of our students, our institutions, and our wider culture(s)—and to continue to do so as we work with these industrial (and other) influences that may want to restrict the manifestation of such a critical pedagogy. We may recall the heretical tradition of the university system within the European tradition and its often troubled relationship with church and state as exemplary of such a history of resistance. Liberation theology offers a more current example of such interior/exterior forms of critical and transformative engagement.[2] There is therefore no ivory tower separation from the political and social impact of our collective actions as educators and researchers. Instead, the awareness of this synthetic and discursive relationship between pedagogy and power requires us to develop a vision that "celebrates not what is but what could be, that looks beyond the immediate to the future . . . [t]his is a call for a concrete utopianism" (Giroux, 1983). Although it is impossible within the limits of this text to argue more conclusively, I suggest that critical pedagogy represents an imperative within the history of ideas and is therefore an a priori condition to current and future pedagogical praxis.

Currently our premiere institutions of higher education are essentially corporations engaged in a public discourse on the past, present, and future of our collective cultural identities and what constitutes our individual relationships to these forces. Through the actions of academics on the front lines, these institutions of higher learning police ideas in many ways and often serve discreet but powerful ideological masters that work against the goal of a critical pedagogy. This practice is indeed an outgrowth of Paulo Freire's (1996) banking concept of education, in which "apart from inquiry, apart from the praxis, individuals cannot be truly human" (p. 53). We must remain very vigilant in the scrutiny of changes incurred through market forces or other economic forces that are concerned with short-term savings rather than the systemic costs of such initiatives.

The current conditions occasioned by the meteoric rise of the field of information technology within late capitalist societies offer perhaps the most profound exemplar of such power relations to the world of academia. Within this context, as academics we must engage the issues that will best serve the larger project that we attend—namely, our role as critical pedagogues charged with the past, present, and future representation and articulation of ideas, institutions, and cultural production as we act as agents of change within our spheres of influence. Our relationship to technology is therefore inextricably linked to the project of growing or

---

[2]Strictly speaking, liberation theology should be understood as a family of theologies—including the Latin American, Black, and feminist varieties. For more on liberation theology, visit http://home.earthlink.net/~ronrhodes/Liberation.html

limiting critical consciousness. As Freire (1993) states more succinctly, "[t]here is a dialectical relationship between the material world that generates the ideas and the ideas that can influence the world by which they are generated" (p. 67). Within this context, it is clear that institutions of higher education must engage in imagining futures that serve the needs of the social, political, and economic conditions of the present and near future. Yet this imagination requires a dynamic vision and a rapid revision of curricula in a historically unprecedented scale and scope. As Freire (1995) argues in *Pedagogy of Hope,* such a vision first requires the necessary political act of dreaming, which produces historical change through our agency, because "there is no change without dream, as there is no dream without hope" (pp. 90–91). Where should our technologies take us, our ideas, and our cultural traditions? Technology only has agency when we abdicate our own because our technologies—like our metaphors—are the vehicles of our dreams rather than the destination of our journeys. Yet, as art needs metaphors to communicate, humans need technology to survive. We might even conjecture that our emergent condition as human may indeed be tied to our relationship with tools as a means of enabling what we consider humanity. We therefore must imagine the future(s) we want these vehicles to take us to as well as making sure the equipment is well maintained for the duration of this ongoing travel. This is particularly the case given that technologies currently threaten our sustained existence as well as question the nature of what is human in fundamental ways.

New media technologies, such as the Internet, CD-ROM and DVD-ROM, digital audio and video, and various forms of interactive multimedia technologies, are shaping our world and worldview at an unprecedented scale.[3] These new media are assuming an ecological force engaged in a dynamic dance between natural and synthetic systems. Given this context, I would like to suggest an obvious principle: that new media technologies should ultimately enrich rather than depreciate the quality of our personal and collective lives. This principle must be a precondition for the emergence of a critical pedagogy that engages these problems in a way that neither collapses the complexity of the problems such changes embody nor celebrates the inherent supply-side needs they predicate on higher education.

For us to attempt to realize this goal, we first must be able to envision the systemic implications of new media technology and content on public consciousness. We also must be able to distinguish reality from the illusions offered through utopian visions of emerging technologies. Despite the hype, it is clear that new media technological and social developments in the last 2 decades of this century have indeed presented our global community with an ontological shift. This ontological shift plays an affective role in the formation of individual and collective identities, in the way we think of existence and the nature of reality, and, by extension, inaugurates a new reality and new ways of thinking about multiple realities. Onto-

---

[3]New media is used as singular, rather than plural, because media alone would point to its integrative and synthetic relationship to other media as well as support the popular usage of the term.

logical questions such as "What is the nature of existence" must be reframed in this generation to "what is the nature of existence today, in this context, and under what related conditions," given the technological developments of this century and their formative influence on our individual and collective consciousness.

The ontological nature of new media technology is in process and will remain so for generations until our physical bodies begin to adapt to life in cyberspace (virtual worlds) and our creative cognitive abilities rise to the level of our capacity for making tools. Simply stated, our technologies evolve fast, our minds not quite as fast, and our bodies at a much slower rate. It is certainly too early to tell whether these changes are and will be evolutionary or devolutionary in a qualitative sense, but it is likely that human life on the planet will never be of the same scale or enjoy the historical cultural stability of place, lineage, and shared belief systems as those of previous generations. Despite the seductions of a futurist utopian vision, we should not be tempted to relinquish our primary world (the physical world) to the possibilities presented by virtual worlds. Rather than abandoning this complex, contradictory, problematic, and splendid primary world to the thin construction of less-complicated virtual cosmologies, we should work toward the integration of these worlds based on a shared goal. I hold that the most compelling world exists somewhere betwixt and between the two in a synthetic and symbiotic relationship. It is in these virtual/physical places and spaces—these new media interfaces—that the transformative possibilities of new media exist as well as the most promising pedagogical sites for our responsible activities as academics and public intellectuals exist.

These new media interfaces are sites of cultural work that provide markers as to how new media will enhance or limit our lives. The discoveries that occur at these interfaces between the physical and the virtual worlds will have tremendous impact on how we shape our conceptions of the social, moral, and ethical realities of our personal and collective future. The depth of the changes brought on by this ontological shift calls for a shared set of pedagogical beliefs or principles through which new media innovations may be critiqued, assessed, and evaluated. Given such a project for higher education, the following are some general observations that academics should engage on a foundational level within their own disciplines as a starting point to this crucial process of pedagogical transformation:

- New media is a part of our global ecology. The impact of information technologies shapes our environment and our relation to it in ways that would have been unimaginable to previous generations. The illusion that new media technologies somehow operate outside of the limitations of our environment must be dispelled to more fully understand the systemic relationship of new media to other sets of systems. The mind-body and human-nature binaristic splits (among others) must be abandoned for models that recognize the dynamic relationship of things to their systemic contexts. This position requires a systems view of the physical and cognitive world(s), as well as our personal and collective forms of consciousness about the possibilities of worlds within worlds. We might consider such an approach a form of green media, which suggests that how our technologies are produced is connected to how they are used, for what purposes, and under what type of sustainable conditions. We must therefore individually and collectively be held accountable for the implications of this process on our collective global environment.

- New media technologies and content shape consciousness and identity. New media technologies currently provide an increasingly important technological matrix for information dissemination and identity engineering in late capitalistic societies and societies under capitalistic cultural and ideological siege. These technologies and content play an affective role in the formation of personal, familial, communal, national, and cultural identities. It is imperative that cultural producers and consumers (including academics) be critically aware of the power that these technologies exert, the motives underlying their use, and the nature of their appropriate and inappropriate integration within contexts relevant to the formation of personal and collective identities. Although new media offers new possibilities and sites for transforming the power relationships of information production and dissemination, it simultaneously provides new means of reproducing dominant subject positions and power relations. Within identity politics, new media can function to both enhance, restrict, or enhance and restrict, the growth of critical consciousness.

- New media is symbiotically linked to aesthetics (the philosophy of the senses). Our present new media ontological shift provides a rupture with our previous perceptual relationship to the world. New media technology and content harnesses human desire through aesthetic means and produces new forms of aesthetic experiences. A new media ontology is aesthetic by nature because aesthetics is concerned with how ideas are formed, shared, and contested through the matrix of the senses. New media is therefore a new synthetic form of art making and is subject to aesthetic discourse. However, new media presents new challenges for negotiating meaning through sensory input by providing new types of experiences and forms of art production and consumption. In light of the new media ontology, the field of aesthetics should return to its original breadth of meaning as the branch of philosophy dealing with the senses. This definition must also be restated in light of the political dimension of new media aesthetics.

- New media facilitates new forms of reproduction and synthesis. New media allows for the synthesis and reproduction of previous technologies and content and introduces new forms of communication (including new art forms) that require a critical method grounded in a theory that can accommodate complexity and contradiction without being reduced to relativism. The project of developing a theory and criticism of new media must therefore respond to a number of factors to meet these new demands. These synthetic relationships require the field of new media to be linked in symbiotic relationships with other more traditional forms in subtle although important ways, rather than being seen as a replacement for previous technologies. It is important to note that there has been significant growth in the volume of print media since the rise of the Internet and CD-ROM distribution of various texts. Perhaps most significant, digital files lose no quality through reproduction, which changes the scale and scope of their distribution as well as present problems of authorship, ownership, and other related ethical concerns.

- New media functions as a pedagogy. New media technologies and content must be assessed according to how they assist in the qualitative enhancement of human life. Although the use of digital educational resources, simulations, gaming models, interactive communications, and other applications of new media technologies may offer a variety of benefits to education, they must also be evaluated carefully regarding their content and the complicity of their instrumentalist uses in schools and other pedagogical sites. The potential benefits of new media include access to enormous databases and other content that will dwarf the offerings in even the best school libraries, interaction with other learners and learning communities from a diverse range of cultures and classes, empowerment brought about by actively constructing learning materials and environments through hypertextual and hypermedia webs, and the ability to communicate electronically with individuals and communities that would be inaccessible under other conditions. Although these transformative possibilities offer new ways of teaching and learning, they also raise some critical questions for pedagogical consideration. One of the key limiting factors of such possible uses of new media technologies is the economic ownership of information and the development of shared resources that are meant to have the widest accessibility possible. A new media pedagogy should therefore be held to the goals of a critical pedagogy, which strives to share information in the most generous and equitable manner while maintaining a critical relationship to the content available through such resources.

- New media influences the construction of our personal and collective conceptions of reality. New media can threaten the integrity of human experience by privileging virtual and asynchronous interactions over the physically present interactions in real time and space. Our relationship to our world requires metaphysical grounding to structure the nature of our experience. The proliferation of virtual environments, simulations of primary world experiences, and multiple forms of identity made possible through new media communications tend to confound such a grounding. Although our personal conceptions of reality and realities will indeed remain varied, our collective metaphysical grounding should exhibit patterns of a shared vision of what the public policy of new media technologies is, should, and will be.

- New media develops faster than our ability to adapt to the conditions it occasions. In contrast to the time line of our exponential technological evolution over the past 2 centuries, our bodies—and to a lesser degree our minds—are ancient and require much longer periods to adapt to our contemporary needs. For example, many of us today require the simulation of work through exercise to avoid physical atrophy and to reduce the impact of stress on our physical and mental health. In this sense we are already cyborgs in many ways, and our dependence on machines to amplify our physical limitations and to accelerate our evolution is growing at an alarming rate (on a metaphysical scale). I believe we should become more comfortable with—or rediscover—our ancient physical, spiritual, and social needs and be suspicious of any attempts to digitally disrupt these analog rhythms in any sweeping or totalizing manner. We are analog beings who require the continuity of analogic life to maintain our physical, emotional, and psychological health.

• New media benefits some cultures and populations more than others. As has been pointed out in several contexts, the illusion of new media as a radically distributed form of democracy and the myth of total access on a global scale must be given a virtual reality check. In *What Will Be* (1998), Michael Dertouzous, director of MIT's Laboratory for Computer Science, home of the World Wide Web Consortium, warns us that "left to its own devices, the Information Marketplace [the Internet, World Wide Web, and Intranets] will increase the gap between rich and poor countries and rich and poor people" (p. 241). Similarly, the dominant culture emerging as the primary economic—and by extension, social—force in cyberspace speaks a particular language (indigenous to Silicon Valley and its environs) and shares a particular futurist ideology. This critique is not meant to suggest that cyberspace and other sites of new media technology cannot be used in more transformative ways but rather to indicate the current state of affairs. We should therefore work toward a sense of community in cyberspace that more closely resembles the composite diversity of our global population and their cultures, rather than the elitist representation offered through contemporary conditions.

• New media is a force of both liberation and oppression. Unlike the context of Freire's *Pedagogy of the Oppressed* (1996), where one could more clearly distinguish the oppressed from the oppressor, in our contemporary global society the oppressed and the oppressors participate in a dynamic system of power relations with less distinguishable borders. Given our contemporary situation in the United States in particular, most of our citizenry are both complicitous and compliant as the oppressors and oppressed. This complicity extends to the producers and consumers of new media. Although the technologies of new media offer new possibilities for uses that can serve to rupture oppressive conditions, they can also impose new forms of oppression. We must therefore try to distinguish the power relations present in particular new media texts and work to provide ruptures in more oppressive forms of new media ideologies based on shared visions for the development of critical consciousness. On an economic level, the open source movement is a clear manifestation of such a project.[4]

• New media innovations and ideologies must always be assessed in light of how they enable critical consciousness. As a complex and dynamic system of

---

[4]For more on the open source movement, see Tari Lin Fanderclai's chapter in this volume.

communication, new media technologies and content should be evaluated in light of their transformative potential, rather than the most recent speed rating of a microprocessor, the cool graphics available in a new computer game or digital film, or utopian predictions that romanticize new work conditions in cyberspace. In the same way that new media is not one thing but a number of elements that as a dynamic system embody more complexity that any one new media text, the complexity of the new media ontology should not be considered equivalent to a collapse of meaning. We may therefore use some of these considerations as first steps in assessing the impact and relevance of new media technologies within the context of our own disciplines and subsequent personal moral and ethical dimensions.

New media technologies and content therefore exert a global economic and social force, shaping human consciousness and identity in fundamental ways. As a complex aesthetic system, new media provides for new methods of reproduction and synthesis—a melding of previous art and communication forms into a new hybrid of old and new, popular and elite, and representing a broad sweep of taste cultures. Because we learn through mediated experiences, new media functions pedagogically, influencing our personal and collective conceptions of reality and meaning. Because our bodies and minds evolve much slower than our technologies, new media is developing faster than our abilities to adapt to the changing conditions it presents. New media benefits some cultures more than others and is therefore a force of liberation and oppression. On the economic borders of new media culture, the plugged-out simply don't exist. Participation in new media culture is a privilege that the vast majority of the world does not enjoy. It is clear in this context why the lure of the digital world offers us a final illusory solution to the enlightenment project: denial. Homelessness, disease, malnutrition, poverty, and other undesirable realities of our primary world will not likely be programmed into our virtual worlds. Given this context, innovations and ideologies must always be assessed in light of how they enable the previously stated project of a widening and deepening critical consciousness.

Any text that attempts to prognosticate the future implications of technology runs the risk of over- or underdetermining the complexity of conditions at work based on the current state of affairs. Just as some predicted that radio would eradicate ignorance and television was to be the great educational vehicle for the future, I fear the Internet may ultimately be subject to similar limiting factors driven by the cold calculus of money. We might note how popular hate radio and professional wrestling are at present—as is pornography on the Internet. Despite these limiting factors, new media offers much more freedom to produce content, rather than simply consume, as in the case of radio and television, and is producing radical new forms of communication that will continue to find transformative uses.

Although I take a decidedly critical look at new media, I believe that the ontological shift occasioned by new media technologies offers profound possibilities for transformative and positive changes in our personal and collective lives. I believe that the project of developing a vision for a promising and sustainable future is a required condition for negotiating the appropriate relationships for technological growth in praxis for the next century and beyond. This goal will help us determine where our tools and technologies should take us and why it is important to be there. I hope that we may begin to see technology as a vehicle that can help us travel to such a location, rather than considering our technologies the destination of our individual and collective journeys.

Our technologies are the vehicles for our metaphors, dreams, and nightmares. The question I would like to propose is "where do we want these technological vehicles to take us, and why?" Our vision for the future, or lack thereof, will have profound consequences on the quality of human and other forms of life on this planet. Our challenge for the present and the future is to develop a vision that can be shared by diverse global cultures. This vision will be shaped through aesthetic means using new media technologies, and it is political in nature. I therefore believe that the project of developing a progressive vision for a more personally rewarding, socially equitable, and ecologically sustainable future should be a central concern for research in the field of new media and the pragmatic application of information technologies with higher education. Let us hope we build a future in which we can live with more stability and sustainability and that our technologies serve such a goal. Our institutions of higher education play an important role in these matters and therefore require academics to ask some stark pedagogical questions about which goals our technologies should serve. I hope we ask these questions with forethought and answer them from our hearts and minds. Imagined futures often produce very real outcomes.

17

Critical and Dynamic Literacy in the Computer Classroom: Bridging the Gap Between School Literacy and Workplace Literacy

Paul J. Morris II
Pittsburg State University

I came to graduate school late in life from a job in television broadcasting. I worked as a broadcast engineer in radio and television for nearly 20 years, and my knowledge of computers has grown out of that job. Indeed, I bought my first PC in 1986 because many of the machines I operated were run by computers, and I needed to keep up with current trends in communication technology. After I received my PhD it was, in part, my computer experience that helped me to obtain several job interviews. During one on-campus interview at a large Midwestern university, I was invited to attend a computer conference. The conference was small, but some important people in computer-assisted instruction (CAI) presented. Along with the CAI experts, a contributing editor of *Wired* and another academic who had just written an anticomputer book presented. The conference presentations were interesting enough; however, the concluding, town meeting discussion turned into a free-for-all, in which the academic presenters felt obliged to defend the use of CAI in the classroom, and the computer-phobic academic alternately attacked and was attacked by the audience. The *Wired* editor, who was seated near me in the back of the room, seemed bored with it all. As a former broadcast engineer in radio and television, I was somewhat annoyed with the conversation. It was probably my understanding of computers as a practical tool that prompted me to ask, "Shouldn't we, as teachers, take advantage of any tool that engages our students?" The CAI specialists inferred that my question was not very pedagogically sound, and the computeraphobe, who had never owned a computer in his life, dismissed me with a witty aphorism. "Well," he said. "My mother used to put honey on my oatmeal so I would eat it." My question, obviously irrelevant to the more engrossing theoretical discussion I had just interrupted, was basically ignored. For me, their responses not only minimized the value of computers as intrinsic and extrinsic motivators for literacy but downplayed the practical benefits of simply learning how to use a computer. They also seemed to be dismissing the connection between the literacy skills we teach and what students actually do on the job or in their communities after they finish school.

I relate this story not to diminish the importance of pedagogical theory, especially as it relates to CAI, but to highlight the differences I have observed between how academics perceive computers and how business and industry see them. For us to ignore the practical applications of computers as literacy motivators and as an important prerequisite for obtaining a position in the workplace seems imprudent; on the other hand, it seems equally unwise to design school curriculums that are based solely on the needs of business and industry or uninformed by theories supported by credible literacy research. My purpose in this chapter is to try to find ways to bridge this gap between theory and practice, between what we teach in the classroom and what people actually do on the job, between schoolhouse literacy and workplace literacy. Specifically, I will discuss how the computer classroom can help span these gaps. My experience with computers both in the communication industry and in the academy gives me some insight into how both sides use computers, and it may be a useful position for a bridge builder. Before I propose some practical bridge-building strategies, I would like to say something about literacy and computers and then offer my own multifarious definition for the word *literacy*.

I am not about to suggest that it is possible (nor even advisable) for schools to become so specialized that they could actually teach the literacy skills necessary for any particular job. For our students the academy is very much a part of the real world and very much like a real job. Although we should not teach for the workplace, we should pay attention to how people use their literacy skills in the workplace. I agree with the following from workplace writing researcher Rachel Spilka (1993):

> Professional writing specialists, both in the academia and the workplace, . . . need to agree that professional writing pedagogy in academia needs to be influenced, somehow, by those practices preferred and accepted in workplace cultures, while those practices, in turn need to be influenced, somehow, by that knowledge preferred and accepted by academia. (p. 210)

I would extend this reciprocal relationship a bit to include, as others have done before me, composition specialists and composition pedagogy in general. Literacy teachers in colleges and secondary schools would benefit from the kind cooperative exchange of ideas that Spilka is suggesting here. Further, the sort of research that Peter Sands encourages in his chapter in this volume also would be greatly enhanced by fostering closer relationships between writing professionals in academia and in the workplace.

In *The New Literacy: Moving Beyond the 3Rs* (1996), Stephen Tchudi and I discuss the problems with traditional notions and narrow definitions of literacy, offer a more complex definition, and then urge schools to take a comprehensive and firsthand approach to literacy education. The primary research for the book consists of interviews with more than 30 people, children and adults from all walks of life. We asked these people to reveal their literacy stories—how they became literate, what they learned in school, and how they adapt their literacy skills to what they do in the workplace. In many of the interviews, we discovered that school literacy and workplace literacy complement each other, though often in indirect ways. Two brief examples from the book will illustrate what I mean. One woman, who majored in English literature in college, worked for a ship building company and read complicated contracts, deciphering them for her boss. Although she believed her degree in literature was impractical, it was obvious that the critical literacy skills she had honed on Shakespeare and Milton transferred easily to marketable job skills. In another interview, a Paiute reservation police officer, who had recently returned to college, disclosed that he learned how to write on the job. His simple and direct ("just the facts, ma'am") style of writing served him very well on the force, but in college he enrolled in a remedial composition class because he thought his writing was not good enough for college. Literacy stories such as these suggest that often there is no clear connection between how schools teach literacy and what literate citizens actually do in their communities.

Although I can understand the allure of wanting to connect all school writing to the literacy skills needed to find a job, it seems pedagogically unsound and perhaps even dangerous for colleges to attempt the kind of radical restructuring of freshman composition curriculums that Elizabeth Tebeaux and others have suggested. Tebeaux (1996) believes the following:

> freshman composition courses could be adjusted to harmonize with rather than counter the paradigms used in nonacademic writing courses: emphasis on richer contexts (case studies), deemphasis on the essay, introduction to document design, elimination of expressive discourse, and inclusion of writing, such as instructions. (p. 51)

Tebeaux's emphasis on richer contexts is laudable; however, her deemphasis on essay and expressive writing assumes that the techniques in this type of writing—for example, narration and self-reflection—are antithetical to the writing skills needed in technical fields. (Ironically, Tebeaux quotes Peter Elbow on the page immediately preceding the previous quote. Elbow, of course, began his teaching career at MIT.) And though it may not be Tebeaux's main objective, one could interpret her "rethink[ing] of freshman composition courses" as a not so subtle attempt to mold students into the roles that she believes are best for them in a capitalistic society. As Jami Carlacio points out in her chapter of this collection, the "rhetoric of technology" may be a bit too intimately connected to the "rhetoric of democracy" in our society. Academic attitudes about business and industry affect how our students relate to the workplace. It therefore may be just as important to teach the critical literacy skills necessary to question certain practices in business and industry as it is to prepare students to be good workers.

I recently taught a summer course in technical writing, and I found the students in the class to be much more engaged than my 1st- or 2nd-year composition students. One reason for this engagement may be their age difference (freshman and sophomores rather than juniors and seniors), but it may also have to do with the fact that the writing in the class seems to have a clearer connection to writing in the workplace. To reproduce this kind of engagement in other writing classes, literacy learning has to be connected to the literacy skills that people need to become workers, teachers, activists, community leaders, and so on. Schools that fail to make this connection or bridge this gap may be condemning their students to classes or curriculums that have no clear relationship to student literacy needs. As illustrated by some of the people Tchudi and I interviewed, literacy learning in this mode can become a hit-or-miss proposition in which teachers cannot possibly know what is helping students and what is not, and students see little correlation between what they are learning in school and what actually happens in the community or on the job. Because the computer is such a vital tool in the community and the workplace, the computer classroom offers vast opportunities for students to engage in literacy tasks with authentic contexts.

Not having been brought up in the academy, I often find myself looking askance at academic ideas and theories that I believe are unrealistic or impractical. I have never heard a businessperson, an automobile technician, or an engineer argue theoretically about the practicality of using a computer in the workplace. In the television industry, computers are considered an indispensable tool for the operation and maintenance of any station. From the control room to the studio, from the editing bays to the newsroom, nearly every aspect of the operation depends on computers. Auto mechanics use computers to diagnose problems in cars; doctors use them to diagnose diseases; accountants use them to do our taxes. Most people in business and industry take it for granted that they would not be competitive without computers, and obviously they want their employees to be computer literate. The computer industry itself is the largest growth industry in America. Recent statistical information about employment opportunities for college graduates suggests that the major employer of liberal arts graduates are companies involved in or with computer technology. In fact, a recent article in *The Journal of the Midwest Modern Language Association* (Hellekson, 1998) was written by an individual who quit her teaching fellowship at the University of Kansas to pursue a job at a printing company (an industry now very much dependent on computer technology). "In short," she says, "because of all the research strategies, teaching skills and writing and grammar skills that I possess as a result of years of English, I have found a job that exploits them and values them more than the academy does" (p. 4). Her use of the terms *values* and *exploits* to describe the expertise that she brings to her new job helps to illustrate just how disengaged schoolhouse literacy is from workplace literacy. Paradoxically, it is because she feels her skills have been exploited and undervalued by the academy that she has decided to take a job in the private sector. It would be interesting to know how many of the skills she perfected while studying English in college as a student and a teacher were enhanced in the computer classroom.

At the 1998 Convention of the Conference on College Composition and Communication (CCCC), Cynthia Selfe, the keynote speaker, urged the members "to pay attention to how technology is now inextricably linked to literacy and literacy education in this country." Selfe's main concerns in "Technology and Literacy: A Story about the Perils of Not Paying Attention" are that schools may not be considering the implications of combining technology and literacy and that people of certain socioeconomic status may be left behind in our society. She suggests that if schools do not start thinking about how technology is linked to literacy, then they risk exacerbating the social and educational discrimination that already exists in our society. She also warns that large-scale literacy projects, such as the one advanced by President Clinton and based on expanding computer technology in the schools, are not a panacea for complex social problems in America. For Selfe, paying attention to the link between technology and literacy is not simply deciding whether we should or should not use computers in the classroom. Instead, paying attention means schools and teachers must resist narrow, politically driven solutions to literacy problems in our society. It means literacy teachers should quit avoiding technology and make it as much their business as reading and writing. Finally, it means doing the kinds of scholarship and research that several authors in this book have encouraged and using that research to inform classroom practice.

The address was originally posted in draft form on the Internet, and I urge anyone who has a stake in education to read it. She invites comments at the end of her address, and I responded on two fronts. First, I questioned Selfe's concerns about social inequities and the computer classroom. I cited a survey conducted by the Tomás Rivera Policy Institute in Chicago. According to the Institute's most recent statistics, 33.7% of Hispanic households own computers, up from 13% in 1994 (TRPI, 2001). With the advent of cheaper technology, computers may someday be as prevalent in people's homes as televisions are today. Second, although I agree that teachers should discuss the implications of new technologies in the classroom, I'm not convinced that teachers are rushing madly into the computer classroom sans purpose or theory. As computer coordinator and composition specialist for the English Department at Pittsburg State University, I give workshops on computers and literacy that are attended by mostly graduate assistants, part-time faculty, and newer full-time faculty. Most of the colleague's in my department are similar to the teachers Selfe describes in her article. They are either not interested in computers, fearful of new hardware and software, or resistant to new teaching strategies. However, those who do attend the workshops are more than willing to discuss the theoretical implications of what they are doing in the computer classroom. Like me, they realize that until academics can recognize and accept the computer as an indispensable classroom tool, students will leave school ill-prepared, lacking the critical and dynamic literacy skills necessary to compete in the workplace.

The terms *critical* and *dynamic* are taken from *The New Literacy* (1996). Throughout the book, Tchudi and I use the terms critical literacy and dynamic literacy to describe more specifically what children and adults actually do as literate citizens. Critical literacy is the ability to move beyond literal meanings; to interpret texts, films, video, computers, and the Internet; and to use writing (films, video, computers, and the Internet) not only to record facts but also to analyze, interpret, and explain. Mike Rose (1989) refers to critical literacy as "framing an argument or taking someone else's argument apart, systematically inspecting a document, an issue, or an event, synthesizing different points of view, [and] applying a theory to disparate phenomena" (p. 188). If a person discusses a film or the Internet, argues over political issues, analyzes complicated scientific data, or interprets a law brief or contract, he or she is engaged in critical literacy. Writing an essay, poem, story, novel, scientific paper, or law brief and making a film, video, or Web page, likewise, are aspects of critical literacy. Any metacognitive activity wherein individuals reflect on their own thinking processes can also be considered within the range of critical literacy. Much of what I describe here as critical literacy is what curriculum designers and school administrators like to refer to as critical-thinking skills. These skills, they say, are what we are trying to impart to our students. My sense is that we do not actually teach critical-thinking skills. Students are critical thinkers long before they arrive at school (see Bissex, 1980; Donaldson, 1978; Harste, Woodward, & Burke, 1984); they simply are not thinking critically about the subjects that we in the academy value. Of course, teachers can enhance critical skills in students by helping them to adapt the thinking skills they already possess to the tasks at hand.

Dynamic literacy goes beyond text to include the skills and abilities that are likened to them (e.g., science literacy, historical literacy, computer literacy, baseball literacy, plumbing literacy). Because it encompasses both basic and critical literacy, dynamic literacy can be seen as using these other properties of literacy in practical situations. However, as we discovered in our interviews, people who are dynamically literate can be adept at manipulating language within their respective discourse communities yet not be able to read or write well. For example, one man who once owned a butcher shop could neither read nor write, so to advertise his meats he used to borrow the advertisements from a large supermarket and substitute his own prices. Dynamically literate individuals exhibit characteristics derived directly from the literal definition of dynamic. This word comes from the French *dynamique*, which in turn is derived from the Greek word *dunamikos*, meaning "powerful," and *dunasthai*, meaning "to be able." Recognizing the power of language as a tool to transform their lives, the dynamically literate are able to change the way they use words to suit their own purposes. Whether using texts, computers, radio telescopes, or the spoken word, they can effectively integrate their use of language with their need to belong to and to operate within a given community.

Along with critical and dynamic literacy, Tchudi and I included basic literacy (Morris & Tchudi, 1996), more closely associated with literacy in the 3Rs. We do not separate these three definitions, nor do we perceive them as hierarchical. Instead, we see them as three rings, all of which overlap in their connections to the other rings of literacy. Because the people we talked to did not use one single basic literacy to operate in the world, we used these three terms to help us understand what we were discovering in our interviews. An understanding of the 3Rs may have allowed our interviewees to encode and decode language symbols, but their critical literacy helped them to reflect on how and why they used language, and their dynamic literacy helped them to apply both their basic and critical language skills to specific jobs or tasks. Dynamic literacy seems to be more associated with workplace literacy in that it requires that a person be able to adapt specific language skills to a particular job or discourse community. The computer classroom, with its emphasis on task, publication, collaboration, as well as individual initiative, becomes the ideal place for creating assignments that enhance both dynamic and critical literacy. So, with these definitions in mind, I would like to turn now to some classroom activities that may help bridge the gap between schoolhouse literacy and workplace literacy as well as literacy theory and classroom practice.

It is possible to turn classrooms (traditional or computer) into literacy laboratories, and this can be accomplished through the collaborative learning techniques widely supported by researchers such as Mike Rose (1989), Russell Hunt (1989), Susan Jarratt (1991), and many others. In *Lives on the Boundary*, Rose repeatedly recounts how the underprepared students in his remedial classes work in groups to help each other "generate meaning and make connections" (p. 145). Hunt suggests making writing assignments "more instrumental" or "real" by "creat[ing] situations in which student writing—and the teacher's own—is read for its meaning, for what it has to say, rather than as an example of a student theme to be 'assessed' or 'evaluated' " (p. 95). Teachers and students collaborate as partners in this nondirective approach to literacy learning. Although adopting a more sophistic approach, Jarratt (1991) states her vision of a composition course:

> [I envision a] composition course in which students argue about the ethical implications of discourse on a wide range of subjects and, in so doing, come to identify their personal interests with others, understand those interests as implicated in a larger social setting, and advance them in a public voice. (p. 116)

Jarratt sees critical and dynamic literacy growing out of the conflict inherent in disparate voices. The collaborative techniques that these experts support have helped to shape the peer response groups (for writing and reading assignments), the collaborative writing assignments, and the small- and large-group class discussions that many teachers now use in their classrooms. Further, they are all useful for creating a sense of community.

But how does a teacher connect literacy to the larger community outside of the academy? A letter to the editor is an obvious example of an authentic literacy task that gives students a chance to see their opinions printed in the daily newspaper. In the computer classroom, this assignment can be accomplished via e-mail, and because most newspapers now have their own Web sites, they solicit e-mail letters. Another assignment I developed asks two students to collaborate in the writing of an argumentive essay. Each student takes a position, pro or con, with respect to some community issue. Besides the more common issues, such as gun control or the death penalty, students in my class have chosen to debate whether our college should have a campus pub and whether our state should raise the age requirement for a driver's license to 18 years. There is also an oral component to this assignment in which the students debate their topic in front of the class and then invite comments from their classmates. Summaries of the arguments can be given as PowerPoint presentations. To prepare for this assignment, I encourage students to use community-based resources, such as interviews with experts in and out of the academy. When I can, I bring experts into the classroom to talk about how they use language in their disciplines. Of course, students are encouraged to take advantage of resources on the Internet. Through these and other literacy activities, I attempt to show my students how the practical use of language can enhance their lives and improve their understanding of themselves and the world around them. I want my students to experience communication with others in an environment that, as closely as possible, represents what happens to them in their own communities outside of school. Through my collaborative assignment, which employs not only collaboration but traditional, computer, and community-based research, I am trying to foster critical and dynamic literacy in my students.

Unfortunately, collaboration in the classroom often presents many challenges to the traditional curriculum. Collaborative activities take up too much time and are difficult to organize. For my argumentative assignment to be effective, my students need a minimum of 8 weeks, or half a semester, to complete their papers. Students, indoctrinated by the media, past schooling, and in a society that privileges individual achievement over group effort, are often resistant or see little value in collaborative learning projects. Because most of the students in my classes work and go to school, they complain that it is too difficult to meet with their partners outside of class. Finally, collaborative projects are difficult to assess. I recently submitted some of my students' collaborative papers for the year-end assessment of our Introduction to Research Writing course. A month later, the freshman composition director told me that the instrument they used to assess critical thinking skills (a holistic reading that used a grading scale of 1 to 6) was set up to evaluate individual writers and not writers in collaboration. Certainly part of what needs to be done to bridge the gap between school and workplace literacy is to convince some of our peers of the significance of collaborative writing in the school and the workplace.

In "Bridging the Gap: Scenic Motives for Collaborative Writing in Workplace and School," James R. Reither (1993) argues:

> for a view of collaborative writing that encompasses far more than literal writing-together, on the ground that an expanded notion of writing-as-collaboration will help teachers see more clearly the kinds of changes they need to make in their classrooms ... I do suggest that making classrooms in some ways more like collaborative workplaces will help students understand better what writing is and does. (p. 197)

Reither is trying to convince teachers to make real connections between school-house and workplace literacy, urging them to turn their classrooms into literacy laboratories where students learn to collaborate not as individual, solitary writers but as a community of literacy learners and researchers. To accommodate this type of learning environment, Reither suggests "three major changes in the class-room" (p. 202). First, the students themselves must conceive of research projects linked directly to their fields of study: "As far as possible, it must be students, not teachers, who conceptualize the project and the process" (p. 202). By allowing students to choose topics and projects from their fields of study, teachers can adapt the critical literacy skills that the academy values to the dynamic literacy skills of the workplace. Second, teachers must redefine their roles, acting as men-tors and managers rather than "delivery systems" (p. 204) for information. The advantage of this approach to classroom management is twofold: It resembles, as closely as possible, the role an employee has with his or her boss in the workplace, and it frees the teacher to work more closely with students individually or in groups. Third, Reither states:

> Teachers must help students recast their own roles in the classroom: from people who come to class to listen to lectures, read textbooks, and write term papers and ex-ams to people who come to class having done some research outside of class, ready to pool their findings with those of others, evaluate the present status of their devel-oping knowledge, and identify gaps in knowing that need to be filled. (p. 205)

Acknowledging students as capable knowledge makers already in possession of certain critical and dynamic literacy skills, Reither urges teachers to treat them as active rather than passive agents in the education process. The teacher's role then becomes something akin to what Tchudi refers to as a teacher-mentor, in which the teacher helps to expand on the literacy experiences that students get through a range of media, while realizing also "that the range of experiences must originate with student needs and desires to integrate their own lives" (Morris & Tchudi, 1996, p. 215).

Tchudi put his teacher-mentor theory to work recently, devising an innovative classroom-workplace connection in one of his graduate classes at the University of Nevada, Reno (UNR). Students in his class put together "a book commemorating more than 80 years of Frandsen history" (Pike, 1998, p. 9). Frandsen Humanities is the old English and Foreign Languages building on the UNR campus, and they wrote the book to commemorate the renovation of the building. From drafting to revising, from page layout to publishing, the students were in charge of the project, while Tchudi acted as their mentor and managing editor. "We made $2,000 for the department on this project," he told me. That a collaborative classroom literacy project might actually pay for itself is always helpful when trying to convince administrators of the validity of classroom-workplace connections. Another benefit of the project was the development of a new class, Professional Editing and Publishing, which is team taught and began in the fall of 1998. According to one of the English professors involved in the project, "It's one of the most practical classes you can take in the English department" (Pike, 1998, p. 9). Unlike some of the examples I mentioned earlier in which the literacy skills taught in schools remain disconnected from those needed in the workplace, in this type of literacy classroom there is a direct connection between what is happening in the classroom and what happens on the job. Besides the obvious practical advantages of teaching professional editing skills to students who will soon be in the job market, graduate level classes such as this one help future literacy teachers to integrate the practical aspects of language use with the language theory they are learning in their other classes.

Classroom publishing enhances both critical and dynamic literacy because it offers opportunities for students to write for authentic purposes and to publish their work for audiences beyond their classmates or teacher. It also offers an engaging and accessible avenue for bridging the gap between school literacy and workplace literacy. There are a number of books to help teachers take advantage of not only conventional publishing of books, pamphlets, or newsletters but also Web page design and multimedia presentations. In *Classroom Publishing: A Practical Guide to Enhancing Student Literacy*, for example, Laurie King and Dennis Stovall (1992) have written a "how-to and here's-how-they-did-it" (p. xi) book for publishing in the classroom. This text contains step-by-step help for literacy teachers who want to use publishing to "*demonstrate* [original emphasis] to their students that language functions to communicate" (p. 3). Along with the how-to information, King and Stovall supply teacher testimonials about the breadth and effectiveness of classroom publishing. For example, English teacher Pat Graff, at La Cueva High School in New Mexico, has developed a publishing project where his students produce children's books. In a booklet he wrote about the project, Graff says, "All the elements of a successful program are in this unit: writing to a real audience, dealing with the community, and developing creativity and self-esteem while working with one's peers" (p. 7). The success of Graff's program also derives from a clear connection between the authentic, collaborative writing tasks in which his students are engaged and the kinds of literacy skills they will need as adults.

Some educators have taken the classroom-workplace connection beyond classroom publishing projects to actual workplace simulation. Nancy Allen (1996), a professional communications professor at Eastern Michigan University, believes that "the professional writing classroom can provide a bridge between the campus and the workplace by introducing future professionals to the complexities of electronic literacy before they arrive at the office" (p. 217). Allen defends the use of "workplace simulations" in the school settings because they offer students a chance to "focus . . . on understanding rather than efficiency, a goal that academe can indulge but industry must often forego" (p. 227). She describes one simulation in which writing students working in collaboration evaluated poorly written documentation for a software package used to teach calculus, observed users of the program, and questioned a math professor about his use of the program. Later, the students discussed what should be done to improve the manual and sent a proposal to the math department. Although it took the students 9 weeks to finish the proposal (obviously much too long for industry or business proposals), Allen says that the simulation was a success because it "provided students with an opportunity to learn principles of good technical writing as they applied them to a real project" (p. 227). By offering her students authentic contexts for applying the concepts of page design and formatting to a real purpose and audience, Allen connects her literacy classroom to the workplace. And unlike actual internships in the workplace, workplace simulations in the classroom allow students to hone their critical and dynamic literacy skills in a nonthreatening environment.

For example, in similar project but on a larger scale, Allen joined Gregory Wickliff (1997) to design a workplace simulation that involved 61 students from three different universities. In this simulation students "participated as members of a hypothetical software development company. Each university site represented one company department (site 1 = production, site 2 = finance, and site 3 = marketing)" (p. 202). After dividing each site into teams of two to five people, the students were told to collaborate in the design of "the final module of a software program intended for independent insurance companies" (p. 202). The deliverable for the project was a recommendation report for their design. Allen and Wickliff mention several problems with electronic collaboration, writing across disciplines, and their own nondirective approach to the project that occurred during the 6-week task. For example, in an admirable but unwise attempt to be strictly nondirective, one participating instructor refused even to make the normal decisions that a project manager might actually be required to make on the job, and in doing so, the students were forced to devote an inordinate amount of time to insignificant decision-making tasks. But as in her earlier simulation project, Allen believes the benefits of the workplace simulation depended not on how closely it compared to a real workplace situation but on what it taught the students (and the teachers) about the value collaboration and about learning to work through their own problems:

As a result of our participation in this project and analysis of the activities that occurred, we believe strongly that electronic collaboration projects are valuable because they complicate in interesting ways our collaborative writing processes and our relationships to disciplinary cultures. They challenge teachers and students to question definitions and assumptions about learning and writing that have grown transparent through their long acceptance. They also help us recognize ways in which our tools not only influence but mold and press on the writing that we do. (p. 217)

As Allen and Wickliff suggest, the rewards of teaching literacy in more complex ways benefit both students and teachers. Through collaboration, teachers can gain new insights into how writing is used in disciplines other than their own, and students learn, among other things, that the writing process involves much more than just putting words on paper and that there are real purposes for writing beyond the typical academic research paper. Finally, Allen and Wickliff remind us that new technology such as the computer continues to reshape and, I might add, complicate and enhance the way we use language in the classroom and the workplace.

While looking for ways to bridge the gap between work and school, literacy teachers may want to observe what their colleagues in the science, technology, and business departments are doing in their computer classrooms. For example, the Hydrology Department at UNR uses the computer classroom to help future secondary and college teachers learn more about how to teach watershed management. The program they use, called the Watershed Management Simulator, was developed to emphasize "the interdisciplinary aspects of resource management" (p. 2). Through the manipulation of natural rainfall and reservoir capacities, students learn how ineffective watershed management adversely affects not only the environment but also industry and the levels of potable drinking water. Designers of the program claim it helps "students [to] focus on the integration of natural and social sciences" (Higgins, 1989, p. 2). In other kinds of classrooms, computer simulators are used to teach pilots how to fly and to show soldiers how to drive a tank, connecting the classroom directly to what actually might occur on a commercial airliner or in a real battlefield situation. The team that designed the *Watershed Management Simulator* and workplace literacy researchers such as Allen and Wickliff have adapted similar learning techniques to the college classroom. By using authentic tasks to forge connections between literacy and technology and school and work, these kinds of classrooms enhance both critical and dynamic literacy and help students to understand the collaborative aspects of knowledge making and how communities of knowledge makers are interconnected. Further research into the ways computers are used for teaching and learning in other disciplines may help literacy teachers develop new ways to design and structure their own computer classrooms.

We can even draw assignment ideas from our interactions with colleagues in other disciplines. While acting as coordinator for a writing-across-the-curriculum (WAC) assessment project in the Business and Computer Information Systems

(CIS) department at UNR, I was introduced to some excellent assignments dealing with the ethics of the Internet and computer technology in general. Fritz Grupe, a CIS professor who sat on the assessment committee with me, asked his students to respond to ethical dilemmas that dealt with computer technology. The following example is from one of his assignments:

> Assume that you are a professional systems developer. You are approached by a telephone company to develop an expert system that will greatly facilitate the repair of computer switching devices. It is clear to you that the system when implemented will replace many junior and senior maintenance people whose participation is essential in the analysis and design stage. Management has described the new system as an aid to the maintenance people and has downplayed the possible negative outcomes such as the reduction of jobs when repairs are automated.
>
> A. Do you as a contractor have any obligation to inform these people of the possible consequences of their participation in the analysis, design, and development process?
>
> B. If you were one of the people involved, and if you knew the impact the system would have on you personally, would you be obligated to participate honestly and completely in the development of the system?

Given the increasing occurrences of corporate downsizing, this particular ethical dilemma allows students to discuss the moral implications of certain business philosophies. What are your obligations and responsibilities to your employer? To your fellow employees? This exercise also forces students to examine the good as well as the bad effects of new technology. Is capitalism too intimately connected to technology? In this type of assignment, students learn to analyze choices they may have to make in the business world, and they may even help to promote a kind of ethical literacy.[1]

Taking my cue from Grupe, I have designed an undergraduate class in which students would study the rhetoric of the Internet. The goal of this course is to help students see the connections between rhetoric, ethics, and Internet communication. Other more practical goals include showing students how to use the Internet, how to take advantage of different sources, how to find what is suitable for their specific purposes, and how to be more discerning in their selection of information sources. Through analysis of various issues and data presented on the Internet, they would evaluate and discuss the implications of using it as a kind of global town meeting, in which citizens of the world can gather and give voice to their concerns. Students would also be asked to explore the validity of a claim such as this one made by consumer advocate Ralph Nader and quoted in *PC World* (Janu-

---

[1]Most technical writing textbooks devote at least a chapter to ethics. However, despite the important role ethics has played in the history of rhetoric, I don't believe I've ever seen a freshman composition textbook that dealt with anything more serious than plagiarism.

ary, 1996): "The Internet is more than a collection of databases. It's the interaction of a large community of people who see themselves as citizens rather than customers. The political activity on the Internet is a positive resurgence of democracy" (p. 193). If Nader is right, what role should censorship play (or not play) in the global town meeting? And how does a citizen in this global town meeting learn to tell the difference between "good" information and "bad" information? In her keynote address, Selfe talks about the importance of teachers discussing the implications of CAI. In this class, which focuses on the ethics of the Internet as well as how to use it, students will be asked to reflect on their own use of the computer as a communication tool in school, at home, and on the job. By teaching the practical aspects of Internet use along with the ethical concerns, I hope to offer my students the critical literacy skills necessary not only to be good researchers in the workplace but also to be skeptical of the data they find on the information highway.

As computers continue to reshape literacy in schools and its connection to the workplace, researchers such as James Kalmbach (1997b) perceive a need for educators to move beyond the notion of computers as simply tools for writing. According to Kalmbach, the "instrumental" view of computers as tools allowed teachers with little technological savvy and lots of knowledge of writing pedagogy to better adapt to the computer classroom (p. 263). Although he regards this view of computers as a significant step in the evolution of CAI, Kalmbach believes the following:

> [Eventually] faculty need to see themselves as writers creating texts within the spaces of the computer-supported classrooms. . . . We are writers of learning spaces in which students develop strategies for using the intertextuality of software to create texts. When we modify preference settings, change menus, or add aliases to networked resources, we are writing the classroom. (p. 265)

Kalmbach regards computers as writing spaces, a metaphor that takes into account the effect that computer technology has had on writing. In the instrumental view of CAI, the teaching of writing is regarded as being separate from computer technology, whereas in Kalmbach's writing spaces metaphor there is no clear distinction between where the computer technology begins and where the writing ends (and vice versa). As people become more intimately involved with the technology of writing on computers in school and in the workplace, it forces them to view literacy in more complex ways. Therefore, Kalmbach believes teachers must create learning spaces where students are allowed to experiment with language as it coalesces with evolving technology. As testament to the success of this view of literacy learning, Kalmbach offers the work of three graduate students from his college. Two of the students presented electronic theses—one defended a "cyberpunk short story . . . , a hypertext with an elaborate interface and rich multimedia content—graphics, videos, sound, and text," and the other defended a hypertext novel, "whose simple interface disguised a complex underlying linking structure" (p. 261). The third student wrote a *Kids Works II* instruction manual for sixth graders and then tested it with local children (p. 270). These student projects, which combine technology with writing for the specific purpose of creating new kinds of texts, are the result of a new pedagogy that refuses to separate the technology of writing from the technology of computers. Because of the experimental nature of these literacy projects, they offer opportunities for further research into the connections between school and workplace literacy. These literacy projects also present areas where literacy professionals in academia and professionals in the workplace can begin to share the knowledge and practice of their two cultures.

I have presented only a few ways that teachers can use the computer classroom to enhance literacy. Undergraduate and graduate students from across the curriculum could be allowed to do computer presentations along with traditional papers. They could learn how to design their own Web pages, newsletters, books, journals, multimedia presentations, or electronic journals. Graduate seminars can evolve that allow students from a variety of disciplines to observe and analyze the processes and rhetorical strategies involved in each of these tasks. Graduate assistants would then use their knowledge of computers and other communication devices in their own classrooms. In fact, graduate classrooms could become research projects for analyzing the effectiveness of these new approaches to literacy pedagogy in the computer classroom. These projects also present interdisciplinary approaches to literacy. They offer opportunities for faculty, graduate assistants, graduates, and undergraduates from across the university to work together to improve the quality and breadth of student literacy and to help students be better prepared for jobs in business, industry, and the academy.

As I conclude this chapter with all of these creative and innovative student literacy projects, none of which would have been possible without computer technology, it occurs to me that I need look no further for the answer to the question about computers and student engagement that I asked at that Midwestern conference. I have tried to show that teachers, by viewing literacy as a complex, multifaceted process, can help bridge the gap between school literacy and workplace literacy and between theory and practice. In the process I hope I have also shown how computers can be used to engage, motivate, and inspire students. Finally, I would like to reiterate that I am not endorsing a workplace-specific curriculum. Instead, I want to encourage teachers and schools to help students connect what they learn in school to what happens in the workplace by using the computer classroom to generate authentic contexts for literacy learning.

18

Collaborative Research, Collaborative Thinking: Lessons From the Linux Community

Tari Fanderclai
Connections MOO

Recently I was involved in an effort to coordinate a cross-institutional study of collaboration in electronic environments among students in writing classes. Our motivation was this: Research takes a long time, and computer-assisted pedagogy exists in territory that changes daily and overwhelmingly. Technology develops, the Internet grows, access spreads. Information bombards us at a rate that simply was not possible in the print-only world. And every development brings social changes that affect the classroom. By the time a researcher can pilot a study of the pedagogical uses of a particular technology, the technology may well have altered so much as to change the questions that need to be asked or the methods that can be used to answer them. A lone researcher can't keep up. Even a small group of researchers can't keep up. We need more ways to collaborate on a much larger scale.

After a few weeks of discussing this problem in various online forums, six geographically scattered teachers and researchers in computers and writing—James Inman, Catherine Spann, Sharon Cogdill, Barry Maid, Janice Walker, and I—decided to coordinate a cross-institutional study. Our idea was to try to deal with some of those overwhelmingly large questions and fast-moving targets by getting a large number of researchers to contribute data and data analyses to the effort. We chose a research area, defined questions for the study, devised a methodology, and solicited volunteers from various online communities. The researchers would conduct the study in their classes, collect data, and report their conclusions. The coordinators would then write an overall report based on the findings presented by the various researchers.

We also encouraged the participants to reuse their data for projects and publications of their own. Eventually, we hoped to provide a space where researchers could share everything they produced relevant to our research topic: background literature, alternative methodologies, raw data, initial conclusions, complete papers. Researchers using the site would be able to examine and contribute to stages of the research project matching their interests and expertise, without necessarily working through an entire instance of the study on their own. For example, an expert in research methods might contribute to the methodology; in response, an expert in data analysis might pick up raw data from the site and post an analysis.

Our theory was that in a short time, a community of researchers working in this way should be able to generate a great deal of material relating to a particular research question. By encouraging people to contribute in the areas they know best and are most interested in, such a community should be able to generate higher quality and more timely research than an individual or small group.

As might be expected, our pilot effort produced very little data to apply to answering our research questions and a great deal of firsthand knowledge of the difficulties of coordinating a large-scale collaborative effort. We need a clearer model of a process for conducting large-scale collaborative projects, and we need to learn more about the essential elements and the kinds of attitudes that make large volunteer efforts work.

Our experiences and our long-term goals drew me to investigate the largest and most successful volunteer collaboration I know of: the Linux development community. As a result of my preliminary investigations, not only do I want to put the model to use in a project I'm involved in, but I also believe that all of us who work in areas where large collaborations would be valuable should examine the Linux model very carefully and begin to put its lessons into practice. There are many parallels between the problems we need to solve and the problems the Linux community has already overcome. I believe we could learn a great deal about how to develop our own collaborative research model by beginning with the principles the Linux community has already discovered. I present here a summary of many of those parallels, some suggestions for ways we could begin to put the principles of the Linux model into practice, and reasons why I think the model just might work.

THE LINUX DEVELOPMENT EFFORT

Linux is an open-source operating system that provides a Unix operating environment for desktop computers. Briefly, open-source software is software that is distributed free, along with its source code. Licensing for open-source software permits modifications and derivative works and may not discriminate against the use of the software by any persons or groups, or in any field of endeavor. The complete Open Source Definition, drafted by Bruce Perens and revised with the help of the Debian Linux developers, is available at http://www.opensource.org

Many people are familiar with open-source software because they regularly use such well-known instances as the Apache Web server, the Netscape Communicator Web browser, and, of course, the Linux operating system. A growing number of academics are also interested in the concept of providing their products—course materials, research papers, and the like—to the rest of the community under terms similar in spirit to open-source software licensing.

But the most interesting aspect of the open-source movement for a research community should be the cooperative, collaborative model of software development that has evolved, and the canonical example is the wildly successful Linux effort. Created in 1991 by Linus Torvalds, Linux has been developed by a loosely coordinated, open community of volunteer hackers all contributing code, bug reports, and bug fixes via the Internet. In a paper titled "The Cathedral and the Bazaar," Eric Raymond (1999a) undertakes to explain the success of the Linux model of software development.

In his paper, Raymond contrasts the traditional model of software development with the style Torvalds uses for the Linux project. Raymond characterizes the traditional model as the cathedral style, in which "the most important software (operating systems and really large tools like Emacs) needed to be built like cathedrals, carefully crafted by individual wizards or small bands of mages working in splendid isolation, with no beta to be released before its time." Of the Linux model, he says the following:

> Linus Torvalds's style of development—release early and often, delegate everything you can, be open to the point of promiscuity—came as a surprise. No quiet, reverent cathedral-building here—rather, the Linux community seemed to resemble a great babbling bazaar of differing agendas and approaches (aptly symbolized by the Linux archive sites, who'd take submissions from anyone) out of which a coherent and stable system could seemingly emerge only by a succession of miracles.

Raymond set out, he says, "to understand why the Linux world not only didn't fly apart in confusion but seemed to go from strength to strength at a speed barely imaginable to cathedral-builders" (1999a).

Raymond began his study of the Linux model of development by participating in the Linux development community. Then he conducted one of his own projects, the development of the mail client Fetchmail, in the bazaar style he had observed. His conclusions from these experiences highlight key factors of a highly successful kind of collaborative thinking and working, one that may well be generalizable to research communities.

BARRIERS TO LARGE-SCALE COLLABORATION IN THE HUMANITIES

The Internet has removed what should be the chief barriers to large-scale collaborative efforts: time and distance. We can communicate instantly with researchers in practically any field. We can share documents as fast as we can write them. It should be a short step, then, to using these capabilities to facilitate the collaboration so many of us recognize as necessary.

But our research traditions and reward systems can be barriers to collaboration. Humanists tend to value individual products and great minds; the greatest rewards are for individual works. When work is coauthored, it is often understood that a first author is present, who gets the main credit; often the point of a collaboration is for the lead researcher or writer to get the work done, while giving the collaborators some experience in an area in which they are novices and learners. Student milestone works such as dissertations are never done collaboratively; the student must prove that he or she is capable of producing a large-scale work alone. Establishing oneself in the field requires carving out a territory of one's own and building up a store of personal intellectual property. As a result, the research procedures that we are familiar with often favor working alone.

Raymond might say that humanists traditionally work in the cathedral style, building our projects and papers alone or within a small, closed group of collaborators. There is certainly nothing wrong with cathedral building; one could hardly argue with the vast body of valuable work it has produced. But many of the research problems currently facing us in the area of computer-assisted education are simply too large, the changes too rapid for researchers working alone to make much headway. The bazaar style of development offers a new way of thinking and acting collaboratively that could serve us well.

COLLABORATING, BAZAAR STYLE

The Linux style of development overturns the notion that good software can only come from a developer or small group of developers working in isolation, probably in a commercial setting, releasing their products only when they are bug-free and as nearly perfect as possible. Linux releases are frequent so that developers and users can help test and submit bug reports and fixes. Thousands of developers, testers, and users contribute to Linux; certainly they do not all have contact with the central coordinators. They need not be experts on the entire Linux system—in fact, it's unlikely that anyone could be because the system is vast. Contributions are accepted from anyone who has the skills to work on a particular area. The contributors are volunteers, motivated not by money or direct career advancement but by the chance to work on interesting problems and to help improve software that answers their needs. They believe in sharing code and information and solutions to problems rather than in hoarding them, for when the work already done is available to everyone, each new volunteer can use his or her time to contribute something new.

And they have produced a high-quality operating system at a speed that seems impossible. The sum of this collaboration is truly greater than its parts, and creating something that transcends what an individual could produce on his or her own is surely the reason to collaborate on any project.

Eric Raymond's (1999a) analysis of the Linux development style offers more than a method for conducting a software development project. By asking himself why the method works, Raymond uncovers principles of collaboration and ways of thinking collaboratively. These principles lie behind the Linux community's ability to attract and keep volunteers, make efficient use of all the willing hands available, and produce quality work, all without the various financial and career-enhancement motivations that accompany corporate or institutional collaborations.

It should be possible to apply many of Raymond's conclusions to our own large-scale collaborations. I propose using them to examine ways of thinking and acting that we might need to revise or develop to work collaboratively across disciplinary and institutional boundaries and to begin to build our own models for large-scale collaboration.

BAZAAR-STYLE THINKING AND ACTING

"Every good work of software starts by scratching a developer's personal itch" (Raymond, 1999a). Certainly this is an overgeneralization, and yet, as a rule of thumb, it's important. The person who needs a solution to a problem has an intimate understanding of that problem and of the complexities that need to inform the solution and is highly motivated to find or create a well-designed solution. If a developer has a need for a piece of software to perform a particular task, that need forms the motivation and informs the design of the tool. But so often, Raymond observes, "software developers spend their days grinding away for pay at tools they neither need nor love."

As teachers, we speak often of the problem of getting students to invest themselves in their work, and we spend many hours looking for ways to help them find problems they are genuinely interested in investigating. We know that people produce their best work when a problem takes hold of them and demands their attention, when they aren't just working to complete an assignment.

Yet, driven by the need to produce and publish and present, we often assign ourselves projects that we aren't all that interested in. Such projects are not usually our best work; many of them go unfinished. No large-scale collaboration effort will continue without a coordinator or group of coordinators who are personally fascinated by the project. A large collaboration needs a coordinator who will be driven by his or her fascination with the subject matter to follow through and who knows how to attract interested and talented people to the project.

And so, Raymond advises, "To solve an interesting problem, start by finding a problem that is interesting to you."

"Provided the development coordinator has a medium at least as good as the Internet, and knows how to lead without coercion, many heads are inevitably better than one" (Raymond, 1999a). Although open-source developers believe strongly in cooperation and code sharing, it is important to recognize that for many of them, this is not a moral stance. In other words, they don't necessarily work the way they do because they believe that open source is good and proprietary software is evil (Raymond, e.g., notes that neither he nor Linus Torvalds believes that building closed-source software is morally wrong). Rather, they work the way they do because the Linux style of development works so much better, so much faster, than the cathedral style.

This is largely because of the number of heads—in other words, the size of the pool of talent—that an open-source project can bring together to work on a problem. No single corporation could assemble all the people or pay for all the skilled time that goes into Linux development. But with hundreds of people who all have instant access to the code that's already been written and the knowledge already assembled, each volunteer can build on previous work and help to move the project forward.

Like cathedral-building software developers, individual researchers and small groups of researchers are at a disadvantage. Even if an individual researcher had all the skills necessary for a large-scale project, the speed at which such a project must move to produce timely results is beyond the capacity of one person, or even a small group.

It's key that an open-source community has access to all the work its members have already done and all the work that's in progress. Because of our emphasis on traditional means of publication and the values we place on an individual's intellectual property, we usually don't see others' work until it appears in print. There is a delay of several months between the completion of an article and its publication to the community, then another delay while the next researcher builds on the work and publishes, and so on. Imagine, though, the speed at which a research project could progress, given a large group of researchers working on various areas of a problem, and sharing their findings as they work. And I don't mean simply posting their final papers to the Internet rather than offering them to a print journal. I mean something like what Raymond expresses in the next few lessons from the Linux community.

"Release early. Release often. And listen to your customers" (Raymond, 1999a). Early versions of software are bound to be flawed and buggy. In the traditional model of software development, any version of software released to users must be as bug-free as possible. That's not a bad goal, particularly when users are customers who expect to pay for a software product, load it onto their desktops, and begin using it. The drawback is, of course, if there are bugs, users often wait several months for a release that corrects them. New versions with new features take even longer; users looking forward to new functionality in the product may wait for a year or more.

The bazaar approach to development, however, is to release as often as possible; in the early stages of Linux development, releases sometimes happened daily. Users get bug fixes and new features as fast as they are available. Developers are motivated by seeing the steady improvement and progress in their work.

Further, the Linux community listens to its users, and frequent releases mean frequent feedback. Users spot bugs and flaws and are invested in reporting them; further, when the development community makes clear that user feedback is valued, those users are willing to contribute to the overall effort by making many kinds of suggestions for improvement.

Like cathedral-building software developers, researchers in the humanities are used to releasing (publishing) work that is as bug-free as possible. One does not, after all, offer a rough draft to a journal.

But imagine a research project where all of the source materials—the methods, the raw data, the rough drafts, and the tentative conclusions—are released via the Internet as soon as they are created. Users—community members who are interested in using the findings, either by applying them in their classrooms or by building on them in their own research—would have immediate access to the information and could provide feedback, suggestions, and bug fixes. They wouldn't have to wait for a print publication to get the information they need. Researchers could get feedback at every stage of the process, rather than working in isolation with no responses until a final publication comes out.

"Treating your users as co-developers is your least-hassle route to rapid code improvement and effective debugging" (Raymond, 1999a). Many software users are software developers themselves, and an open-source project can make especially effective use of these users. Because they can see the source code, they can find problems and suggest fixes, and in a bazaar style effort, they do. Because Linux meets a need they have, they want the software to work. Because the Linux community values feedback from users and rewards them by using that feedback, and using it almost instantly, those users are willing to spend the time to go beyond making simple bug reports to studying the code itself and suggesting possible solutions to problems. Treated as valued members of the development community, the users behave like members of the development community. They become members of the development community, saving endless debugging time and speeding the improvement process.

The users of a bazaar-style collaborative research project would be similarly valuable. Rather than picturing the rest of the community as out there waiting patiently for our polished and published findings, we need to start thinking of them as the collaborators they could become. Most of them are also researchers. Not only could they provide feedback as they examine and try to apply a research project's findings, but also they could look at the source—the raw data, the methodology, the methods—helping to identify and resolve problems in the research process. If such users are treated as coresearchers, as valued members of the project, they would no doubt respond as many Linux users do and continue to participate in the community, speeding the project along.

"Given a large enough beta-tester and co-developer base, almost every problem will be characterized quickly and the fix obvious to someone" (Raymond, 1999a).

In other words, Raymond says, "Given enough eyeballs, all bugs are shallow." He names this Linus's Law and explains that it may well be "the core difference" between the cathedral and bazaar styles of programming. He explains it this way:

> In the cathedral-builder view of programming, bugs and development problems are tricky, insidious, deep phenomena. It takes months of scrutiny by a dedicated few to develop confidence that you've winkled them all out. Thus the long release intervals, and the inevitable disappointment when long-awaited releases are not perfect. . . .
>
> In the bazaar view, on the other hand, you assume that bugs are generally shallow phenomena—or, at least, that they turn shallow pretty quickly when exposed to a thousand eager co-developers pounding on every single new release. Accordingly you release often in order to get more corrections, and as a beneficial side effect you have less to lose if an occasional botch gets out the door.

With many users stressing the software in different ways, many more bugs are found in a short period of time. And with many developers available to study the problems, the chances are high that someone will find the problem easy to solve. Although this might seem obvious, neither cathedral-building software development nor traditional individual or small-group research collaborations are able to take advantage of Linus's Law. They don't have enough collaborators, and they cannot show early versions of their work to anyone outside the group. Just as it usually takes cathedral builders a long time to find and fix bugs in their software, it may take a closed group of researchers a long time to spot flaws in their methods and their conclusions and to find solutions to those problems.

Working in isolation suggests that only an individual or a small group understands a problem and can find solutions. Yet, like closed-source software development, a closed research project is often simply a race to publish first, because in fact many people have perspectives on the problem and ideas for solutions. And so we have a duplication of effort, necessitated by our tradition of valuing individuals by their collection of publications and intellectual property. In truth, we know that whatever research project we take on could as well be taken on by someone else in the community. We don't work on problems in isolation because we believe that we're the only ones who can solve them—we work on closed projects because our traditional value system says that we have to solve them first to establish our value as researchers. We're afraid if we let slip any hint of what we're working on, another researcher will steal our ideas and beat us to publication.

How much more useful it would be to the community, how much more efficient, if researchers would open their projects to participation from all interested members of the community. Instead of competing with other researchers, each group wasting time as it struggles in isolation with the same problems, open collaborations could pool their resources, saving precious time and taking advantage of the many perspectives and skill sets available. Instead of competing for credit and thereby slowing the progress of the whole community, we need to share the work and the credit, recognizing the value of every contribution and every contributor.

To make such a process work requires an attitude shift. In his paper "How to Become a Hacker," Raymond (1999b) says:

> To behave like a hacker, you have to believe that the thinking time of other hackers is precious—so much so that it's almost a moral duty for you to share information, solve problems and then give the solutions away just so other hackers can solve *new* problems instead of having to perpetually re-address old ones. (1999a)

It's that attitude that makes bazaar style development work. We need to cultivate a similar attitude as researchers if we want collaborations to work, and we need to make collaborations work if we're going to tackle the research questions facing the humanities in the area of computer-assisted education.

PUTTING IT INTO PRACTICE

We are, of course, left with the problem of our traditional reward system and the ways it can interfere with collaborative thinking and acting. It seems to me, though, that the best way to create change may be simply to begin bazaar-style research projects.

The project would have a coordinator or group of coordinators who are invested in the particular research question, interested in studying the problem and finding solutions. The coordinators would need to have the researcher attitude—that is, they would need to believe that all members of the community can be useful collaborators and that the time of every researcher is too valuable to waste in covering old ground. They would need to believe in the value of making rapid progress through shared effort, and believe in it strongly enough to counter any misgivings they or their participants might have about not being able to claim any particular part of the project as their personal intellectual property.

The coordinators would outline a starting place, perhaps by defining a problem and designing a research methodology. They might choose to pilot the initial methods with a small group of volunteer researchers before opening the project completely. Most open-source software projects also begin with an early version of the software already in place; a large collaboration needs a starting place, a design with which to attract potential participants. The coordinators would recognize that what they set out is only the beginning, to be modified and adapted by the researchers who will use it, and build a base that will adapt to the changing needs of the project.

After refining their initial efforts, the coordinators and the initial participants could publicize their efforts to the community, inviting everyone interested to join, stressing the goals and benefits of bazaar-style collaboration and presenting the community with a well-designed yet flexible starting place.

Once the project is opened up, a central Web site for the project could provide a place for contributors to contribute data, conclusions, alternative methods, and new research questions. Researchers could specialize—some might want to test and refine methods; others could collect data; still others could study accumulations of data, write up conclusions, and suggest new directions. All materials could be available to the entire community as soon as they come into existence, and community members could make immediate use of the data, methods, and conclusions, both in their classroom practices and in further research projects. Those users could provide feedback for the researchers based on their experiences with the materials.

Led by coordinators who are devoted to the idea of open collaboration and willing to study available models for making it work, the results of such projects could well speak for themselves. Bazaar-style collaborative research could move us forward because it recognizes what our current great-mind-worshipping, intellectual-property-hoarding, publication-counting, print-enslaved culture does not.

First, isolated researchers cannot hope to make significant progress investigating the enormous and rapidly changing questions we face as we try to understand and develop the role of computing in humanities education. Like the Linux community, we need to develop ways of collaborating that involve as many people as possible so that we can work as quickly and efficiently as possible.

Second, a research project is never finished. Our research questions are never going to get final answers, and we stunt our own growth when we behave as though we can take hold of a problem, solve it, publish a paper, and dust off our hands. Each new development in technology changes our questions and the possible answers. The goal of Linux development is not to finish Linux but to continuously improve it, keeping up with changes in technology and meeting the evolving needs of its users.

Our research goals must be similarly pragmatic—and similarly visionary.

19

Current and Future Research in the Production and Analysis of Electronic Text in the Humanities: Bridging Our Own "Two Cultures" With Integrated, Empirical Studies

Peter Sands
University of Wisconsin-Milwaukee

My title refers to a famous disagreement between C. P. Snow and F. R. Leavis, in which the two scholars battled publicly over what Snow had identified as the "two cultures" of science and the humanities and which strand would be given necessary precedence in the future of the human race.[1] It is hardly worth pointing out that their debate was never fully resolved. Given current understandings of epistemology, scholars in the humanities at least would surely insist on multiplicities of intellectual and other cultures rather than a binary. But even though prominent scholars have pointed to their inutility and lamented their effects on the next generation of scholars, gaps between different kinds of researchers in the humanities are very real. Recent public exchanges have focused on the state of research in the production of electronic texts by humanities scholars and classrooms, with proponents weighing in from all points of the spectrum on the relative merits of the body of research being produced by or available to scholars. Vigorous debates still take place about the roles of humans and machines in the production of text even with the visibly inexorable development of electronic publishing, the growing body of computer-assisted content analysis of linguistic corpora, the ubiquity of personal computers, and the body of personal and scholarly narratives about their use. There are arguments on three levels: "should we or shouldn't we,"

---

[1]C. P. Snow, *The Two Cultures and the Scientific Revolution.* New York: Cambridge University Press, 1959. F. R. Leavis, *Two cultures? The significance of C. P. Snow* [with a new preface for the American reader and an essay on Sir Charles Snow's Rede lecture by Michael Yudkin]. New York: Pantheon, 1963.

"what meaning do we assign to this phenomenon," and "what's next." Like any meme, these discussions reflect an idea spreading on its own throughout the culture and the discipline: that new research initiatives and ideas are needed. They also reflect the difficult tension between a desire for positive changes and a rejection of existing paradigms. In this chapter, I will focus briefly on the specific case of current research in computer-mediated composition teaching, then suggest one direction researchers in the humanities in general need to pursue vigorously: the empirical. Computers and composition is a subdiscipline well positioned for prominence: Electronic discourse is primarily written discourse; the field has as its object of study both that discourse and effective use of emerging technologies in teaching and research, and the field itself is nascent—it can still be shaped.

Empirical research holds the key to a newly invigorated scholarly program integrated with our teaching. It also provides the strongest foundation from which to build new theoretical models for understanding the nature, amount, and reasons for changes in writing, reading, and learning attendant upon the introduction of new technologies. Whether we accept the newness or difference of work in computer-mediated environments—and I do—or we choose to view student texts produced with computers as retaining essential similarity to those produced with other writing technologies—pens, pencils, chalk—we must acknowledge the fait accompli that the computer is the composing tool of choice both inside and outside the academy. Whether we should is no longer a relevant question; what we should, how we should, and what happens when we do are the questions of today. Only a research program grounded in empirical methods[2] and guided by interpretive judgments will provide adequate information over time to create a framework for understanding. I specify empirical here and have in mind to some extent quantitative rather than qualitative methods, even though that is a troubled ground in the humanities more than in any other disciplinary area. As Linda Flower (1997) and others have noted, the general perception in the humanities is that empirical means "counting" or some such, rather than "observing and interpreting." At some level, a somewhat uncharitable argument can even be made that the real split in research on electronic text production and consumption is between empirical methods (including both quantitative and qualitative) and nonempirical methods (including both anecdotal and other forms of reasoning not based on close and systematic observation or research that draws heavily on the methods of library-based research rather than systematic observation). This is a distinction along the lines of that drawn by Mary Sue MacNealy (1999), who says a major difference between library-based and empirical research is that:

> Empirical researchers emphasize a systematic research methodology by reporting details of the research method along with the findings, whereas library-based re-

---

[2]This is not unlike what I take Stephen North to have meant in his call for an integration of practitioner lore and more rigorous investigations in *The Making of Knowledge in Composition: Portrait of an Emerging Field*. Portsmouth, NH: Boynton, 1987. See especially pp. 363–375.

searchers rarely report details of their research methods, nor do they emphasize the systematic nature of their method. A second important difference between the two methods is that library-based researchers track down and analyze other people's ideas and observations, which have been previously published, whereas empirical researchers directly observe and analyze (often using measurements) actual activities, products, and other phenomena. (p. 8)

This description represents a good deal of what is published in computers and composition, as others have found and as I will outline very briefly in this chapter. MacNealy comments on the two cultures issue by concluding that

Distressing is the tendency of some empirical researchers to value quantitative research over qualitative or vice versa. These attitudes impoverish the field of studies in writing, and if these attitudes persist, they will hinder the development of the field into a fully accredited academic discipline. (p. 6)

Fortunately, computers provide unparalleled opportunity for quantitative empirical research at the level of the text that has heretofore been simply impossible. Fortunately, too, computers have demonstrable uses in handling qualitative research methods—to such an extent that there is a miniindustry producing software specifically for qualitative research methods, such as the analysis of large bodies of written text and notes.

Linda Flower (1997), in "Observation-Based Theory Building," outlines the reasons why effective theorizing must be built on effective empirical investigation—that the one, abstract, interpretive, speculative, must be grounded in the other: local, specific, enumerative. As she writes, "Research . . . begins in a problem . . . that leads to inquiry and ends with an explanatory account" (p. 163). According to Flower, for a theory "to be an accountable interpretation of social cognitive action, it must be built on a fine-grained, richly specified vision of the process in question" (p. 170). To get to such a theory, the researcher must build "from the union of two sources of evidence: . . . in part from an intuition or an argument and in part from the complementary evidence of close, systematic observation and data" (p. 172). But that data can only "provide the foundation for interpretation," she cautions (p. 174). Empirical methods, properly applied, add validity to the researcher's assertion that she or he has observed patterns and that the interpretive act—the theorizing—is justified. It is precisely this continuum from close observation at the level of the text to theorizing of meaning that is needed in research in computer-mediated composition.

I am not alone in calling for more empirical research in electronic text production, consumption, and analysis in the humanities. This call is especially found in computers and composition, where empirical methods have curiously not taken much hold. Patricia Sullivan and James Porter (1997) write that "[s]everal works have evaluated the uses of various technologies in the teaching of writing, but none of these treatments has centrally focused on issues of methodology in the study of computer writing" (p. xv). They add, referring to their graduate students, that "too often we see them aligning themselves with what they see as one of the two available prestige camps in the field—Theory or Empirical Research ... [which] underappreciate [each other and] neglect practice" (p. xv). Sullivan and Porter portray this as a net loss, one which leads to a limited worldview and similarly limited fields of inquiry. Similarly, in "The Current Nature of Hypertext Research in Computers and Composition Studies: An Historical Perspective," Scott Lloyd DeWitt (1996) asserts that only a new empiricism will lead to improvements in computer-mediated pedagogies (p. 79). DeWitt's summary of current research clearly shows the dominance of anecdotalism over empiricism in analysis of student use of hypertext, leaving a clear path for the next generation of research design. Following his advice, not just for work on hypertext but the full range of computer-mediated textual practices, research studies such as Susan Herring's (1994) work on gender could be replicated and extended in larger databases, with better controls over the data, allowing researchers to gather information about gender, regional language, variation of language over time, and even about the inroads particular terms make, both geographically and temporally. Such distributed inquiries translate into an enormous body of interpretable data: about possible correlations between physical locations and given variables or about the spread of new terminology and concepts over time and across physical distances. It means that researchers could map variables in a larger data set than is normally possible given actual human limitations, with all that such an act entails—including the ability to ask questions and pose answers that have not been imagined in the past and cannot be imagined yet.

Existing studies that combine qualitative and quantitative measures, though, rarely do significant quantitative work other than through survey data and limited text analysis. For example, in "Computer-Mediated Communication in the Undergraduate Writing Classroom: A Study of the Relationship of Online Discourse and Classroom Discourse in Two Writing Classes," Robert P. Yagelski and Jeffrey T. Grabill (1998) limit their quantitative investigation largely to the length and number of messages students wrote and to less-relevant quantitative evidence, such as the amount of time each spent in the others' class. Yagelski and Gabrill define their study by using two different, nonspecific coding schemes to attempt a content analysis of student writing for each class, rather than an analysis of specific words or phrases. In the same issue of *Computers and Composition,* Patrick Slattery and Rosemary Kowalski (1998) focus their quantitative work on survey data that gathered information about writing process and composing behavior to make their argument that students using word processors learn a variety of ways to conceptualize the written word. Neither study reports or interprets significant quantitative information about the actual texts students wrote: their choice of metaphor, shifts in vocabulary, complexity of sentences, and so on. They instead enumerate relatively superficial—important, to be sure, but surface-level—information about the broad contours of the body of data.

Others, of course, have seen that the text-processing power of computers can be harnessed and concur in finding that too little has been done to explore doing so. Brian Huot (1996), in "Computers and Assessment: Understanding Two Technologies," refers to "the limited literature available on using computers to assess student writing" (p. 232). Huot notes that "using computer programs that count words and other textual features could tell us a great deal about what readers are paying attention to in various contexts" (p. 239) and, following a summary of potential uses of computers in assessment that ranges from mechanical textual analysis to Web-based portfolio archives, concludes that "we have much to gain by harnessing the enormous power in these two technologies and much to lose if we do not" (p. 241). The published research in computers and composition is not as broad generally as the published research in other areas of humanities computing; conversely, other areas, such as the analysis of literary texts or of historical documents, or of large corpora in linguistics, have largely focused on quantitative analyses.

In fact, whereas most research in computers and composition is not quantitative and does not even use existing software for text or content analysis, this is not the case in all disciplines. Huot (1996) notes in his article on computer assessment that the bulk of the available literature comes from outside our discipline (p. 232). A recent article by William Evans (1996) in *The Social Science Computer Review,* for example, confirms that advances in computer-aided content analysis— including advances linked to theoretical innovations that "would be very difficult to pursue without computer-supported protocols"—are of increasing interest in other humanistic and social science disciplines (pp. 271–272). William Evans reviews 16 software programs and 86 separate studies on computer-aided content analysis. Several recent book-length publications for historians similarly point to the possible uses of computers in research, including uses easily applicable in other humanistic and social science disciplines, such as text analysis.[3]

Without going so far afield other than to gather models and expertise, existing kinds of studies could in fact be enhanced, replicated, and otherwise performed across a wide range of class, gender, and geographic groups, as well as performed over time on large electronic textual corpora. For example, Christopher Holcomb's analysis (1997) of the formal features of his students' texts in *Inter-Change* discussions at Texas A&M University takes a close look at the specific words students wrote to provide a nuanced and complex picture of his own classroom. Founded as it is on close reading of the actual texts, Holcomb's conclusion that the rhetoric of democracy and egalitarianism in computer-mediated discussions oversimplifies the complex reality of those discussions is far more convinc-

---

[3]See Peter Denley and Deian Hopkin, *History and Computing* (Manchester, UK: Manchester University Press, 1987); M. J. Lewis and Roger Lloyd-Jones, *Using Computers in History: A Practical Guide* (New York: Routledge, 1996); Evan Mawdsley and Thomas Munk, *Computer for Historians* (Manchester, UK: Manchester University Press, 1993); Daniel I. Greenstein, *A Historian's Guide to Computing* (New York: Oxford University Press, 1994).

ing than a host of anecdotal studies. But were such a study to be conducted on a larger textual corpora, scholars could likely begin to formulate better theoretical understanding of humor and other linguistic features in online environments.

Research projects such as these are not without their problems. A deep-seated suspicion of empirical work among many composition and literary researchers may stand in the way of acceptance (Flower, 1997, p. 170). Even where that suspicion is absent, compositionists do not always have the needed background and training to handle large statistical projects and may not have the needed background to spot flaws in their own research designs. Additionally, tagging texts produced through synchronous discussion for retrieval and analysis by software will be at best a daunting task.[4] But it is not an insurmountable task. Project participants will have to agree at the outset to maintain careful file-naming and storage protocols, to work with ASCII text files wherever possible, and to work with standardized software as much as possible. Additionally, researchers will have to choose which features of their texts will be tagged, how they will be tagged, and the time frame for completion. In a large, multiple-site, longitudinal study, a central processing office will be needed to provide continuity and error checking, and that will have to be funded through grants or other institutional support. Finally, legal issues surrounding the copyright of student writings, research findings from the database, and other complications will have to be worked out, as much in advance as possible. But only when computer-mediated pedagogy begins to produce better empirical scholarship outside the dissertation will it also enter fully into what Holdstein and Selfe (1990) called the "second generation" of critical, skeptical, "investigation beyond (but not necessarily without) classroom practice" (p. 1).

Full disclosure: Behind the kind of research I advocate here is my belief that the computer can be both a site of resistance and an entryway to other sites of resistance, that it can and should be put to use in the service of a liberatory pedagogy, and that scholars and teachers of liberatory pedagogies must resituate themselves within the powerfully enabling discourses of empirical research and argumentation too often woefully absent from their recent work.[5] Additionally, because of the situatedness of experience and the impossibility of successfully isolating all possible variables in research on writing, I believe that small-scale, local research projects existing independently of but connected to larger research projects might successfully create pedagogies that closely resemble tactics of guerrilla insurgency as advocated by Peter McLaren (1996; pp. 160–162) for radical educational change.[6] Sullivan and Porter (1997) suggest this as well with their argument "that the study of writing-with-the-computer requires a situationally

---

[4]The Text Encoding Initiative uses SGML; see http://www.uic.edu/orgs/tei/index.html

[5]See, for example, Flower 170*ff*.

[6]Peter McLaren (Spring, 1996), "Paulo Freire and the Academy: A Challenge from the U.S. Left." *Cultural Critique*, 151–184.

sensitive approach to research" (p. xvi) or a "situated reflexivity" (p. 186) and that "computer writing technology indeed can potentially be a tool of resistance and transformation—if we can figure out what situated uses of it make that possible" (p. xvi). Accordingly, I present here not a fully formed research plan or instructions that can simply be transferred, templatelike from here to there but a call for one particular kind of research, supplemented by examples that can provide more empirical data usable for theory building and can be used in a liberatory pedagogy.

This is an auspicious time for such projects. Though there are dangers of miscommunication, of intellectual disagreement, of revelations of ignorance, of time ill-spent, they can be minimized. Connecting the ways computers are used in both classrooms and research settings to create a classroom-based research setting both more rigorous and more productive of usable, replicable, and extensible research projects would help the evolution of computers from their twin uses as word-processing devices and statistical workhorses into key elements in the new information space of the university. Digital texts that themselves in part define the information space of the emerging classroom offer the possibility of integrating the research binary I describe. The best computers and composition research already involves close reading of the texts students and others produce in collaborative writing environments such as Daedalus *InterChange* or a MOO;[7] other projects have concentrated on statistical information gathered through survey instruments such as the Daly-Miller Writing Apprehension Survey.[8] Imagine, though, a study of metaphor in student writing that examines not a single class's output over a semester but looks at 100,000 papers written during a 5-year period in several countries, or of sentence length, or verb phrasing, or transitional phrases.

Such careful, empirical studies have in the past been limited by the number of cases researchers could fairly examine. The same problem has held true for empirical investigation into literary and historical texts: Researchers are human, days are short, texts are long. But it is not difficult to imagine a large-scale project that takes the transcripts generated by three classes using synchronous conferencing via MOO, Daedalus Integrated Writing Environment, or any similar tool, tags those transcripts in Standard Generalized Markup Language (SGML),[9] and applies text-analysis software tools to the transcripts. This is clearly a direction that researchers will be moving in anyway. In their recent research, published as *Electronic Discourse: Linguistic Individuals in Virtual Space*, Boyd H. Davis and Jeutonne P. Brewer (1997) read electronic student writing culled from nearly 10

---

[7]Multiple-user dimension, Object Oriented = MOO. These text-based virtual-reality environments are in wide use for the teaching of writing, although little is known outside the communities of writing instructors and online role-playing games, their other main use.

[8]See John A. Daly and Michael D. Miller, "The Empirical Development of an Instrument to Measure Writing Apprehension" (*RTE*, Winter 1975, 242–256); and Wendy Bishop, "Qualitative Evaluation and the Conversational Writing Classroom" (*JTE*, 1989, 267–285).

[9]SGML is the international standard markup language for describing documents to machines.

years of computer conferencing associated with their linguistics classes (from 1989 to the mid-1990s). Davis and Brewer, whose work I stumbled upon while revising this essay, tagged the "corpus of interactive mainframe conference discourse" and studied the texts for patterns, "finding it to combine features of both the written text that it is and the oral text that it reads like" (p. 157). This is an assessment uttered in countless conference papers and professional conversations, but Brewer and Davis give it solidity through their careful, computer-assisted analysis of a well-defined textual corpus:

> Electronic discourse, like any other use of language performed by people in interaction for the purpose of making or sharing meaning, is replete with formula: the repetition of words and phrases, the replication of patterns in titling, the emulation of strategies such as rhetorical questions. (p. 165)

But it differs from other forms in its combination of characteristics of oral and written symbol use and in its ready availability for markup, analysis, and interpretation with the assistance of computer tools. Its emergence as a prominent means of communication demands careful analysis and opens new research fields (and new possibilities for crossing the sometimes arbitrary boundaries between fields). These texts have linguistic features similar to Old English or other early written texts: They are orthographically irregular, they exist to some degree independent of their original context, and they have characteristics of both oral and written speech. With reference to audience reception and multiple authorship issues, Billie Wahlstrom and Chris Scruton (1997) of the University of Minnesota have already used similar issues in antiquarian textual studies to suggest lessons for researchers in technical communication and document design,[10] but there is much more to be mined in technical expertise alone with regard to the tagging and analyzing of such irregular texts. Researchers in Europe have for some years now applied computer-assisted text analysis tools to medieval texts; the project I describe would draw on that empirical research in creating a framework for similar analysis of the corpus of synchronous conferencing texts.

Let us further suppose that the texts are drawn from classes identifiably different from one another: economically, geographically, linguistically, perhaps even temporally (e.g., adult vs. traditional learners). And let us further suppose that the classes engaged in debate over the same texts and the same issues because the participating teachers coordinated their efforts. If each class met twice a week, and conducted two 30-min conferences around the texts, generating, say, 20 pages of text each session, then they would produce 840 pages of text for analysis in a 14-week semester. At this point, the text-analysis tools developed for the study of large corpora of literary texts become obviously useful in generating empirical

---

[10]Billie Wahlstrom and Chris Scruton (1997), "Constructing Texts/Understanding Texts: Lessons from Antiquity and the Middle Ages," *Computers and Composition, 14*(3), 311–328.

data about the linguistic features of the student-produced text, and careful mining of the data will yield raw information that can be correlated with other observations to form an interpretive framework for those texts. Interpreting that data would require the special training and research interests of compositionists and ethnographers, marrying the two sides of the existing binary.

Let's look further at possible projects that bring the text-analysis tools of the literary and historical scholar to bear on electronic texts. These tools are best known in primitive forms in the grammar- and spell-checking software that ships with WordPerfect or MS-Word. Less well known are Editor 5, a more powerful text-analysis tool distributed by the Modern Language Association of America (MLA), WordCruncher, and TACT: Text-Analysis Computing Tools, which has long been programmed and distributed by the Computing in the Humanities Group at University of Toronto. TACT has also recently been distributed, with a manual and CD-ROM, by the MLA.[11] WordCruncher and TACT are concordance and full-text retrieval systems that allow scholars to perform complex analyses on large corpora of texts. TACT, for example, contains 16 programs that create concordances, graph the physical distribution of words throughout a text or texts, compare word appearance and usage across texts, and so on. "*TACT* transforms texts specifically into word and anagram lists, concordances, distributions, node-collocate tables, and maximal-phrase indexes" among other possible uses (Lancashire, 1996, p. 138). These tools are broadly scaled and automated versions of the skill sets that professional editors and proofreaders bring to their jobs, with added ability to organize and present in interpretable forms data about patterns so large and diffuse that human readers cannot reasonably be expected to discover them.

A common feature to this software is searching for linguistic units or markers such as word frequency, or the distance between occurrences of specific phrases, or, in the better programs, the context in which certain words or phrases appear. As with Flower's (1997) dictum regarding data—"All data can do is provide the foundation for interpretation"—we can say up front that such studies are of little use unless they are also married to some form of interpretive analysis. Critical skepticism toward quantitative studies "rests on an important fact: no text-analysis tool ever produces conclusions" (Lancashire, 1996, p. 138). This is perfectly in keeping as well with the underpinning of Sullivan and Porter's *Opening Spaces* (1997), which insists on the necessity of human interpretation and abstract reasoning applied to the empirical data gathered in research. Simply put, text transformations done by computer need interpretation by scholars. Eric Johnson (1993) of Dakota State University advocates using text-analysis programs in graduate humanities computing classes to "make it easy to answer a series of questions that otherwise can be answered only by intuition, guess, or uncom-

---

[11]Ian Lancashire et al., *Using TACT with Electronic Texts: A Guide to Text-Analysis Computing Tools, version 2.1 for MS-DOS and PC DOS.* New York: MLA, 1996.

monly mind-numbing research,"[12] but he also insists that the text-analysis programs only provide starting points for the interpretive work of reading. This is a little bit of pointing out the obvious. But so ingrained in our culture is the perception that computers are infallible, so ingrained in our culture is the notion that automation of mundane tasks is equivalent to having actually performed them (nothing quite alienates one from labor like a computer that transparently performs the time-consuming tasks of syllable counting or concordance generating), that it bears stating explicitly. Teachers roundly excoriate student reliance on spelling- and grammar-checkers, for example, because, even though the software is notoriously incapable of natural-language processing, its suggestions are often treated as dictates, followed without question. As computers come closer to the ability to process language in a manner indistinguishable from human language processing, these tendencies are surely going to be exacerbated. Perhaps not to the extent imagined by Norbert Wiener (1948), the father of cybernetics, who foresaw the potential for a new slave-labor market through the connection of human with machine bodies,[13] but certainly to a degree that requires ethical teachers and researchers to frame their work in a manner that emphasizes the "creative, highly intuitive activity" that humans can bring and machines cannot.[14]

A very basic example of text-analysis work points in the general direction I seek. In "Using the Eyes of the PC to Teach Revision," Ed Klonoski (1994) describes a method for using the grammar- and spell-checking functions of a word processor not to merely perform those tasks but to teach students to interpret the data critically and make informed decisions about writing. Klonoski teaches students to use style or grammar software in the word processor to create search patterns for particular verbs, verb phrases, nouns, noun phrases, and other grammatical features that the student is learning to control. This teaches students to use the software actively and in a limited way, rather than letting students uncritically accept erroneous advice, "always emphasizing that the computer suggests, but the writer decides" (p. 77).

Useful as a teaching tool, Klonoski's approach can also be enhanced for use as a research tool by adding to his search, consider, revise strategies. Most word processors have some form of version control or document comparison feature; teaching students a simple protocol for saving drafts with sequenced file names, then using the computer to compare the documents and produce a third, combined, document automatically marked to show all changes will create a set of documents researchers can use to track both global and local revisions. By combining this data with other more traditional research instruments, investigators could quickly create a body of information that would move beyond the anecdotal

---

[12]Eric Johnson (1993), "Electronic Shakespeare: Making Texts Compute," *Computer-Assisted Research Forum, 1*(3), 1–3.

[13]Norbert Wiener, *Cybernetics.* Cambridge, MA: MIT, 1947.

[14]James P. Hogan, *Mind Matters: Exploring the World of Artificial Intelligence.* New York: Ballantine, 1997, p. 358.

and local to the specific, empirical, and global. What's more, one can easily imagine a research project across a number of institutions, using standard software, simple protocols for naming and storing files, a sequence of shared assignments and teaching strategies, and shared survey instruments. Such a project would in a short time create a searchable, electronic archive of data scholars could mine in ways similar to—but much more powerful than—the extensive research Lunsford and Connors (1995) performed for their editions of *The St. Martin's Handbook*.[15]

Before looking at the implications for research by professional scholars, it is instructive to look at ways such research can—and should—be integrated in the classroom itself. Imagine a semester's work done by a student in a computer lab or in a class where the teacher has required students to do their writing—including journaling and other informal writing—via computers. At the conclusion of that semester, the student has a digitized corpora of texts for analysis with TACT or other tools. Students could begin by examining concerns and comments identified by instructors during the semester. Guided by the reading of their instructors and peers, the software will provide students with page and line references for frequently repeated phrases or mechanical errors or other textual features. On the one hand, students could then use that analysis to study, practice and remove those errors. But on the other hand, students could also use that analysis to make an informed case regarding the shifts and growth in their writing over a semester—marrying the usual metacognitive document that introduces a portfolio with a sophisticated, fact-based linguistic analysis. In addition to giving students impartial information about their writing, and empowering students through reflective analysis of the data, the software gives students and instructors a common, neutral starting point for discussion of the writing, removing it from the contentious perception of subjectivity with which student-teacher portfolio conferences sometimes begin.

Now imagine the same student learning to perform those analyses on a body of texts that is produced over, say, 4 to 6 years. We now have a self-directed and generated longitudinal study of a writer's development that could easily be used in the creation of a learning record or portfolio covering the writing and thinking of the student's entire baccalaureate career and giving raw data that could be used to demonstrate successful learning. Students could analyze their own texts for evidence of changes in their use of language. If, as John Swales (1990) claims, following Kate Ronald (1988) and others, that among the chief goals of education in disciplined and disciplinary language use is to teach students to "sound right" for their discipline, such an exercise would provide students with closely examined evidence from their own writing. Swales concludes that to increase the possibility of students learning to write in their disciplines, they have to be taught to perceive

---

[15]Connors and Lunsford analyzed 21,000 student essays marked by instructors, "eventually identifying the twenty most common error patterns facing college writers today" (front matter). *The St. Martin's Handbook*, 3rd ed. New York: St. Martin's, 1995.

their disciplines' own peculiarities of language use; they have to have rhetorical awareness generated by "a perceived rationale for the communicative behavior" (p. 234). They develop that perceived rationale by developing a rhetorical awareness that "usefully develops in participants an increasing control of the metalanguage (negotiation of knowledge claims, self-citation, metadiscourse, etc.) which, in turn, provides a perspective for critiquing their own writing and that of others" (p. 215). A teacher engaged in longitudinal computer-aided analyses with students would be working at a level virtually guaranteed to impress upon the student information about language use that too often remains an abstraction divorced from the student's context. Recursing the analysis over a period of years would give the student continual opportunity to reflect on learning from a relatively neutral starting point. A researcher with access to a database of similar portfolios and reflective analyses could literally mine it for insights for decades.

Now imagine a course in which students engage in four kinds of writing: personal, informal journaling; formal essay or paper writing in multiple drafts; synchronous conferencing for classroom discussion; and asynchronous e-mail conferencing. Applying textual analysis tools to the four kinds of writing separately and as a group, students not only could form and express judgments about tone, register, diction, and change over time but also would have at their disposal textual evidence necessary for making academic arguments about their writing. If the analysis were extended by members of the class to the texts that the class produced together, a group project on social construction of knowledge would indeed be powerfully aided by the textual evidence alone. The known consequence of taking their own language apart and studying it closely is demystification of writing processes and validation of student identity as scholars of their own experience. This soft result cannot be discounted as a desirable pedagogical outcome of such hard quantitative work.

Finally, imagine a course in which the tools of textual analysis are first demonstrated on the course syllabus, then all the syllabi the instructor has written, then on all the syllabi and course descriptions that the department provides, then on as many electronic syllabi and course descriptions and catalog pages as the university provides. Such a class would provide students with raw data for comparative analysis of the language used to describe students, learning processes, and other key terms in pedagogy, opening the door both for a radical pedagogical awareness of the act of education and for a text-based analysis of the language the university itself uses to "invent the student," if I may invert Bartholomae's phrase (1985). One can easily imagine pairing such powerful empirical tools with Freirean cultural analysis activities or the text-based critical pedagogy Ira Shor (1987, 1996) describes. Raising the critical consciousness of students would be best accomplished by tying political analysis of hegemony and ideology to such powerful statistical tools that aid in identifying patterns. It is not difficult to imagine scholarly research that builds on both the reporting of what happens in classroom such as those I describe and the more empirical research on student writing.

Because textual analysis software provides such a large amount of uninterpreted data—literally thousands of pages sometimes—many teachers and scholars do not use it: It is easy to see from the vantage point of the experienced academic how likely it is that there will be much separating of wheat from chaff in the aftermath of compiling the raw data. Evidence suggests, though, as described by Eric Johnson (1998) in "The World Wide Web, Computers, and Teaching Literature," that students trained in textual interpretation can profit from the textual analysis of large corpora:

> There were significant patterns and rhythms of words and sentences that were impossible to notice by simply reading texts, but they could be discovered with computer programs such as those used in the class: they could determine a range of surface details, matters of symbolism and theme, and, sometimes, how an author's craft achieves particular effects. They had learned to ask questions about texts that would not have occurred to them prior to the course [online].

I favor this hermeneutic approach, interpretive and meta-aware, that is grounded in detailed observation and recording, a grounded approach to a research-based development of rhetorical awareness in student writers.

One could suggest that students are not capable of distinguishing good data from bad or that they will likely be seduced by the powerful myth of objectivity and accuracy that surrounds statistical information in our culture. But I cannot think of a better case for the necessity of teaching empirical analysis of texts in our classes. Only the introduction of such research will expose students to the idea that empirical research is open to hermeneutic analysis—indeed, that such analysis is necessary, particularly in a society increasingly glutted with digital texts of widely varying accuracy and usefulness. Close reading and textual analysis are increasingly becoming survival skills. If, as Linda Flowers (1997) says, "Fine-grained, observational theories can encourage the rhetoric of exploration and construction" (p. 170), then that is what is needed, for such a rhetoric, grounded in close analysis, could move research in computer-mediated communication beyond the practitioner lore characterizing our current publications and into a broader awareness and application of the full range of theoretical, empirical, and interpretive approaches made possible by humanities computing.

20

Imaging Florida: A Model Interdisciplinary Collaboration by the Florida Research Ensemble

John Craig Freeman
Emerson College

CraigCam: Hey Ron
CraigCam: Your sound is breaking up.....

Vannevar Bush, inventor of some of the earliest, most powerful computer technology in the 1930s, and leader of the effort to build the first atomic bomb, wrote a cornerstone article for the July 1945 issue of *Atlantic Monthly*. In it, he attempted to lay out a roadmap for shifting the attentions of the scientific community back toward peacetime activities. In addition to conceptualizing the vast apparatus of the Internet, including hyperlinked databases, date compression as well as collaborative and remote access to those databases, he predicted the advent of a contraption he called the memex, nothing less than the personal computer itself:

> Consider a future device for individual use, which is a sort of mechanized private file and library. It needs a name, and to coin one at random, "memex" will do. A memex is a device in which an individual stores all his books, records, and communications, and which is mechanized so that it may be consulted with exceeding speed and flexibility. It is an enlarged intimate supplement to his memory.
>
> It consists of a desk, and while it can presumably be operated from a distance, it is primarily the piece of furniture at which he works. On the top are slanting translucent screens, on which material can be projected for convenient reading. There is a keyboard, and sets of buttons and levers. Otherwise it looks like an ordinary desk. (p. 41)

Vannevar Bush was responding to the dilemma of the burgeoning volume of specialized knowledge that science was producing and the inefficiency by which this record could be accessed and consulted. Over a half century later, clearly, the memex technology has arrived. Although this technology is increasingly affordable, efficient, and reliable, the forms by which knowledge flows through it are just beginning to emerge. Vannevar Bush had thoughts about this as well.

> Our ineptitude in getting at the record is largely caused by the artificiality of systems of indexing. When data of any sort are placed in storage, they are filed alphabetically or numerically, and information is found (when it is) by tracing it down from subclass to subclass. It can be in only one place, unless duplicates are used; one has to have rules as to which path will locate it, and the rules are cumbersome. Having found one item, moreover, one has to emerge from the system and re-enter on a new path.
>
> The human mind does not work that way. It operates by association. With one item in its grasp, it snaps instantly to the next that is suggested by the association of thoughts, in accordance with some intricate web of trails carried by the cells of the brain. It has the characteristics, of course: trails that are not frequently followed are prone to fade, items are not fully permanent, memory is transitory. Yet the speed of action, the intricacy of trails, the detail of mental pictures, is awe-inspiring beyond all else in nature.
>
> Man cannot hope fully to duplicate this mental process artificially, but he certainly ought to be able to learn from it. In minor ways he may even improve, for his records have relative permanency. The first idea, however, to be drawn from the analogy concerns selection. Selection by association, rather than by indexing, may yet be mechanized. One cannot hope thus to equal the speed and flexibility with which the mind

follows an associative trail, but it should be possible to beat the mind decisively in regard to the permanence and clarity of the item resurrected from storage. (pp. 39–41)

As an artist, it is these associative trails and the form that they might take that I find most engaging. I am interested not only in the role that hypermedia might play in the advancement of scientific knowledge to aid the scientist in the laboratory, but I am also interested in the role that it might play in the advancement of poetic wisdom.

Let me point out one more important idea that Vannevar Bush had before I proceed:

> There is a growing mountain of research. But there is increased evidence that we are being bogged down today as specialization extends. The investigator is staggered by the findings and conclusions of thousands of other workers—conclusions that he cannot find time to grasp, much less to remember, as they appear. Yet specialization becomes increasingly necessary for progress, and the effort to bridge between disciplines is correspondingly superficial. (pp. 29–30)

Here is the call and the justification for interdisciplinary collaboration. Now we have the tools, and the forms are emerging to accommodate it.

IMAGING FLORIDA

Imaging Florida was the first of an ongoing series of collaborative interdisciplinary projects known collectively as Imaging Place. It was produced throughout the late 1990s as part of the Florida Research Ensemble (FRE) collaborations at the University of Florida. The group was brought together by a common interest in electronic technologies and a desire to explore new methods to deliver our respective research and creative production to audiences beyond predominate cultural and educational institutions. We realized that the new technology required a new form with which to represent our disciplinary insights, and even a new methodology of research, to break from the limits that were isolating our disciplinary knowledge and practices from the other disciplines and from the public at large. The FRE included myself, specializing in documentary photography/digital video virtual reality, writer and cultural theorist Greg Ulmer, media and performance artist Barbara Jo Revelle, architect and urban planner William Tilson and the late Gordon Bleach.

Our name reflected some of the features of the new direction that we set for ourselves:

- Florida (not just that we lived in Florida but that our local setting as well as the universal abstractions of our various fields provide the basis for our work)
- Research (that the nature of our methodologies should not be assumed but should be examined critically)

- Ensemble (that we work collaboratively rather than exclusively as individuals and that the methods of inquiry peculiar to art making and theory are tested for their applicability to practical states of affairs outside the academy and the gallery or museum)

During the formation of FRE, I was experimenting with early forms of the Imaging Place method at the Castillo de San Marcos in Saint Augustine and the 19th century sugar plantation ruins along the east coast of Florida. Both are rich in history, metaphor, and contemporary political relevance. The goal of the Imaging Place method is to document sites of cultural significance, which for political, social, economic, or environmental reasons are under duress, at risk of destruction, or undergoing substantial changes. This includes historic sites as well as sites of living culture, which are being displaced by globalization and the collapse of industrial modernism.

The Castillo de San Marcos, built 1672–1695, served primarily as an outpost of the Spanish Empire, guarding St. Augustine, the first permanent European settlement in the continental United States and also protecting the sea route for treasure ships returning to Spain. I was attracted by the geometry of the architecture, which made it manageable to organize a systematic virtual reality shoot throughout the site.

In the 1830s through the 1840s, molasses milled at the sugar plantations was shipped to the Indies for the manufacture of rum. The mills were destroyed during the Second Seminole War by Native Americans angered by the exploitation of their land. Disputes over land use relating to sugar production continues to be one of the most explosive issues in Florida to this day.

I worked with sequences of still images and a simple rollover interactive interface. If you rolled the mouse to the right of the image, the computer would quickly load the next image in a rightward pan. Rolling left would pan left and rolling up would sequence images forward into the virtual space. Rolling down would sequence the images backward.

As a group, FRE began the Imaging Florida project by looking at the entire state, but gradually we narrowed our focus to the Miami River. At a mere 5 miles long, the Miami River cuts right through the center of downtown Miami. Once a major hub for U.S.–Caribbean trade, the river port and its culture has slowly given way to the mechanized container shipping that was developed throughout the 1970s and 1980s, at the Port of Miami, just beyond the mouth of the river on Biscayne Bay. Today the river is torn between a wide array of diverse factions and interests. These factions include the small-scale Haitian shipping businesses, which are too poor to operate out of the Port of Miami, and the developers, who want to raze the place and put in condos, shopping malls, restaurants, and a river walk. The Coast Guard try to keep the place safe, but they can not speak Creole, and the drug dealers take advantage of the river's tradition of contraband trading. When you include the neighborhood folks who just happen to live there, you have a dynamic mix that makes traditional forms of documentary media, such as still photography, film, and video, problematic if not inadequate.

Imaging Florida has been exhibited in various forms both nationally and internationally, including at the Centro de la Imagen in Mexico City; the Johnson Museum of Art in Ithaca, New York; the Orlando Museum of Art; the Jacksonville Museum of Contemporary Art; and Ambrosino Gallery in Miami. The Virtual Florida CD-ROM is included in the groundbreaking traveling exhibition Contact Zones: The Art of CD-ROM, curated by Timothy Murray of Cornell University. It has been presented at professional conferences including the 1998 Southeastern College Art Conference, the 1998 Computers in Writing national conference, the 1998 Society for Photographic Education national conference, and the 1999 College Art Association national conference.

---

ron: can you hear me at all?

What follows is a montage cross-section of the Imaging Florida collaboration in action.

---

ron: can you hear me at all?
CraigCam: yeah, sure

It includes excerpts from all of the different technologies that we used on daily basis. Each technology has its own set of strengths and weaknesses, which I hope becomes apparent as you read. I was careful not to strip too much of the chaos out, because chaos seems to be one of the defining characteristics of interdisciplinary collaboration.

---

ron: can you hear me at all?
CraigCam: yeah, sure
CraigCam: how is Paris?
ron: I just downloaded this, trying to get it to work
CraigCam: iVisit is great software.

```
Subject: Interdisciplinary Collaboration
   Date: Fri, 5 June 1999
   From: "John (Craig) Freeman" <freeman@uml.edu>
     To: Florida-l <florida@nwe.ufl.edu>
```

Hello FRE and others,
I sent my first draft of the article for James Inman's book on Interdisciplinary Collaborations in the Humanities and his response was positive. He has asked me to fine tune some structural elements and I wanted to get some feedback from this group.

Inman wrote:
> Audience. Readers of this book will need some help from you in order to navigate successfully the different genres they will see in your chapter. At the outset, for example, you might indicate that your collaborative method is moving beyond print media as apparatus, this allowing you to then explain to readers that they will see a mix of communicative genres, each suggested by electronic media (the emails, the chat, etc.). If you say up front that the point is for the reading to be a little uncomfortable, a little challenging in print, then I think readers will get up for the challenge and be excited about what you've put together.

> Also with audience, the other aspect of helping the reader along is taking us with you, as you weave in and out of postliteracy theory and the texts that re/present it. With the chats, for example, you could either explain their significance up front, or you might want to spend some time discussing each transcript, explaining its specific value and lessons. Another way of thinking about this revision suggestion is that readers will need an interface—they are operating mostly in a paradigm that calls on them to expect paragraphs and formal arguments and prose, which you do give them sometimes, but not all the time. I understand that there's an underlying rationale/methodology to the way you've woven the various texts together, and making that more visible will enable readers to come with you—to duck in and out of multiple media, to understand what postliteracy means

through their reading (interpretation) and their reading (grammar-induced).

Craig wrote:

> OK, although I completely understand this need for explanation, I think it represents one of the fundamental obstacles to moving postliterate forms into the public realm. Let me explain. Ulmer and I have been talking about a new form of writing that would be more like curation; take an idea from here, an idea from there, fill in the gaps and assemble it in montage form. What I did for the chapter was to take a cross section of materials that have been produced around the Imaging Florida project (collaboration in action.) They include email exchanges from this list and its archive, telecam chat, transcriptions from meetings and written materials from the Imaging Florida web site. I cut them up and put the pieces back together in a kind of mosaic montage. One of the most important lessons that I have learned from Ulmer's work with electracy is that it allows for an increased reliance on poetic association making possible forms of knowledge exchange that are structured much more like human thought. An example of electracy would be effective email skills, or more abstractly, the capacity to focus on several communication tasks simultaneously. Here is the logical argument part; explication undermines the poetics of association. It takes the responsibility of the connections out of the audience's hands. To explain a poem is to strip it of its associative power. Associative power resides in the reader's ability to internalize the poem's emotive evocations based on her or his lived experience. The problem is that people are used to explanation. Although it is easy enough to explain my motives, as I did above, it goes against my judgment as an artist and my commitment to theory by practice. In the case of this article, I am attempting to express the complex idea of postliteracy by writing it in a postliterate form. I could, for instance, just include this post in the article, which would expose my motivations for the Imaging Florida project, postliteracy theory and the article itself.

More later. Craig

---

ron: Craig, walk me through the audio setup
CraigCam: I wanted to tell you about the Virtual Reality method that I have been developing for Imaging Florida
CraigCam: communication is best by typing
CraigCam: sound is good to get people's attention, then type
CraigCam: especially via modem
CraigCam: the way that you are talking on the phone to someone in Paris and me, here via telecam, at the same time is very postliterate
ron: yes, hold on

```
Subject: Interdisciplinary Collaboration
   Date: Wed, 12 May 1999
   From: "John (Craig) Freeman" <freeman@ufl.edu>
     To: Florida-l <florida@nwe.ufl.edu>
```

Hello FRE and other Floridians,
I wanted to open a discussion about Interdisciplinary Collaboration.

Why is it necessary to collaborate across disciplines?

Craig

THE PROBLEM TOUR

Imaging Florida questions the opposition between pure and applied research by means of a virtual tourism. Our virtual tourism adopts the tourist attraction as an interface metaphor to bring citizens into relation with arts and letters experience in a practical context. Like any interface metaphor, tourism is something already familiar to the users and will help guide his or her encounter with something less familiar or even unknown to them.

Some of the places most attractive to tourists are places associated with problems, trouble, or even disaster. Places such as the sewers of Paris, the red-light districts of various metropolises, battlefields, or famous crime scenes are popular with visitors, complete with a live or recorded commentary explaining how the place works or what happened there.

A recent example of this phenomenon is the Cunanan Trail Guide Maps, which sprang up in Miami the day after the Versace murder. Tourism, as a problem or a set of interlocking problems, includes this property of Florida as attractive not only to tourists but also to retirees, immigrants, the homeless, drug dealers, unemployed workers, and serial killers. Columnist for the *Miami Herald* and fiction writer Carl Hiaasen titled his column dealing with Andrew Cunanan, "Florida: A Real Slime Magnet."

> **ron:** OK, I'm back
> **CraigCam:** I have been thinking that the methods I have been developing for Imaging Florida might be of interest for your work in Europe.
> **ron:** go on
> **CraigCam:** although , I have borrowed freely from the traditions of documentary still photography and filmmaking, the Imaging Place method departs from those traditions by using the emerging nonlinear narrative structures made possible by new interactive media technologies and telecommunication apparatuses.
> **CraigCam:** the Imaging Place computer program is projected up to nine by twelve feet in a darkened space with a podium and a mouse placed in the center of the space, which allows the audience to navigate throughout the project
> **CraigCam:** Ron, are you still there?
> **CraigCam:** Ron?

Subject: **Re: Interdisciplinary Collaboration**
Date: Thu, 13 May 1999
From: Greg Ulmer <gulmer@nwe.ufl.edu>
  To: Florida-l <florida@nwe.ufl.edu>

Craig wrote:

> 1) Why is it necessary to collaborate across disciplines?

Disciplines have a history, of course, and their formation is relative to historical circumstances. My approach to such questions is through grammatology—the history and theory of writing—with its notion of the "apparatus." Literacy is an apparatus (a social machine) that includes not only technologies of language (culminating in print) but also institutionalization and identity formation. The modern university is the most recent institutionalization of literacy, part of a tradition going back to the first school—Plato's Academy, founded as part of the Greek appropriation of alphabetic writing.

Writing as a technology is a prosthesis of analysis: it augments the analytical powers of human thought tremendously. Plato is credited with inventing the first concept, and his PHAEDRUS is credited as the first discourse on method in the Western tradition. The dialogue form (invented as the institutional practice for his school) was the vehicle for dialectic, the seed from which the mighty oak of science eventually grew. The point of this history is that university disciplines are part of the apparatus of literacy, relative to it, and not absolute.

Actually the fit between the world and the disciplinary manner of knowing the world is not all that good. The perfection of literacy in print produced an information explosion that schooling never has responded to completely. Specialization is the institutional form of analysis, and it guarantees its own inadequacy by exacerbating the information overload (producing ever greater bodies of "literature" about increasingly fragmented parts of what is).

At this moment of transition from one apparatus to another (from literacy to what I call electracy) academics in many disciplines are beginning to question the limitations of their fields. The problem is that each discipline produces something akin to a "nationalism." Each has its own socialization processes, its own discourse (or even language), creating identification in the deep sense described in psychoanalysis. Disciplinary relations, within this analogy, often resembles events in the Balkans.

Best,
Greg

> **ron:** Sorry I got cut off
> **CraigCam:** No problem
> **CraigCam:** when activated by the click of a mouse, the project leads the user from global satellite perspectives to virtual reality scenes on the ground.
> **CraigCam:** the virtual reality scenes are lens-based representations of real places, shot in the field with a combination of panoramic still photography and digital video.

    Subject: **FRE Meeting**
       Date: Fri, 14 June 1998
       From: Greg Ulmer <gulmer@nwe.ufl.edu>
         To: Florida-l <florida@nwe.ufl.edu>

Ulmer wrote:
>FRE had a great meeting, discussing plans for the problem tour of the Miami River (virtually). The structure emerging at this point starts as follows (logically):

1. following Barthes in S/Z, cutting out a rectangle ZONE on the map of Miami—specifically the Miami river and environs—and CON-SULTING it the way a diviner or soothsayer consults (__contemplatio__) a section of sky upon which has been traced an imaginary rectangle (or square), to read the flights of birds that cross it.

2. Barbara Jo Revelle enters this zone, in order to register its attunement: spirit of place.

3. Syncretism, hybridize: a new mode of consultation emerges out of critical theory and Afro-Caribbean divination.

4. At different scales, micro to macrocosm, an isotopy forms, homologous across dimensions, linking: Barbara Jo's dreams, The media (fantasy) representations of the zone (river), Haitian Vodun divination (spirit possession) methods Poststructural metaphysics.

5. The Miami River virtually forms a category in a new metaphysics of electracy. Metaphysics (demystified) amounts to: the system of classification practiced within the apparatus of a civilization.

6. In poststructural theory, the Miami River is a cryptophore (perhaps). Juxtaposition: Western empirical problem solving (literacy) postliterate (secondary) magic (electracy).

What does this sequence mean in practice?

1. Craig plots the physical place. This place is the floor plan (map) of a memory theater.

2. What the theater remembers (the mnemonic walk) is:

a. (Tilson) the administrative city political legal information network imposed on the physical place. Regulation, a policy: BOATS MUST BE SEAWORTHY.

b. (Revelle) personal individual subjective emotional inquiry recording the place (anecdotes, photos, conversations, dreams, memories, found objects, scraps.)

c. (Ulmer) poststructural discourse network: info on syncretism, on divination methods, on metaphysics as category formation, on complexity theory (form: how to reason using the Miami River holistically as a unit of meaning).

INTERFACE: the tour (collaborative). A linking principle that makes the category intelligible. Our point is that a complex scene such as the Miami River is intelligible and coheres strictly within the prosthesis of digital technology. Similarly, concepts exist strictly within the prosthesis of alphabetic literacy. A concept is to literacy - what the Miami River Zone (virtual) is to electracy

A note, with reference to taking the Miami River as a zone, like that which the soothsayer draws on the sky, to trace the flights of birds that cross the space for purposes of divination. Our divination, our consulting, is with this zone in Miami. We don't need to take the zone too strictly, too literally, sometimes the "flights" or traces will be on site, such as the boatload of Haitians that made a dash for freedom. Other times it might be contextual, such as the Punk Band "Dead German Tourist" (not allowed to play within the city limits of Miami, for being in such bad taste).

Today the paper (the source of our bird tracks) provided another trace: "MIAMI—So much yellow tape was used at a police shooting in Miami recently that all four corners of an intersection of two city streets were roped off by multiple lines, making it look like a boxing ring. Yellow tape blocked traffic on streets nine blocks away from the shooting scene. 'It was one of the most heavily taped scenes I've seen in a long time," Miami Police Lt. Bill Schwartz said.".

We have a bright shiny archived discussion list we could use to brainstorm, think as we write. The use of email as a medium for collaboration is one of my primary interests in Imaging Florida. I don't understand why you folks will not use email. Let's discuss that :-)

---

ron: Sorry I got cut off
CraigCam: The user can navigate throughout the immersive virtual space viewing video clips
ron: oh, I see. The Aerial and satellite images define the divination zone that Ulmer was talking about
CraigCam: exactly
CraigCam: rather than the linear structures of the novel or cinema, this new form allows the story to unfold in a meandering labyrinth of discovery and associations through nonlinear media montage

Subject: RE: FRE Meeting
Date: Sat, 15 June 1998
From: John (Craig) Freeman <freeman.ufl.edu>
To: Florida-1 <florida@nwe.ufl.edu>

Forgive my lurking. I have been in the thick of processing the materials that I brought back from the Miami River trip. Bill and I looked at this stuff last week but I would like to go over some of my progress this week. The trip was very successful. I got Barbara Jo moved in to the Miami River Inn for the next five weeks and she hit the payment running. Grateful to escape bureaucratic bludgeoning at the University, no doubt. The tug trip that Bill lined up with Captain Rick Reed was epic. Miguel and Alberto joined BJ and I. We tugged a 160' freighter out of the river. I shot a sequence of 120 stills, one every minute for the duration of the trip. I will be using these images as the main navigational device for the virtual tour of the river.

After the trip we went to a little seafood place on the river called Garcia's for $4 fish sandwiches and cold beer and discussed the trip and other related issues such as the history of the building of Interstate 95 through Overtown. Miguel is very knowledgeable about all of this. Apparently Overtown used to be a thriving cultural center compared by many to Harlem. When the interstate came through it devastated the community. Notice that there were no on or off ramps built in Overtown.

The next day I returned to try to navigate around the river by car, a daunting task, and shot some Virtual Reality nodes. At lunch time I found myself at Garcia's again but this time there was a news crew there interviewing the 'power lunch' types about the pollution on the river. When I asked the reporter if she had interviewed any one working on the river she pointed at the river and said "it's a trash heap, it's a trash heap," which of course it is. The concern on the river is that the developers will be successful in framing the issue in such a way that it will allow city authorities to condemn it and turned into another shopping mall. I got the newscast on tape that evening. It will be useful in demonstrating how the media chooses the problem.

Another problem in the news is the spread of pfiesteria, a deadly, poorly understood microorganism that has been causing massive fish kills from the Chesapeake Bay to South Florida. The fish develop bloody wounds and then swim in circles for a few days before they die. The neurotoxin produced by the bacteria is known to affect humans. Does anyone know any more about this. Watch the sushi folks.

Sorry to drone. Oh, I also got the aerial images of the river at the county building. The interface for the interactive hypermedia version of Imaging Florida will start from a global perspective using this aerial imagery. The user will navigate by zooming in to the ground level photographic virtual reality nodes connecting the abstraction of the theory with the visceral experience on the ground with.

More later, Craig

CraigCam: Hi Hiroshi, what are you up to?
Hiroshi: just hanging out, what 'bout you?
CraigCam: I was just telling Ron about the new work. Here, connect to him
ron: Hiroshi?
Hiroshi: hey Ron
CraigCam: Hiroshi is in Tokyo
CraigCam: anyway, I am attempting to apply postliteracy theory to the advent of new forms of digital media art
Hiroshi: postliteracy?
CraigCam: although attention spans are getting shorter they are capable of concentrating on many different things at once
ron: so why do you say that our attention spans are getting shorter?
CraigCam: my student's attention spans have shortened measurably
CraigCam: it is the effect of media culture
CraigCam: kids are growing up on a steady diet of remote control TV and video games
CraigCam: the result is a physical change in the human brain
ron: I suppose this is a valid point, but its not to say we don't want to develop specialized expertise anymore
CraigCam: although it is short attention, it is very intense...
CraigCam: capable of assimilating and associating vast amounts of multi-layered information in burst
Hiroshi: maybe it is because with all this technology we are searching for a more than we can get
CraigCam: or the other way around
CraigCam: we create the tools to accommodate the changing cognition
CraigCam: It used to be adequate to store and disseminate information in books and organize the books in libraries
CraigCam: but now you could not build a library large enough to store all of the information that has accumulated
ron: I thought you said that the tools created that condition?
CraigCam: there is just too much information so that we have to create new forms of information generation, dissemination and storage
CraigCam: telecam technology had to happen now
CraigCam: we need new ways of communicating and distributing information, a cybernetic extension of our memory
Hiroshi: I think that because there is more access and therefore more information
CraigCam: yes, chicken and egg
CraigCam: as the amount of human knowledge increases so does the need for new forms
ron: gota go, it is 2:00 a.m. here
CraigCam: C YA
Hiroshi: we have created the need in our search for our reasons for living... that's where all this stems from
Hiroshi: that's what I reason anyway
CraigCam: I'll be back in a sec
CraigCam: I need to read to this email.

MURDOCK'S MACHINE SHOP

Barbara Jo: I'm inside Murdock's Diesel Machine Shop, which is like a big open warehouse, near the 7th Avenue Bridge. I came here to look for Winston, the Jamaican man with the burned boat, whom I met yesterday at the restaurant near the boat with all the old bicycles. He is not here but they say he will be coming in a minute. I try to make a tape of the Haitian women going through bins of old clothes and tying them into bundles to ship to Haiti. In the building next door, they are violently opposed and afraid—holding their hands over their faces—so I make a tape in Winston's shop. I wander around making shots of the density—diesel engine parts all over the floor, a table with dominos on it, a filthy bathroom with a very old, white Great Dane who appears to be dead—curled around the base of the toilet. It must be 98 degrees in here.

Russell is sweeping. He has beautiful bones, luminous black skin, he won't look at me. Someone says he's a Vietnam vet, which is hard to believe—he looks too young. He tells me what company he was in but mumbles. It isn't clear on the tape. There is reggae playing on the radio, very loud, and a man moving engines around with a tractor.

Mike: I'm the guy that runs around and gets all the parts. My boss here, he's well established on the river.

George: I transport all the finished engines, get em outta here.

Barbara Jo: Those are the green ones?

George: Yeah, green is to show they are done. Ready to go. We paint em.

Tyrone: I'm the general manager—I do the paperwork, and I'm Winston's brother. Winston was in a terrible accident. He went to Haiti, and while he was over there he had a terrible accident—broke his head, broke his neck—and there is no medical attention over there. Don't wanna be in Haiti when you almost gonna die, no way. So he come back and I come over to help out, and here I am.

Barbara Jo: Why'd he get into an accident? What kind of accident?

Tyrone: Well, he can tell you about that—if he wants to. We'll leave that for him to tell.

Spencer: Winston done alot of work for me over the past 20 years. I'm his best customer. I don't use nobody else—I don't have no breakdowns. I'm from Bimini. See this shirt—Bimini. I come for general cargo. I got an 80-foot cargo boat and I go back and forth from Bimini and Cat Key and Haiti. But right now they all stop. It's what they call the Caribbean Code. They try to get you to follow all their damn rules. They give you all these things you gotta fix on your boat—the boat that just come in the river perfectly safe for crossing for many years. I been waiting for them fuckers to call me back for 3 day, it's you gotta have chain. I never had no accident, no problem, no incident, no nothing. Over 200 trips going back and forth. The ferryboat that goes to Fish Island sank two boats already right across from the Coast Guard. Me, I never run down nobody, cause nobody no problem, nothing. But I can't go out. They say I can't go out.

Barbara Jo: So why? Why do they stop your boat?

Spencer: The Caribbean Code—they got this big book of new rules and shit you supposed to have but the main thing is to get the little man out of business. What the Coast Guard fail to understand is that all those bicycles and refrigerators, cars and mattress—they help to clean up South Florida. Every boat come to buy stuff, not to steal. They need it in their country. Customs come with a dog this trip—make me take off my tank covers, drain my oil, drain my water, they don't find nothing. Now I got water in my generator from those fuckers. That what I'm against. Everybody up against this shit.

Winston comes in, smiles, but doesn't particularly acknowledge my presence—just joins the conversation.

Winston: I seen it all - the whole plan for the river, up at Mrs. Brickle's Boat Yard. They gonna close her down—you know—up there on the 2nd Avenue Bridge. They shot *Miami Vice* on her property. She got this map of how they plan to make a shopping center, like another Bayside—with shops and fancy little restaurants. They got this whole big plan. Only boats that gonna be on the river be the tour boats, water taxis, and like that—taking the tourist tootin' all round. River walks and tours about history and shit.

Barbara Jo: So why don't you all—all the people on the working river, I mean—you make your living from the river, get together and fight the yuppie gentrification thing? Explain to the Coast Guard about the problem—the double bind. How, if you can't leave the river to deliver your cargo, you can't get the money to fix your boats the way they want.

Winston: Well, it's mostly Haitians. The Haitian is difficult. They face the problem—they don't face the problem—they go around the problem. Coast Guard tell them not to bring their wood boats in here. So they just sit there on their boat, waiting for some message from their gods about what to do or something, or they get together, get a bigger boat—steel boat—instead of show the Coast Guard where they're off base. But then the big boats get seized, too, if they have Haitian flags. The other day a boat come in here with a Haitian flag—big boat—U.S. Navy-built boat, and they seized it. Said it was unsafe. U.S.-Navy built boat. I mean, hello, so we got boats going up the river now that is too big—get stuck in there at low tide—tip right over on there sides and dump all the cargo. But that's the kind of solution the Haitians think up. Don't go at it head on. If the big boat is from Cuba, now that's a different story, or say the Bahamas or someplace Americans like. Then they can come in, go out, no problem.

Spencer: I know this—somebody is getting some kickbacks. Big boats give big kickbacks. They just want the little man out. If 10 Haitians come in here, they send them back. If 20 Cubans come in here, it's oh, hello, can we help you? with a big escort and immunity or some shit. What they call it?

Winston: OK, you wanna know what's really going on—turn off the tape. Here is what's really going on. Between about 1979 and, say, 1988, Cubans running drugs, that's what. Now Haiti do it. Cuba got too hot. Everybody know Haiti do it now. Nobody looking at Haiti that's why. All the drugs go through Haiti now. They can't catch the Haitians.

Barbara Jo: Why?

Winston: The Americans look at the Haitians like they're dumb and stupid. But they are very, very smart people. But the American people think they are stupid—that's how they class them. That's why they can't catch the Haitians—they too smart; they are so dumb they're too smart. Do you know what voodoo is?

Barbara Jo: Yeah, Haitians' version of an African religion.

Winston: Yeah, well everybody there have the voodoo, so they can bring the shit in. No problem. The other day, there's a boat—a wood boat—they can't bring wood boat no more. A wood boat come right down the river past the customs, the Coast Guard, the bridge guys even put up the damn bridge for them. They dropped off 150 Haitians right in the river.

Barbara Jo: Oh, yeah, the Rose Marie Express—it was at night though, wasn't it?

Winston: No—in the middle of the day—they caught less than half of them.

Barbara Jo: Was that because of voodoo, you think?

Winston: Well, you can believe whatever you want but how come they come down the river like they was invisible and half the people disappear into the city if they don't have something working for them? That's more what the newspaper tells it. Them blowing right by all the Coast Guard—they have satellites on you and all. How they do that is what I wonder.

Barbara Jo: Yeah, I wonder, how would that work?

Winston: You tell me. They got some power with them, I know that much. Then 3, 4 months ago, you have a Haitians boat come in here. It had over 400 people on it. The Coast Guard got after it before it got to the mouth of the river—way out in deep water. You know what the Coast Guard did? They took their steel boat and they rammed into that little wood boat—with all them people crowded into it with babies and crying and old people—they called it shouldering.

Spencer: Now that human life right there. They have no respect for human life. They say they wanna protect human life, so they make up all these rules how your boat isn't safe so to protect you. But that's not it. They send the boats back half the time—so how they say they care about if it seaworthy? Send them back—3, 4-day back—with all the people with no food if they want, no water.

Winston: I'm the one who built the engine for that boat—that one they rammed. That was a good boat. They just hate the Haitians, any goddamn Haitian boat they just hate it. Big, little. Why only Haitian boats they send back? They hate the Black man that's why, and they afraid of him.

Spencer: My Daddy run this trip over 50 years. Twenty years I been working for them—coming in here—but now they got this Caribbean Code and I got to suffer. Captain Brown was my daddy. They can plant something on your boat too—they done that to the Haitians plenty times. Captain Brown never need to run no drugs—he had money—he come in here for 50 years—never had no incidents. Everybody say that the wooden boats are dangerous, but that just show they don't know about boats. If a wooden boat gets a hole, you can put a towel in, clothes or a mattress, you can plug it. If a fiberglass boat cracks, it has to be totally dry to fix it. Out to sea, it just go down. You can't plug it less its totally dry. It's just the Haitian boat they want—nothing about dangerous. Wood boat is just as safe as steel boat, fiberglass boat. Steel boat you have to weld on panels—wood boat you just plug it—take a rag—easy. A lot people running the boats don't know a thing about it. Gotta call Coast Guard. These Haitian people build boats without a blueprint or nothing—from scratch—and beautiful, smart. I would go with them in a storm over any of these big fancy boats. Haitians come in here without radio, compass, anything. Coast Guard say you have to have six lights, backup lights; they got one, two, they never have no accident, no problem. They know what they're doing.

Winston: One thing—they only speak Creole. They're supposed to have one guy on the boat speaks English, but he can't read English, and they give him this long list of fix this, fix that, and what you think the guy knows what to do, what to go do? And how they supposed to get the 50 thousand dollars to fix the wench and buy some more lifeboats and all that. Less they can take the cargo back to Haiti to get some money. But no, the cargo just sits and rots by the river—gets stolen. Boat goes back empty or gets seized or just gets stuck here trying to figure out how to get up the money—so there's not no mystery how they get into drugs there, right?

Spencer: They just take out all the garbage the city dump anyway. They take it all back home. They smart people, see? There used to be a big line of them taking all this junk back to Haiti, but that's all over now. Nobody to take out the trash no more. Common sense, my old man use to tell me—common sense. You don't gotta have no PhD; you gotta have common sense. They got common sense—Haitian people.

Winston: Common sense or voodoo, something gets them across the sea. But the American people, they don't get it—they think the Haitians is diseased or something, think they gonna get AIDS. The Haitian boats—I been on the boats when the Coast Guard come on—afraid to touch 'em. Think they're dirty—they'll get sick to just be on the boat. Their idea is to take the boats up on the banks and crush 'em. I seen it. They just take them up and crush 'em and haul it all away. Beautiful boats all painted up nice inside—bright colors—crush 'em.

```
Subject: Re: Interdisciplinary Collaboration
   Date: Fri, 14 May 1999
   From: Ron Kenley <rk@wanadoo.fr>
     To: florida@nwe.ufl.edu
```

Craig wrote:
> 1) Why is it necessary to collaborate across disciplines?

a. Houses and Cities:
Writing as an architect, it has been clear to me for some time that there are hardly any buildings worth their salt which do not represent a successful collaboration across disciplines around the pretext of shelter (for families, communities, and cultures). The assemblage of these shelters in a social organization sense, is yet another aspect of collaboration, showing explicitly and wonderfully the limits of any agreement. In other words, cities refuse to just collaborate, as their inhabitants get on with their own lives. So, places for negotiation are continually reinvented in the form of streets, public transport, theatres, libraries, squares, reused abandoned abattoirs or warehouses, and so on.

b. Transforming Partnerships:

In Europe, all calls for European Union sponsored research ask for partnerships to be established. They have to be diagonal, both in terms of the disciplines engaged and geographically. But what is really at stake is not the declaration of collaboration, but the necessary synergy for any project. Of course, this is partly circumstantial, but I think it is significant that circumstances include the common questions, which different disciplines address in diverse ways. From my angle and place, any collaborative environment must recognize that the institution of collaboration needs to be reinvented at the same time with the establishment of collaborative setups. A term that seems of interest is trans-disciplinary collaboration. As I understand it, this is set up to expect a transformation of the work that emerges, a mutation due to the participation of representatives from a number of disciplines. The New Consultancy (Gregory Ulmer and FRE) is based on problem or project defined expertise operating through testimonial in conditions of emerAgency (http://www.elf.ufl.edu/). I find this a pretty happy definition of transformative practice, a synergetic collaboration among disciplines engaged in the common set of problems.

Greetings from Paris,
Ron Kenley

Cathy: dang that is so weird
Cathy: I can see you
Cathy: perfectly
Hiroshi -> Cathy: Do you have a cam?
CraigCam: I'm back
Hiroshi: cool
Hiroshi: so you were saying?
CraigCam: contemporary culture, is undergoing one of the most substantial transitions in history. This transition manifests itself as a paradigm shift marked by the collapse of industrial modernism and the rise of information technology in its wake
Cathy: whoa hey, did you guys see what that pervert was doing?
Hiroshi -> Cathy: Just blacklist him and he won't bother you any more
Hiroshi: does this have to do with the differences between nomadic, agrarian, industrial and technological cultures.
Cathy: what
CraigCam: yes, in hunter gatherer cultures the story, the song and the ritual were all mnemonic devices, memorization strategies for passing on important information from generation to generation.
CraigCam: knowledge was limited by the amount any one person could remember and the story was easier to remember if it rhymed.
Cathy: Huh
CraigCam: as cultures settled into agrarian civilization, as food began to accumulate creating surplus which led to accounting and then to writing, which led to literacy, the need to remember the whole story dissolved
Hiroshi: they just needed to remember how to read and write

CraigCam: literacy itself is a mnemonic device, data compression for human thought and experience
CraigCam: although literacy was more efficient and better in many ways it didn't come without its cost
CraigCam: the intimacy, community connection and the multidirectional dissemination of knowledge through group discourse were sacrificed in favor of the ability to accumulate knowledge.
Hiroshi: I see, and as culture industrialized and literacy spread worldwide, through the use of the printing press, the amount of human knowledge expanded exponentially
CraigCam: Right, and now in the throes and turmoil of the collapse of modernism it is no longer adequate to organize knowledge only in fixed linear documents that start on page one and end some number of pages later
CraigCam: there is simply too much knowledge and the need for more
CraigCam: furthermore, our brains don't work with linear logic
CraigCam: instead our thoughts, like our accumulated knowledge in a postliterate world, works through poetic association, moving fluidly from one disparate idea to another
Hiroshi: like the hyperlink on the World Wide Web
Cathy: Interesting, really
CraigCam -> Cathy: I can't see you
Cathy: bummer
Cathy: can anyone else see me??
Cathy: I'm American
Cathy: can ya see me yet??
CraigCam -> Cathy: Not yet
Cathy: Shit
Cathy: Ops I mean... shoot
CraigCam -> Cathy: Bang - Bang

```
    Subject: Sinthome
       Date: Wed, 19 May 1999
       From: Greg Ulmer <gulmer@nwe.ufl.edu>
         To: florida@nwe.ufl.edu
```

I'm finishing up some research prior to writing a draft for Miami/Miautre book based on the FRE's Imaging Florida work. The testimonial method is now located within the tradition of the "epiphany," from Baudelaire through Joyce to Lacan. One of Lacan's last Seminars was devoted to Joyce, particularly The Portrait of the Artist as a Young Man, in which he worked out the notion of the "sinthome." The sinthome (archaic spelling of "symptom") supplies the 4th ring of the Borromean knot (the topology of Real, Symbolic, Imaginary). The 4th ring that holds together the other 3 (even after the cut that makes them fall apart) is constituted by the Name-of-the-Father, and manifests itself as the style of the artist (and its logic is that of the pure signifier, passing through languages via the homophone).

Barbara Jo Revelle works well with this theory. Her experience in the chora of Miautre, tuning the testimonial practice, is an epiphany (the experience of the passage from the uncanny to sa-tori). I already noted the coincidence of "Barbara" as the name of the first modality of syllogism (in Medieval classifications), the prototypical one (All men are mortal, etc.). Revelle is equally interesting in French and English, including its very confirmation of the legitimacy of working with secularized revelation (the epiphany) as the relay for the testimonial consultation.

The further point is that the popcycle/mystory generator may be understood as an "epiphanizer"—an epiphany machine. The experi-ment of Miami/Miautre is that of a collective mystory (not his-tory, or historiography, but chorography). Chorography is "collective mystory." A "machine" is needed for an epiphany in this context since no one person exactly "experiences" the reve-lation (each has his or her own local feeling). In composing the Miautre popcycle, Barbara-Bill-Craig-Greg are attuned. To what extent may this attunement (Heidegger) be effective as a cate-gory for others to use?

Best,
Greg

CraigCam: we live in the most visually literate culture in the history of the world. In fact it could be argued that we are the first fully mature media culture

CraigCam: we read far more images than words in a given day, even if a picture is not worth a thousand words and even if 98% of it is corporate garbage

CraigCam: with every hour that children spend in front of video games and with every click of the remote control our attention spans are compressing even as they are shortening

CraigCam: the result is a new type of cognition

CraigCam: I often hear it said that the computer is just another tool, like a paint-brush or a sledgehammer

CraigCam: this is like telling Gutenburg that the mechanically reproduced book was not fundamentally different from the hand-scribed manuscript

CraigCam: instead, consider technology, as Ulmer does, as cybernetic prosthetic apparatus

CraigCam: cybernetic, in that it has the potential to expand biological capabilities, particularly in the case of memory

CraigCam: prosthetic, in that we wear it like an artificial limb or a pair of contact lenses

CraigCam: apparatus, in that any one technology exist only in relation to all other technologies.

Richard: hello.

CraigCam: hey

CraigCam: what's up

Richard: I was just about to sign off and sleep

INVENTING NEW FORMS

Sergei Eisenstein, one of the filmmaker-theorists who contributed most to the invention of cinema, used the workings of the attractions at fairgrounds, sideshows, and circuses as the point of departure for his concept of montage. The challenge Eisenstein faced in the 1920s was how to extend the new medium of silent film to express increasingly complex political ideas to a largely illiterate population. Prior to Eisenstein's work the question of what to do with the new technology was largely unanswered. Often, the film camera was simply pointed at the subject and shot throughout the duration of an event. The resulting film was played back as a single cut, at the same duration as it was shot. Thomas Edison's early filmmaking contained many examples of this form of the single cut film. For example, he made a short film to demonstrate the danger of his competitors' use of alternating current electricity. The camera was set up in front of an elephant, and electrodes were strapped to the animal's legs. A switch was pulled and the beast began to shake, then smolder, until it fell over dead. End of film.

Strolling down the arcade at the circus sideshow, Eisenstein was stuck by the juxtaposition of seemingly unrelated packages of meaning. From the sword swallower to the chicken lady to the tattooed man, the whole of the experience had greater meaning than the sum of its parts. And, like film, it all unfolded over time. This experience became the model for Eisenstein's invention of the cut and paste film-editing techniques we associate with modern cinema. Today we take it for granted that if we see a video of a woman turning to look out a window followed by a scene of a car accident, the meaning that is being conveyed is that the woman witnessed the car accident. We suspend our disbelief, even though the two events were shot at different times and in different places. Studying the formal mechanisms through which popular entertainment held the attention of people, Eisenstein learned how to compose screen space and time in montage.

> **Richard:** your picture is very clear - why?
> **CraigCam:** I use a video camera with video card not a webcam
> **Richard:** You are also sending more frames per second

How do we recognize postliteracy when we see it? Although it is early to be taking this question on with any certainty, there are some characteristics that seem to have emerged.

> **Richard:** sounds like it is starting to rain
> **CraigCam:** where?
> **Richard:** no packet losses
> **Richard:** this is Wisconsin
> **Richard:** round trip time is 2 seconds
> **Richard:** what is your location?
> **CraigCam:** Florida, no rain
> **Richard:** you have your audio blocked
> **Richard:** I have seen no packet losses therefore audio should be clear
> **Richard:** 80 percent of the time the audio is very good
> **CraigCam:** better and better

Montage

Given our new ability to assimilate multiple systems of meanings simultaneously, postliterate forms will undoubtedly be a noisy fragmented proposition.

> **Richard:** my nephew just bought a camera so he can talk with Germans
> **Richard:** he is going to school there next fall
> **CraigCam:** I am trying to figure out how to integrate this technology with photographic VR
> **Richard:** is the VR virtual reality?
> **CraigCam:** yes
> **Richard:** what do you mean by integrating it?
> **CraigCam:** I am building software interfaces that start from satellite images and you can zoom in to ground level
> **Richard:** I like to talk with people when I sit down to eat breakfast. People from around the world
> **Richard:** that is what telecam technology has been useful for to me.
> **CraigCam:** breakfast with Europeans, I use it to collaborate

Association Not Explication

Because the new forms that will emerge will inevitably press the bounds of human communication, we will probably see a move away from the shackles of explanation and a move toward the poetic associations of raw human thought.

> **CraigCam:** I want to make it so you can meet people via telecam when you get to the ground level virtual reality nodes
> **Richard:** are you always above or can you look horizontally?
> **CraigCam:** when you get to the ground you can look in all directions
> **Richard:** then we will have to give our longitude and latitude to the program
> **Richard:** some of the router computers have longitude and latitude info
> **CraigCam:** well the spaces are real places but users will not need to be tied to the place
> **CraigCam:** for instance I could be exploring the Miami River and run into you doing the same
> **CraigCam:** does that make sense
> **Richard:** not really, I am trying to think of what it is good for
> **Richard:** I can think of military uses for this
> **CraigCam:** scary
> **CraigCam:** well, it is an artwork
> **CraigCam:** "good for" may not be relevant

Interdisciplinary Collaboration

One of the most poignant aspects of the collapse of modernism was the monocular vision of disciplinary expertise. Everyone was so focused on their particular practice that few had the generalized vision to see the whole belief system crumbling. We will probably not move back to generalization; we have gone too far for that. Holistic perceptions will emerge from disciplinary experts using technology to bring esoteric understanding together.

> **CraigCam:** I go out on location and take a series of panorama still photographs
> **CraigCam:** the sill images are stitched together into one long continuous pan
> **CraigCam:** then pans are linked with hot spots
> **Richard:** the hot spots are .....?
> **CraigCam:** the hot spots are hyperlinks that connect one panoramic node to another
> **Richard:** so then you click on a hot spot and that takes you to another panorama?
> **CraigCam:** yes
> **Richard:** really quite amazing

Virtual Community

Postliterate forms will probably move away from the masterwork of individual genius toward a multidirectional exchange, evoking the community of oral traditions. However, the virtual community will form through common interest, not physical proximity.

> **Richard:** now the round trip time is down to 10 seconds
> **Richard:** so I think video data gets plugged up in my computer then slowly gets sent out
> **Richard:** the round trip time is down to 4.5 seconds
> **Richard:** it could be that something in my computer has been corrupted and if I shut down and reboot it would be fixed
> **CraigCam:** What is your Network setting
> **Richard:** 80 and 80

Process Not Product

The phenomenon of the art object as a commodity to be possessed, bought, and sold, paralleled the development of industrial modernism. Prior to the advent of the canvas, painting was done in permanent site-specific spaces, such as cathedrals or caves. As we move into an information technology paradigm, we will probably experience a shift toward an emphasis on the process, not the product. Postliterate cultural forms will probably develop in a version with no clear finish, a constant state of becoming.

```
CraigCam: are you on a modem
Richard: yes a modem.
Richard: Maximum upload rate for a modem is 28.8
Richard: I will go to sleep now
Richard: I will then reboot and experiment more tomorrow morning
Richard: the round trip time is going slowly up
Richard: its that rate cap that is the problem
Richard: the internet may be too busy right now?
CraigCam: probably
Richard: pacific time is 11:00 p.m.
Richard: the kids should be in bed
Richard: kids play some kind of games and use up the bandwidth in the evening
Richard: but by now most should be asleep
CraigCam: are you kidding
Richard: my kids are asleep
Richard: only the vampires are awake this late
CraigCam: yes, I think that is what we should be concerned about
Richard: I am just trying to understand why the rate cap is so low
Richard: I have to get some sleep - my eyes are beginning to get blurry
Richard: so good night and I will look for you again
Richard: out of here
```

Our commitment to inventing a new form of media invokes Eisenstein's precedent for learning from an entertainment practice how to author in a new medium. We hope to build on Eisenstein's insights and extend them into new media, helping to create a new language and form.

```
Subject: Imaging Niagara
   Date: Fri, 08 Dec 2000
   From: "John (Craig) Freeman" <freeman@uml.edu>
     To: Greg Ulmer <gulmer@NWE.UFL.EDU>,
         Barbara Jo Revelle <revelle@ufl.edu>

Hi Greg and Barbara Jo,

I hope all is well. I have been intending to write you for quite
some time now but I have too much to say for email. So I kept
putting it off, thinking that I would just call. I wanted to
bring you up to date on my activities since I left Florida.
```

The Imaging Place work is taking off at a rate that is both exciting and intimidating. Last year in the midst of the cross-country transition I successfully produced and exhibited Imaging Appalachia at the William King Regional Art Center in Abingdon Virginia. I worked in a region about 40 miles across by 200 miles long, which paralleled the Holston River in the western most portion of the state of Virginia and into Tennessee. The town of Abingdon was the last colonial town before the fabled Cumberland Gap, gateway to the western wilderness. The work included some of the drama of strip coal mining and the controversy of tobacco dependent agriculture. It was funded by the Virginia Arts Commission and included a one week residency.

When I got to the University of Massachusetts, I wrote a successful grant for seed funding to begin work in New England and to implement digital video into the method. Last summer I started producing work along the Merrimack River, birthplace of Jack Kerouac and the US industrial revolution.

Let me explain the video part. As you will recall from Imaging Florida, the Virtual Reality scenes were constructed with individual panoramic nodes. The user could navigate through the scene by clicking on the node markers in the aerial images or jump from node to node by clicking on embedded hyperlinks in the panoramic images. I have since constructed a VR media apparatus, which includes a sturdy tripod on wheels with a quick-release panoramic head and matching digital video head. When I shoot, I position the panoramic camera and shoot the stills for a node. I then swap the still camera for the video camera and roll the whole thing to the next node position with the video camera running. I have completed the interface programming so that when the user pans around and clicks on an embedded hyperlink in the scene, it switches from panorama to video and the camera view floats seamlessly to the next node. Now that I have video running in combination with the panorama it opens up the possibility not only for fluid navigation of the scenes but I can also develop rich narrative within the scenes. For instance, imagine that the user is exploring the virtual space and she or he comes across a person in the scene. If the user clicks on that person, a video tape monologue plays. The person could be a community elder telling the oral histories of the place or an academic expert ruminating theoretically. Although this method would work equally well for non-linear fiction I have remained committed to documentary.

Now that I know that the method is technically possible I have begun a concerted effort to fund the production of new work. I will be working with CEPA Gallery in Buffalo this summer, to produce Imaging Niagara as part of the upcoming Paradise in Search of a Future exhibition in Fall 2001. The funding sources include a New York Foundation for the Arts Curatorial Grant. I will be doing the fieldwork through CEPA's artist residency program and we have written a grant to the GUNK Foundation to sponsor a public art component of the virtual reality installation. I will be focusing on the whole Niagara River corridor. The Niagara River (always seems to be a River) is about 23 miles long and rich in spectacle and history. The possible topics include the first hydroelectric power project in the world, the Niagara Falls daredevils, tourism and the devastating remains of Love Canal.

Last week I was awarded a $5,000 grant from the LEF Foundation in Cambridge for work in this area. I will exhibit the first version of Imaging New England in May 2001 at Mobius, as part of the Boston Cyber Art Festival.

In November, with the help of the Umass Research Foundation, I flew down to Washington DC to meet with Jeremey Roberts, Streaming Media Specialist and David Johnston, Director of Technology at the national offices of PBS. I demonstrated some of the work, including Imaging Florida and Imaging Appalachia and we discussed the possibility of developing the Imaging Place method into a new form of interactive broadcast media, somewhere beyond the current World Wide Web and interactive television.

These are just the first responses to a massive query effort that I have underway. It will include programs like World Heritage Center at the United Nations Educational, Scientific and Cultural Organization (UNSCO).

The Three Gorges Dam project on the Yangtze River in China, Illisu Dam project on the Tigris in Turkey and the Upper Nile in Sudan will all submerge some of the most significant sites of human development in the world. Talk about Problem Tour.

Of course, what is missing from the newer work is the theoretical rigor and the poetic juxtapositions that FRE provided. I have successfully integrated a collaborative component to Imaging Appalachia and New England, working with locals and colleagues, but it turns out that what we had going was unique and very difficult to replicate. Furthermore, I feel that we never finished what we started and that we were on the threshold of succeeding in the goal of inventing a new form of media. I think that I needed some time to retreat to some level of isolation in order to focus on bringing the technical aspects of the method up to speed, but I am now ready to go into full-scale production.

So, I wanted to see if there is any interest in a new collaboration.

What I propose would be to devise a strategy up front, with a manageable workload and achievable outcomes on a closed schedule. Although it was appropriate at the time, our method during Imaging Florida was far too open-ended.

I think that if the theory is valid, then we should be able to graft it over any divination zone. The Niagara River would be an excellent test. It is, in many ways, similar to the Miami River. It is short, contested, historic, widely represented within culture, exploited and it is changing.

I think that the premise that you, Greg, need not go to the zone is still valid. I am most interested in spoken word theory. I am imagining small talking-theory-head video windows that float around in space between the VR scenes.

If we were to have a videoconference like the one we did with Ron a few years ago, I could simply drop a tape in the VCR and I could collect every thing I need from you, Greg. I have access to a similar facility up here on the Umass campus. We would simply need to revisit the issues that were raised in Imaging Florida, and allow me to gather it on videotape.

Barbara Jo, if you are interested, would you consider meeting me for a couple of weeks in Niagara this summer?

Let me know what you think.

- -
John (Craig) Freeman
freeman@uml.edu

http://www.uml.edu/Dept/Art/Freeman/freeman

Response

Randall Bass
Georgetown University

In his book *Out of Control* Kevin Kelley (1995) writes

> There are two extreme ways to structure "moreness." At one extreme, you can construct a system as a long string of sequential operations, such as we do in a meandering factory assembly line. . . . At the other far extreme, we find many systems ordered as a patchwork of parallel operations, very much as in the neural network of a brain or in a colony of ants. (p. 21)

In this latter kind of operation, says Kelley, "What emerges from the collective is not a series of critical individual actions but a multitude of simultaneous actions whose collective pattern is far more important. This is the swarm model." Yet, Kelley admits:

> These two poles of the organization of moreness exist only in theory because all systems in real life are mixtures of these two extremes. Some large systems lean to the sequential model (the factory); others lean to the web model (the telephone system). (p. 21)

The models of intellectual work in the humanities implicated in this volume operate along some similar continuum. At one end is the traditional model of the individual scholar, laboring alone, in a system that engenders and rewards individual invention. Problems are best solved by the power of one, writing into and out of conversations with other lone workers. This is the traditional environment

363

that makes the university work environment (intellectual work) one of the least cooperative or collaborative work environments one can imagine. This is the environment driven by what David Damrosch (1995) calls the "norms of alienation and aggression" that are "still enshrined in the university and are the products not of nature but of cultural choices, and archaic ones at that" (p. 106).

At the other end of the continuum, as part of the collective imaginary of this volume, is an image of intellectual work that looks more like a swarm. Here problems are best solved by lots of people working—at the power of ten—on similar and related problems without a clear sense of hierarchy or order. The boundaries between "my work" and "your work" are porous at best; problem solving and experimentation is communal and dynamic. Although it appears to be disorganized and chaotic, such a system is, in Kelley's (1995) terms, a self-organizing system, or what he sometimes calls a vivisystem. Such systems, Kelley argues, can be characterized as having four distinct facets that give them their character:

- The absence of imposed centralized control.
- The autonomous nature of subunits.
- The high connectivity between the subunits.
- The webby nonlinear causality of peers influencing peers. (p. 22)

Based on these criteria, it is pretty clear that faculty work in the humanities is a hybrid version of the two kinds of moreness, clearly possessing some of these qualities but not all. Even in the traditional model of intellectual work, scholars are actively connected to each other as subunits, although at the same time functioning in a system that could be characterized as a "webby nonlinear causality of peers influencing peers." Yet there seems to be something almost contradictory about practices that fuse highly individualized and discrete work products with a webby self-organizing knowledge-making network. From this peculiar combination of the two kinds of moreness that characterize humanities work in the academy, an aggregate portrait of electronic collaboration in the humanities emerges, where there exists tendencies that both accentuate traditional practices and antagonize them.

In each of the essays in this section, the possibilities for electronic collaboration in the humanities implies some already existing tension that defines the humanities and its other. The spectre of otherness (the humanities' other) is usually expressed as an opposition that electronic collaboration in the humanities is suited to overcome. A harvested list of implied dichotomies and tensions from these essays might look like this:

- individual scale/large scale
- theory/empiricism
- classroom literacies/workplace literacies

- the academy/the marketplace
- critical thinking/instrumentality
- explication/association
- disciplinary order/interdisciplinary chaos
- physical world/virtual world
- literacy/postliteracy
- disengagement/engagement

The relationship between these pairings or tensions in the context of electronic collaboration in the humanities is not merely complicated but ironic. One dimension of this irony can be captured by the last pairing—engagement and disengagement—because it signals an ambivalence within the humanities of simultaneously turning toward the world and away from it. We might call this the paradox of engagement, and it occurs to me that these discussions of electronic collaboration in the humanities make this paradox evident and meaningful in new ways.

Consider two important points in this section's essays side by side. The first of these comes in Paul Morris's essay "Critical and Dynamic Literacy in the Computer Classroom: Bridging the Gap Between School Literacy and Workplace Literacy." Morris explores differences in both practices and value systems as they manifest themselves in classroom instruction on the one hand and workplace activity on the other, exploring "the gap between theory and practice . . . between schoolhouse literacy and workplace literacy." In his exploration, Morris makes the distinction between critical literacy (a term he gets from Mike Rose) and dynamic literacy. Critical literacy is related to critical thinking and central to most undergraduate curricula. It is—to put it simply—what we value in the academy. Dynamic literacy is something like critical literacy in action. Dynamic literacy comes from the use of literate skills in practical situations, with a felt sense (perhaps unanalyzed) of the power of language in action. As Morris states:

> Recognizing the power of language as a tool to transform their lives, the dynamically literate are able to change the way they use words to suit their purposes. Whether using texts, computers, radio telescopes, or the spoken word, they can effectively integrate their use of language with their need to belong to and to operate within a given community.

The distinction matters, according to Morris, because of the gap between what the academy values and what the workplace values. According to Morris, a key objective for the future—and the practices of collaboration (electronic and otherwise) in the humanities—is to bridge this gap and find ways to connect critical to dynamic literacies.

Contrast this approach to Timothy Jackson's in his essay "Imagining Future(s): Towards a Critical Pedagogy for Emerging Technologies." Jackson opens with this opposition:

As academics, we are in the consciousness business. This is the end game of our efforts, if we are serious about the impact of ideas onto the body politic as opposed to purely instrumentalist motivations driven by market forces which ultimately produce supply-side pedagogies. This is the essence of what it means to *profess*.

In emphasizing the centrality of critical pedagogies for future uses of electronic technologies in education, Jackson posits an opposition between academic uses of such technologies and resources that engender "empowerment brought about by actively constructing learning materials" and more "instrumentalist uses" complicit with the marketplace. In short, for Morris, instrumentality is what makes critical literacy viable and the humanities valuable; for Jackson, instrumentality is the antithesis of critical thinking and thus a kind of antihumanities.

At some finer grain of conversation Jackson and Morris's positions probably could be easily reconciled; nonetheless their opposition points to a conflict intrinsic to the humanities, one that crucially informs (or deforms) any discussion of electronic collaboration in the humanities. The attraction to the political is often accompanied by a disdain for the practical: workers but not workplace, economics but not marketplace. This bias against use would not be so vexing if it didn't seem to have both traditional and progressive versions. Traditionally, it exists in long-standing claims to the value of liberal education and humanistic conversation but the intellectual insistence that their value in application only endures if it not weighed down with the mundane necessity of proving their applied value; progressively, the bias against use manifests in critical and disciplinary practices that advocate engagement in politics but primarily as theory and intellectual action. Practical action, let alone engagement in practical marketplace activities, is often viewed in opposition to critical or political intellectual action in the academy.

This bias comes in part from the deeply held belief that one cannot be of the world and properly critical of it simultaneously. But that only captures part of the resistance. These inherent divides at the heart of electronic collaboration in the humanities point to the ironic relationship between the nature of community and work in the humanities. In short, traditionally to participate in the community of the humanities (i.e., in the academy) means to engage in solitary, not collaborative, work. Collaborative work—except in ways narrowly conforming to traditional research models—always runs the risk of being outside the norms of community practice. The ill fit between the collaborative and the communal in the humanities is, to say the least, ironic.

I am taking my implicit definition of community here from Phil Agre's (2000) suggestion that there are two competing models that structure the university: the commodity and community models—whose countertendencies are particularly emphasized in the increased technology saturation (and recapitalization) of higher education. In Agre's scheme, the community model sees the university as an "idealized microcosm of society," which upholds and promotes "the norms of collegiality, provides a forum for debate, and maintains structures of democratic governance" (p. 5). The community model regards learning as being as much about immigration and acculturation into communities of practitioners that share ways of seeing and doing work in the world. By contrast, the commodity model sees the university as a competitor in the marketplace, of which educational services are the principal commodity, with students as the customers. Traditional universities, to the extent that they operate on the commodity model, have considered their services as bundled, valuing the package of academic opportunities, social life, and athletics, maybe even security, spirituality, and a nurturing environment, all as a bundle of services. With the advent of the Internet and distributed learning environments, universities increasingly will move more toward practices of unbundling, where learning as a product and outcome is marketable in ways separate from the other parts of the package. Although in the minds of many, the commodity and community models compete in the university environment, they have always existed in tension and will likely continue to do so. Even without the intervention of technologies such as the Internet and practices such as distance learning, commodity and community influences check and balance each other in many realms of higher education—in the development and focus of undergraduate students, in intellectual property policies, in transfer and articulation practices, and in those activities that both encourage and limit collaboration.

One of the key features of the community model is the way the university is designed to promote and manage membership in multiple communities. Agre (2000) puts it:

> It is unclear, and usefully so, where the boundaries of the university community lie. Is each university a community unto itself, or do communities form along disciplinary lines, or do the universities of the world form a single cosmopolitan community? All are clearly true to some degree, and the institution is designed to manage this multiplicity. (p. 6)

When one looks across the current landscape of collaboration practices in the humanities, they are very much shaped by the tensions between commodity and community. In the humanities, the community–commodity split bears an ironic relationship to many of the pairings (terms in tension) listed previously. In the humanities community activity is deeply identified with individualistic, detached, and disengaged work. Similarly, where one might think that collaborative activities (large-scale sharing, knowledge production) would line up on the community side, they are often associated with commodity activities. Arguably, the majority of collaboration in the humanities now takes place primarily in the context of commodity activities: coediting of books, journal issues, textbooks, CD-ROM products, and so on.

Certainly that was most evident in a recent study sponsored by the National Initiative for a Networked Cultural Heritage (NINCH). Known as the Building Blocks Project, the NINCH online survey (2000) engaged more than 100 scholars, teacher, curators, and librarians in reflections on their mode of working with primary and secondary materials and their broader intellectual and professional practices. One section of the questionnaire specifically addressed collaborative practices. The collaboration question reads this way:

> Do you collaborate with others in your research, teaching and public outreach? If so, describe the division of labor? You might consider the following questions in your response:
>
> - If you collaborate, do you and your partner(s) work on the same material?
> - What do you most value about collaboration?
> - What do you find are the obstacles to collaboration?

As the "History" field report summarizes the responses to the collaboration questions, "Many respondents reported working collaboratively on textbooks, journals, websites, and databases. Many fewer reported doing cooperative work with sources or cooperative writing projects." To be fair, the other area of collaboration cited was teaching, where team teaching and the occasional collaborative construction of technology-based curricular materials (usually in the context of external grants) were the only other common venue for collaboration. But teaching, like coauthoring a published work, is a commodity practice, even if in actuality these collaborations demonstrate the viability of hybrids between the commodity and community models. Like the open-source programming model that Fanderclai takes up in her essay in this volume—and as Agre cites as an example in the marketplace where the community model invades a commodity context—the community model tends only to thrive in "those niches where incentives are structured to encourage sharing" (Agre, p. 4). This is seen repeatedly in the NINCH surveys. For example, the "Language and Literature" field report sums it up this way, "the 'lack of institutional commitment to collaborative work [and] general culture of isolation in scholarly work in humanities' were identified as serious obstacles to collaborative work."

This latter link between institutional incentives for sharing and the "culture of isolation in scholarly work" has origins that are as much institutional as epistemological. That is, collaboration in the humanities is often present at the point of production, synthesis, and dissemination (even the NINCH survey groups Teaching with Publication, Public Programming and other forms of Dissemination). But throughout the NINCH survey responses, the scholar's work is largely portrayed as private and individual. One historian in the questionnaire captured this intellectual model clearly:

> Historical research consists of an individual mind engaging directly with documents and then ordering and assessing them. The notion that two people can look at the same materials and come to different conclusions is an accepted feature of historical work—hence all the books on the same topic. This means that collaborative work

may take the form of common data collection but does not lend itself readily to communal analysis.

Another respondent, a librarian/scholar, put the same sentiment even more succinctly: "As a librarian, most of my work is collaborative, as a scholar, most is not." These kinds of responses corroborated the findings of a 1988 study about the work practices of art historians, observing that "truly collaborative scholarly effort was uncommon among art historians" due mostly to "institutional encouragement of narrow specialization and concomitant professional competition" (Blackwell, Beeman, & Reese, 1988, pp. 76–77).

There is however a common thread throughout the NINCH and art history studies of an appreciation of collaboration for its capacity to bring differing dialogic perspectives to bear on a particular problem. This context of collaboration lies somewhere along the continuum of two kinds of moreness, somewhere between the isolation of the individual scholar and a vision of open-source knowledge-making. This vision of collaboration at the heart of the humanities builds closely on the idea of dialogue as a primary context for faculty work and is shaped by the peculiar nature of problems within the humanities.

Dialogue in the humanities (and all its euphemisms and dysphemisms) is a self-organizing system without a center, beginning, or end; it is a disciplined network of interstices and vectors. Of course, the academic conversation has both structure and hierarchy, governed as it is by institutional categories of status, professional protocols of discourse, and the mechanism of filtration, such as peer review. But these are all subordinate features of the larger entity: the ongoing intellectual dialogues themselves. And the ways that humanities faculty structure their professional engagement with multiple dialogues determines the arrangement of their membership in various communities.

Faculty do not just belong to multiple communities at once (as Agre points out) but operate in different communities differently. One categorical distinction with which to understand these differences may be the "two types of work-related networks" explored by John Seely Brown and Paul Duguid (2000) in their book *The Social Life of Information*. They distinguish between networks of practice, made up of people who share knowledge and practices (and often professional identity), and communities of practice, where people engage in common work.

In networks of practice, knowledge is "easily assimilated" and can be shared widely. The Internet's first powerful applications to broad audiences were at the service of networks of practice, inside and outside the academy. Although networks of practice are excellent for their reach, taken together, "such social systems don't take action and produce little knowledge" (Brown & Duguid, 2000, p. 145). Nearly all scholars, curators, librarians, and educators self-consciously belong to one or more networks of practice. Communities of practice, as used by Brown and Duguid (and adapted from work by Lave and Wenger) are "subsections of these larger networks of practice." "Communities of practice . . . are relatively tight-knit groups of people who know each other and work together directly." As part of this work, often carried out with some local, face-to-face interaction, communities of practice are constantly engaged in "negotiation, communication, and coordination"—in other words, activities that far exceed sharing knowledge and best practices (pp. 142–143).

This distinction between networks of practice and communities of practice seems useful to the question of collaboration in the humanities because it seems arguable that only in the latter do people tend to truly collaborate. In communities of practice, as Brown and Duguid (2000) frame it,

> the demands of direct coordination inevitably limit reach. You can only work closely with so many people. On the other hand, reciprocity is strong. People are able to affect one another and the group as a whole directly. Changes can propagate easily. Coordination is tight. Ideas and knowledge may be distributed across the group, not held individually. These groups allow for highly productive and creative work to develop collaboratively. (p. 143)

Every community of practice has its own way of working and its own particular equilibrium between dialogue, difference, and productivity. Here again the NINCH surveys make visible in revealing and fascinating ways common underlying assumptions about work in the humanities. In the survey responses, many general activities of collaboration really belong to networks of practice:

- Collaboration enables people to bring rich resources together for the benefit of a much larger group.
- Most of the materials I share with my colleagues are available online, so it is quite easy to share them.
- I would like to see more collaboration between institutions for data collecting—images from early travelers accounts for example, shared across a range of institutions—along with occasional symposia bringing together teachers and students from the institutions working on similar topics.

In contrast, other more detailed descriptions of collaborative work in the NINCH survey more neatly fit the community of practice definition of shared resources, knowledge, and practices—for example, the following:

> I have collaborated in the past on research projects (both a monograph, and an edited personal narrative), and in teaching. Each collaboration is different. In the monograph, the two principals consulted extensively through the data collection phase, the design of the monograph, and following the mutually agreed upon outline, each wrote each chapter, which were then combined in still more negotiation. This project was based on a statistical database whose assembly we directed, and we each had access to the results of the statistical analysis. In the personal narrative, we divided the translating and editing responsibilities in half, but exchanged and commented on

each other's work . . . The great benefit of such collaboration is an ongoing dialogue and the stimulation of an immediate sounding board for ideas as well as the need to clarify and defend positions if they don't agree with those of the collaborator. Collaboration is very stimulating and can even be exhilarating.

Yet another response puts collaboration in the context of a dialogue not only among individual scholars but also among intellectual or ideological positions:

Historical scholarship remains wedded to a model of the independent researcher and thus it does not normally encourage collaborative efforts. Of course, historians are also notoriously independently minded in their research agendas and their interpretations of source material and thus it is often difficult to achieve true collaboration. In my experience, it seems that collaboration occurs more often as a product of opposition to an established orthodoxy or a certain historiographic position rather than as a result of a common agenda.

The irony here, simply put, is that in the traditional scholarly structures of the humanities, communities of practice are often based on the antithesis of a communal model. Collaboration in the humanities is more likely to take place at the point of dissemination (or reproduction) of knowledge than at the point of research or creation. The reason that true collaborative problem solving (knowledge production) is so rare in the humanities is not because institutional design militates against it (in many ways it doesn't) but because of the absence of soluvable problems and the bias against use.

The visions of collaboration in these essays nearly all spring from the attempt to identify, and in some cases redraw, the contours of solvable or authentic work in the humanities. In a pedagogical context, Paul Morris explores a variety of collaborative learning techniques that depend on teaching students to work together to generate meaning and to complete assignments that are more instrumental or real. Collaboration in these classroom assignments becomes a way of bridging the divides between instrumental and critical, theoretical and empirical. Through pedagogical design, collaboration leads to knowledge making for students through an amalgam of dialogue, instrumental authenticity, and the simulation of tight-knit communities of practice in the classroom.

Interestingly, the model is nearly identical to John Freeman's discussion of the Imaging Florida project conducted by the Florida Research Ensemble. In the case of the Florida Research Ensemble, "The group was brought together by a common interest in electronic technologies and a desire to explore new methods to deliver our respective research and creative production to audiences beyond predominate cultural and educational institutions." The Imaging Florida project brought together practitioners from ostensibly different fields, or networks of practice: documentary photography/digital video, cultural theory, media and performance art, architecture and urban planning. The group functions collaboratively through latent shared assumptions about culture and representation made manifest and functional through an authentic set of problems—specifically, the documentation of cultural places:

> The goal of the Imaging Place method is to document sites of cultural significance, which for political, social, economic or environmental reasons are under duress, at risk of destruction or undergoing substantial changes. This includes historic sites as well as sites of living culture which are being displaced by globalization and the collapse of industrial modernism.

Their capacity to work as a community of practice (a tight-knit group of people engaged in a knowledge-based project) despite belonging to traditionally different networks of practice (loose alliances by professional identity, defined by common practices) depends on their ability to work with, through, and inside multiple technologies (collaboration and representation) and to communicate through what Freeman calls postliterate forms. These postliterate forms are as much dependent on the nature of the technologies as on the rhetorics they use, open to the logic of associative argument and poetic meaning in documentary contexts. The work of the Florida Research Ensemble, as described in the Freeman essay—just like the various critical and collaborative practices in the others—all portray new electronic technologies as a means for bridging traditional divides without resolving their underlying tensions. But can these potentially transformative practices—linked to new technologies, rhetorics, and community practices—have their transformative effect without a fundamental (and highly unlikely) transformation in the notions of work and problems in the humanities?

This is the question that kept nagging at me as I read and reread Tari Fanderclai's analogy of the Linux programming community as a model for how large groups (or communities) of researchers might "learn to collaborate on a much larger scale than ever before." Fanderclai admits that in the humanities "our research traditions and reward systems can be barriers to collaboration. Humanists tend to value individual products and great minds; the greatest rewards are for individual works." But at the same time, Fanderclai points out (citing Eric Raymond) that what is key to sharing in the Linux community is the incentive to share. "The contributors are volunteers, motivated not by money or direct career advancement but by the chance to work on interesting problems, helping to improve software that answers their needs." Operating something like a self-organizing system, the bazaar-style research and development model of the open-source community exceeds the efficiency of even the largest in-house staff because it brings together "hundreds of people who all have instant access to the code that's already been written and the knowledge already assembled," and in that way "each volunteer can build on previous work and help to move the project forward." Indeed, the programming model of the open-source community blends the best of both networks of practice (especially in terms of reach, knowledge sharing, rapid dissemination) and communities of practice (the capacity to create new knowledge). Although the "key to sharing in the Linux community is the incentive to share," the incentive to share generates from the prospects of making something work better, more efficiently, practically.

In trying to construct an analogy to the open-source community for scholarly work in the humanities, Fanderclai suggests, "Imagine, though, the speed at which a research project could progress, given a large group of researchers working on various areas of a problem and sharing their findings as they work." In part, the incentive against this way of working is in our own (internalized) sense of value:

> In truth, we know that whatever research project we take on could as well be taken on by someone else in the community. We don't work on problems in isolation because we believe that we're the only ones who can solve them—we work on closed projects because our traditional value system says that we have to solve them first to establish our value as researchers.

Although this is an astute point, the word I keep sticking on is *solve*.

I remember many years ago hearing a senior scholar in English make the earnest assertion that "There is a lot of very important work to do in the 18th century." Only in the academy, and especially in the humanities, could that statement make logical, let alone professional, sense. The dialogue is the project. We only work on problems if we can't solve them. From this context, it is hard to imagine widespread agreement for Damrosch's (1995) following claim:

> [The academy's culture of solitary and alienated work] can be changed if necessary—if, as I believe, more and more topics would benefit from sustained discussion among people with different expertise and perspectives, while relatively fewer topics are still best worked through by single scholars meditating on their favorite authors. (p. 107)

There are some clear examples to tangible projects in the humanities—such as the multiculturalism movement, ethnic studies, and women's studies—that have advanced particular agendas successfully, bridged instrumentality and critical practice, and crossed classroom and community lines in meaningful ways. Potentially, another such exception receiving much attention nationally (indeed internationally) is the scholarship of teaching and learning. This is the context for Peter Sands's essay on bridging theoretical and empirical research in the field of computer-mediated composition. Sands asserts that digital technologies enable researchers who study writing processes to work on bigger problems together, pooling their data and scaling up their questions. Sands argues that written texts—as the object of composition instruction research—are particularly susceptible to the use of computer-based tools for the broad analysis of patterns, both qualitatively and quantitatively. Fundamentally, such technological tools are potentially transformative of the very nature work in the scholarship on teaching and learning because they would allow humanities researchers to engage empirical and quantitative methods (admittedly, traditionally suspect domains in the humanities) for advancing theoretical arguments. Rather than reifying the theory-empiricism split, working with large amounts of systematic data adds "validity to the researcher's assertion that she or he has observed patterns, and that the interpretive act—the theorizing—is justified." Sands concludes, "It is precisely this continuum from close observation at the level of the text to theorizing of meaning that is needed in research in computer-mediated composition."

This new work would require not only a new attitude toward materials but also new skills and methodologies, especially learning how to use the new technologies and their concomitant meta-languages in the context of text-analysis practices. Like the postliterate forms used in the Imaging Florida project, these technology-based meta-languages imply more than technical empowerment: They imply a kind of functional transgression of some of the most basic biases of the humanities, yet in the name of the most central critical, progressive values:

> Behind the kind of research I advocate here is my belief that the computer can both a site of resistance and an entryway to other sites of resistance, that it can and should be put to use in the service of a liberatory pedagogy, and that scholars and teachers of liberatory pedagogies must situate themselves within the powerfully enabling discourses of empirical research and argumentation too often woefully absent from their recent work. . . . Accordingly, I present here not a fully formed research plan or instructions that can simply be transferred, templatelike from here to there, but a call for one particular kind of research, supplemented by examples which can provide more empirical data usable for theory building and can be used in a liberatory pedagogy.

Here, as in the other essays, out of the analog divides comes a vision, as each of the essays in this section holds out some comparable hope: that electronic collaboration in the humanities could be ultimately transformative. The question persists over the relative potential of humanities fields and practices to absorb effects that contradict their most deeply held biases: the solitary nature of communal work, the bias against use, and the abstract nature of insoluble problems. Being structural and philosophical, these phenomena provide causes for the divides, and it is not immediately apparent how electronic technologies by sheer acceleration or multiplication can address them. In fact it may be that the deep structural biases become even more entrenched in a digital age. Greg Ulmer frames it this way in a sobering sentiment in the Freeman essay:

> Actually the fit between the world and the disciplinary manner of knowing the world is not all that good. The perfection of literacy in print produced an information explosion that schooling never has responded to completely. Specialization is this institutional form of analysis, and it guarantees its own inadequacy by exacerbating

the information overload (producing ever greater bodies of "literature" about increasingly fragmented parts of what is).

If the printing press gave rise to excessive specialization, then there's reason for concern that digital information environments will only continue that effect. Given that prospect, it is possible to see electronic collaboration in the humanities not as the future's mainstream way of working but as a set of meta-methods that only work because they work against the grain.

This possibility resembles a suggestion by Pat Hutchings (2001), exploring whether the right home for the scholarship of teaching and learning is disciplinary practice or its own interdisciplinary field. On the one hand, she says, "field matters" and clearly "it follows then that the right 'community of practice' for scholars of teaching and learning is peers in their home discipline." Yet, she concedes, at the same time, "some of the most interesting and pressing issues about our students' learning live in the 'trading zones' between fields," and therefore it might be some discourse as a separate interdisciplinary field that most enlivens the scholarship of teaching and learning. Hutchings also sees there is possibly a third option beyond the choice between disciplinary and cross-disciplinary specialists:

> an aspect of the work of many faculty who undertake it as amateurs—in the highest sense of that word—eager to find occasions and mechanisms to examine their work as teachers, to share their work with colleagues, and to bring their habits, commitments and intelligence as scholars to their work as educators.

What is appealing about Hutchings's third way is that it neither ignores the antagonism (the disciplinary divides) nor proposes a single set of practices that aspires to resolve them. Hutchings's third way wears its irony on its sleeve. And it may be that electronic environments and tools—with their capabilities to change space, time, and scale—are a viable, promising means for doing the same, living the irony at the heart of collaborative work in the humanities. In the end, these essays in this volume seem less collectively to point toward some normative set of practices than to portray new and seductive models for working against and across dominant practices. This does not diminish the power of the vision here. In fact, quite the opposite. Given all the work there is to do in the 21st century—not to mention the 18th—it may be just what the doctors ordered.

Afterword

Anne Ruggles Gere
University of Michigan, Ann Arbor

Item:
Sharing Endnote bibliographies between us will be tricky once we have a document that we're both trying to work on. For now, the guru suggests that we have one master database, maintained most likely by me (since I have only one computer, and I have the guru to help.) In this case, both of us can add references to our respective references libraries as we go along. Once a month we would combine the lists, thus updating the master copy. To do this, you would need to send me your current reference library. I would then import it into mine, updating the master, which I would then send back to you.

Personal communication, September 2001

Item:
Ok gang, it's true that there's a program that will allow each of us to work on the proposal. Using this program, we can pull up a copy of the proposal in our individual offices, make changes, and then download the revised version. It's located in AFS space, and when we meet tomorrow I'll show you how it works. I see two problems with it—Only one person at a time can work on it at any given time and the program only works within the building, so none of us can work on the proposal from home.

Personal communication, August 1999

Item:
We have received your inquiry about listing co-investigators on the application form. It is impossible to list more than one person as Principal Investigator because the computer program currently used by the Division of Research Development and

Administration cannot accommodate more than one name on that line. Please send
us the name of the Principal Investigator for this project.
"Division," University communication, January 1991

So what is left for an afterword to do? The authors included here have laid out a
rich variety of theories, observations, and strategies that illuminate the ways technology shapes and is shaped by collaboration in the humanities. The respondents
for each section pull together the issues raised by individual chapters, showing
how they all contribute to the theme of the section and to the larger questions
raised by the book itself. Surely, dear reader, you do not need more synthesis or
summary. What may be useful is a look toward the future, a discussion of what
this book tells us to consider as we move onward in the quickly changing world of
technology. I think this can be done best through an interrogation of the terms that
remain relatively unexamined within these pages. I mean terms like *collaboration, humanities*, and *community*. But first I want to trouble one of the central assumptions of this collection.

As I read the chapters included here, a dominant, but mostly unarticulated assumption is that technology fosters collaboration among humanists. Coming out
of a paradigm that, as Paoletti, Sies, and Jenkins observe, privileges "the lone humanities scholar whose work is focused on a single period or text or person," humanists are grateful for and happy to celebrate a medium that helps us to work
together. I share this perspective and take real pleasure in the ways technology enables me to collaborate with colleagues. I think back to my first PC, a loaner from
IBM when Big Blue was first attempting to convince professors that computers be-
longed in faculty offices. My friend Bob Abbott, a statistician, somehow arranged
this loan and introduced me to the rudiments of word processing. Subsequently,
Bob and I wrote several publications together, a process made much easier by the
revising capabilities of our new computers. When HBJ Writer was still known as
WANDAH (Writers Aid and Helper), I agreed to serve as a Beta Test Site—even
though I wasn't sure what the term meant. Watching my students use WANDAH,
thinking about how the invention programs could be improved, and sharing insights
with other writing instructors helped me find ways I could improve my own composition classes. Several of the articles that now appear on my list of works published
were composed collaboratively via electronic means. Even this afterword owes its
origins to many e-mail exchanges, electronic attachments, and a much more advanced word-processing program than the one I used in 1982. Still, however, I
want to question whether technology always aids collaboration.

Each of the items listed at the beginning of this chapter represents a moment of
frustration when technology thwarted my desire to collaborate. I am just starting a
new book on Native American teachers, so it seemed like a good time to learn
Endnote, a program that could help me manage the hundreds of citations I am collecting as I read and visit archives. Cari, a graduate student who shares my interest
in Native American teachers, has been working as my research assistant, and

when I mentioned my interest in using Endnote, she said she'd been intending to learn it also, so we agreed to work together. Fortunately, she has a partner who knows more about computers than the two of us combined, and he agreed to serve as our guru. Early on, Cari and I agreed that we would like to build a joint list of references that would incorporate our overlapping reading lists. We assumed that Endnote would enable us to do this smoothly, but as Cari's message indicates, it will be a cumbersome process. The technology will not allow us to collaborate as we would like.

A few years ago several colleagues and I were working on a large proposal for external funding. The narrative alone was more than 50 pages, and then there were all the appendices, charts, and budgets. Each of us brought slightly different expertise to the project, so we agreed to share the writing tasks, which meant that we each needed access to the developing document. When we learned that our division maintains a program that enables several authors to access and work on a single text, we were delighted. We figured that as we got closer to the due date, we could sit at our respective computers and insert new paragraphs, revise sentences, and check references. When Stan checked on the details, however, we quickly realized that our options were more limited. The technology immediately available to us would not support the sort of collaboration we had in mind.

My university, like many research-focused institutions, has a highly efficient and bureaucratized office for processing grant proposals. When a colleague and I worked collaboratively on a proposal several years ago, we asked this office to tell us how we could both be listed on the "Principle Investigator" (PI) line because the university rewards those who serve as PIs. Someone from the Division of Research Development and Administration (DRDA) quickly let us know that our request could not be granted because the program that generates all the internal documents was set up for one name only. Even when we appealed, arguing that neither of us could rightfully be designated Principle Investigator, DRDA remained inflexible. As far as I know, the line still accommodates only one name.

I could offer more examples, but you get the idea. Even though technology has helped me participate in many types of collaboration throughout my academic life, it has also presented obstacles. Programs that will accept only one name on a given line, protocols that work only within the range of a given server, and research tools designed without coauthors in mind all disappoint as much as they support. Corrigan and Gers, two authors included in this collection, make a similar point when they explain that the particular e-mail software used had a negative effect on students' experiences in a collaborative project. Instead of facilitating the collaborations planned by their instructors, the technology made it more difficult. As we look beyond this book to larger issues surrounding electronic collaboration in the humanities, I urge that we acknowledge the frustrations and limitations as well as the capacities and possibilities.

Related to this is the issue of agency. Sure, I am as likely as the next person to utter sentences such as "it doesn't like that" or "it thinks I have the keyboard plugged in" when I'm trying out a new routine on my computer. But I get nervous when I hear statements like "technology will lead us . . ." or "electronic communication requires" because it sounds like the computer is developing a mind of its own. The term *technological determinism* has been used to describe the ways we assign agency to electronic tools, and I think it's useful to keep reminding ourselves that it's dangerous to think that computers can think as we do or that they can determine our behavior, particularly when we are discussing the impact of technology on disciplines where it is less common, particularly when the increasing capacities of technology promise to seduce us. In this regard, Knowles and Hennequin offer a helpful caution when they say, "While we cannot argue that using technology automatically causes collaboration, our classroom experience demonstrates the potential for technology to encourage collaboration." Equally helpful is Jackson's observation that "technology only has agency when we abdicate our own, since our technologies—like our metaphors—are the vehicles of our dreams rather than the destination of our journeys."

Future discussions of electronic collaboration in the humanities need to proceed cautiously by acknowledging drawbacks and avoiding technological determinism. Participants in such discussions would also benefit from careful thinking about their own central terms. Inman, Reed, and Sands note in the preface to this collection:

> The nature of the humanities itself has been threatened with fragmentation beyond reasonable assembly and, at times erasure from educational memory, as its core values have diffused widely and diversely enough that they can no longer be easily recognized.

If we assume this is an accurate rendering of the current situation—and I do—then it seems reasonable that questioning of the term *humanities* needs to be an ongoing project, especially as we consider its relationship to technology. Inman, Reed, and Sands claim that "interdisciplinarity seems to be a hallmark of the contemporary humanities," and chapters like those by Freeman and Rickly demonstrate the truth of this claim. As Freeman shows, technology facilitates cross-disciplinary and, even, international exchanges among scholars. Rickly's examination of computer-mediated communication in writing across the curriculum programs demonstrates how such programs can foster discussions that reach across the boundaries created by individual disciplines.

When Inman, Reed, and Sands enumerate the representatives of the humanities included in this collection, they include the following: literary studies, writing studies, media studies, history of science and technology, women's studies, information science, American studies, literacy studies, technical and professional communication, graphic art and design, and communication studies. This list demonstrates how much the humanities have been permeated by varying versions of cultural studies. It also invites comparison with other lists. For example, the

National Endowment for the Humanities (1998) lists more than 85 disciplines on its application form, dividing many into large categories, such as languages and ethic studies. But this arbiter of the humanities prefaces the list with the following statement: "The listing is not comprehensive and is not meant to define the disciplines of the humanities." One could also consider the lists of departments, programs, or both, given the humanities designation by various colleges and universities. Such lists would differ from one another and from the areas represented in this book. And then there's the issue of methodology. As the chapter by Sands makes clear, "gaps between different kinds of researchers in the humanities are very real." Sands goes so far as to describe the humanities as functioning like the two cultures of science and the humanities described by C. P. Snow and F. R. Leavis many years ago. These varying conceptions of the humanities suggest that our future conversations should include more attention to what we mean when we use this term.

Community is another term that is used frequently in this collection, and it, too, merits questioning. Several of the authors included here consider community, suggesting various meanings and implications. Carlacio, for example, calls easy assumptions about technological communities into question by looking at the economic divide that technology exacerbates. In examining democracy, along with its circulations and capacities for obscuring, Carlacio points to gender inequalities and critiques the techno-literacy rhetoric that assumes technological literacy can bridge socioeconomic differences. Sewell, in contrast, takes a relatively unproblematic view of electronic communities, focusing on features—such as location, shared interests, shared property, interaction, communication, obligation, and emotional connection—that characterize them. Gajjala and Mamidipudi push the discussion of community in a very different direction by considering the ways the modes of production shape communities. In probing the relationship between the online and the everyday, the local and the global, the analogue and the digital, they expose the power differences that mark various communities.

Authors take conflicting positions on the questions of how and if technology shapes community. Trupe argues that technology replicates rather than transforms communities. She writes: "It appears that online communities reproduce dysfunctional group behaviors as well as functional ones, codependencies as well as mentoring, conflicts based on race and class as well as appreciation of multicultural diversity." She also notes that assessment exerts a powerful effect on classroom communities, often devaluing collaboration. Corrigan and Gers take a similar position when they claim that students bring their own literacies to any technologically based project. They may modify them in response to new demands, but the core remains untouched. Fanderclai and McComas, on the other hand, each focus on the ways technology contributes to a community's functioning. For Fanderclai the key lies in open-source software, which provides considerable room for innovation because it is distributed free along with its source code. The Linux community described by Fanderclai developed what is called a bazaar-style community, one in which multiple agendas and approaches circulate at great speed and to great effect. According to McComas, a MOO can create a community that enables transformative learning. In particular, the play that occurs in a MOO creates conditions in which participants feel safe, experience student-centered learning, and develop alternative and often critical insights. Reading about these different perspectives convinces me that we need further conversations about what we mean when we say "community." And then, of course, we'll need to keep thinking about how communities and technologies interact.

Of course the most commonly used term in this collection—*collaboration*—is also the most complicated. This collection is not alone is using this word to mean a variety of things; many scholars in composition studies have taken to using collaboration to connote several different kinds of processes, and this collection echoes that variety. Morris sees collaboration, specifically collaborative writing, as central in the project of closing the gap between school and workplace literacies. For Morris, computer technology offers a way of overcoming student resistance to collaboration. Paoletti, Sies, and Jenkins, for example, describe collaboration as a three-stage process that engages students in discovery, peer editing, and moving beyond the classroom. In addition to describing it as a valuable contribution to student learning, they also point to the value of collaborative scholarship among faculty. For Reed and Formo, collaboration means composing a single text together (even though it means authors supporting one another to produce multiple texts to several others included here), but because of the interactivity and textual innovation enabled by electronic communication, they include their computers as a third collaborator in the model they describe. Inanimates also figure in the way Kibbee and Szylowicz portray collaboration. For them the library, in particular an archival collection, becomes part of the collaborative experience. Individuals and institutions thus become linked in collaboration.

I notice some absences in these discussions of collaboration. For example, the range of participants imagined or described is relatively limited. In my own work, some of the most fruitful collaborations have united university and secondary school faculties on common projects. Technology, including videoconferencing and listservs, allowed us to sustain conversations begun during intense summer workshops, to encourage communication between our two populations of students, and to access archives that informed our teaching. Similarly, I have seen how technology fosters university–community collaborations. Service-learning projects that provide writing assistance to community groups, after-school tutoring projects, and writing workshops are all enhanced by technology. We who value the humanities ignore its extra-academic manifestations at our peril.

Two of the most interesting discussions of collaboration focus on the issue of time. Prell, who describes a technology-assisted service-learning project, considers collaborations between universities and community agencies (still another configuration embraced by this term) and urges that such collaborations need to think of time in more extended terms. She urges that university representatives move away from thinking in terms of semesters and summers to establish long-term and sustainable relationships. Inman looks at collaboration as operating in the past and future as well as the present. In defining collaboration as temporally constructed, he claims that "effective collaboration does not always involve group activity; instead . . . anyone who takes into consideration the past and future in assessing the present indeed practices collaboration by referencing what others have done or will do."

I find these two views of collaboration particularly compelling because it seems to me that discussions of technology need to focus more on issues of time. In a world where we have more available information than we can ever use or absorb, time becomes a crucial factor. As Richard Lanham (1994) once put it, the economy of attention is a central concern in the electronic age. Where shall we focus our minds when there are so many interesting issues competing for our (finite) mental capacities? For humanists this is an especially pressing concern because long spaces of uninterrupted time are essential to the work we do. We regularly exchange money for time—in the form of sabbaticals or course release—because we recognize our own need for time. Yet, we also multitask and move in several directions at once, frequently aided by electronic means. How do time constraints shape the kinds of electronic collaborations we construct? How might we use time differently in these constructions? What does time mean in the context of collaboration? What might it mean? These are only a few of the questions that this book suggests we consider. I expect that our wired community will discuss these and many other issues that emerge when we think about electronic collaboration in the humanities.

Notes on Contributors

Randall Bass is associate professor of English at Georgetown University and director of the American Studies Crossroads Project, which is affiliated with the American Studies Association. His print and electronic publications include *Border Texts: Cultural Readings for Contemporary Writers*, the collaborative *Engines of Inquiry: A Practical Guide for Using Technology to Teach American Culture, Electronic Resources for the Heath Anthology of American Literature*, and *Electronic Archives for the Teaching of American Literatures*. Bass currently takes leadership or collaborative roles with the Epiphany Project, the New Media Classroom Project, and the Center for Electronic Projects in American Cultural Studies at Georgetown University.

T. Lloyd Benson is Walter K. Mattison Associate Professor of History at Furman University, specializing in 19th century America. He has written several essays on antebellum community formation. Since 1998, he has served as editor for the E-DOCS: Historical Documents on the Internet discussion list.

Jami Carlacio currently teaches in the Writing Workshop at Cornell University. Her interests include composition theory and computers and writing; nineteenth-century women's rhetoric and histories of rhetoric; and contemporary theories of rhetoric. She has a chapter (with Alice Gillam) in *Professing Rhetoric* (Lawrence Erlbaum Associates) and an essay in *Rhetoric Review*. Carlacio is currently working on a manuscript on women's rhetorical practices in the nineteenth century.

Dagmar Stuehrk Corrigan is the Writing Center director as well as a lecturer at the University of Houston-Downtown. She is also a discourse studies doctoral student at Texas A&M University.

Mary E. Fakler is an instructor of English at the State University of New York, College at New Paltz. Her field of expertise is in collaborative writing, and with her colleague, Joan E. Perisse, she instituted The Peer Critiquing Computer Project. She has presented her research at numerous national and international conferences.

Tari Lin Fanderclai lives in the Boston area, where she works as a technical communicator and human–computer interaction researcher. She is cofounder and cocoordinator of the Netoric Project and the MOOShop Project, and she's a member of the editorial board for *Kairos: A Journal of Rhetoric, Technology, and Pedagogy*. Fanderclai is also the architect, developer, and administrator of Connections, a MOO devoted to educational endeavors.

Dawn M. Formo, associate professor of literature and writing studies at California State University, San Marcos, serves as the coeditor of *The Writing Instructor,* a digital journal. At CSUSM, Dawn has directed the University Writing Center and the General Education Writing program. Currently, she serves as the associate director for a grant that links high school and university faculty and students to reduce the need for remediation among incoming university students. She deems the project, OWL, a thrilling success. She hopes that high school and university students find the forthcoming online book clubs equally engaging. In addition to her OWL research, she has fun meddling with feminist rhetoric and composition theory.

John Craig Freeman, artist and educator, uses digital technologies to produce documentary virtual reality exhibitions that lead the user from global satellite perspectives to digital video scenes on the ground. As the user explores these virtual spaces, the story of the place unfolds in a meandering, nonlinear media montage. Freeman has adopted strategies of project-based interdisciplinary collaboration. His most recent work, titled "Imaging Place," draws on the expertise of cultural theorists, architects, historians, scientists, and community leaders. His work has been exhibited internationally. He currently runs the digital media art curriculum in the Art Department at the University of Massachusetts, Lowell, and serves as an associate professor. He has focused his academic activities on integrating computer technology and theory of electronic culture into a visual art curriculum and exploring interdisciplinary approaches to education and technology.

Bill Friedheim is a professor of social science at the Borough of Manhattan Community College, where he teaches courses in American history. Friedheim's publications include the books *Freedom's Unfinished Revolution* (New Press, 1996) and *A Web of Connections* (McGraw-Hill, 1998), as well as the Web site for the

Professional Staff Congress of the City University of New York. Friedheim has led a number of widely attended faculty development workshops, his most recent one titled "Crossing Urban Borders: The New Media Classroom."

Radhika Gajjala teaches in the School of Communication Studies at Bowling Green State University. Some of her publications are found in journals such as *Gender and Development, Feminist Media Studies*, and *Works and Days*. Her Web site address is http://www.cyberdiva.org and her e-mail address is radhika@cyberdiva.org.

Anne Ruggles Gere is professor of English and professor of education at the University of Michigan, where she chairs the Joint PhD Program in English and Education. She is a past chair of CCCC and a past president of NCTE. Thanks to patient colleagues and students, she actually uses technology in her teaching and research. Her recent collection of essays, *Making American Literatures in High School and College* (NCTE, 2001), was enhanced by and demonstrates the uses of technology in the humanities.

Simone Gers is department chair for Writing and Humanities at Pima Community College's Desert Vista Campus. She is in the doctoral program in language, reading and culture at the University of Arizona.

M. Wendy Hennequin teaches English at the University of Connecticut. She has published articles in *Literary and Linguistic Computing* and *The Early Drama, Art, and Music Review*.

James A. Inman is assistant professor of English at the University of South Florida, where he directs the writing center and teaches graduate and undergraduate courses in professional and technical writing, rhetoric and composition, and graphic design. Inman's publications include *Taking Flight with OWLs: Examining Electronic Writing Center Work* (Lawrence Erlbaum Associates, 2000; with Donna N. Sewell) and *Computers and Writing: The Cyborg Era* (Lawrence Erlbaum Associates, forthcoming), as well as articles and reviews in various journals. Inman serves as coeditor and copublisher of *Kairos: A Journal of Rhetoric, Technology, and Pedagogy* and as cocoordinator of the Netoric Project, and he serves on the executive board of the International Writing Centers Association.

Timothy Allen Jackson is associate professor of New Media at Ryerson Polytechnic University. His experience covers both theoretical and production concerns of new media design, emerging art forms, and critical theory. Jackson has shown work in several cities in the United States and continues to maintain design clients in New York and Washington, D.C. He has also lectured and been published widely on the subject of new media.

Virginia Scott Jenkins is visiting assistant professor of American Studies at the University of Maryland. Her dissertation was on the American front lawn, and she continues to write on how Americans use this space. Jenkins's current research is on the impact of the importation of bananas on American popular culture and the 1918 Spanish influenza pandemic—what the American public knew and what was published in newspapers and magazines. She is also interested in foodways, historical archaeology, venacular architecture, western history, advertising and consumption, and issues of community formation and design.

Jo Kibbee is the head of reference at the University of Illinois at Urbana-Champaign and associate professor of library administration. She has written on library management issues, comparative librarianship, and library applications of information technology.

Nancy Knowles is assistant professor of English/Writing and the writing coordinator at Eastern Oregon University. She has published articles in *Woolf Studies Annual* and *Literary and Linguistic Computing* and is technical liaison for the Oregon Writing Project at Eastern.

Annapurna Mamidipudi is a field worker in Dakstra, India. She regularly posts to the sa-cyborgs electronic list, which is affiliated with the Spoon Collective. With Radhika Gajjala, Mamidipudi recently published an article, "Cyberfeminism, Technology, and International 'Development'," in *Gender and Development*.

Karen L. McComas, CCC-A/SLP, is associate professor of communications disorders at Marshall University. She received her Bachelor of Arts degree in speech pathology and audiology from Marshall University in May of 1977 and her Master of Arts degree in speech pathology and audiology from Marshall University in August of 1978. McComas worked as a speech-language pathologist in the public school from 1978 to 1986 in Carter County, Kentucky and Lincoln County, West Virginia, and her expertise includes the use of Internet technologies for teaching and clinical purposes. McComas has extensive experience using virtual reality, listserv, e-mail, the World Wide Web, and newsgroups to supplement traditional courses, clinical activities, and online coursework.

Paul "Skip" Morris returned to school at the University of Nevada, Reno (UNR), where he received his PhD in composition and rhetoric after nearly 20 years in the broadcast industry. While at UNR, he co-wrote *The New Literacy: Moving Beyond the 3Rs* with Stephen Tchudi. Morris is now the computer coordinator for Pittsburg State University's English Department in Pittsburg, Kansas. His chapter is the articulation of his desire to synthesize his own practical experience with computers and his research in literacy learning.

Jo B. Paoletti studies apparel design and the history of textile and clothing, and she has spent about 25 years researching and writing about children's clothing in

America. This has given way to an interest in how to do the best possible job of interdisciplinary education. Paoletti holds a joint appointment as an associate professor of American studies and faculty director of College Park Scholars in American cultures, a living-learning program at the University of Maryland in College Park.

Joan E. Perisse is an instructor at both SUNY College at New Paltz and Marist College. After receiving a BA in English and secondary education and an MA in English, her interests expanded to collaborative learning and technology. She codeveloped with her colleague, Mary E. Fakler, many collaborative projects, including the Peer Critiquing Computer Project, which have been presented at conferences both nationally and internationally.

Christina L. Prell is an assistant professor of communication at McDaniel College. Her current research focuses on the use of computer networking technologies for local community purposes. At WMC, Prell teaches courses in computer-mediated-communication, research methods, and mass media.

Cheryl Reed is a recovering academic. She began writing about electronic collaboration as an assistant professor of writing at Pennsylvania State University, Hazleton. Lured by the siren call of applied research, she accepted her current position as research manager for the Telemedicine Project at the Naval Health Research Center, San Diego. She and her colleagues explore how telecommunications technologies can enable remote access to health care for deployed Naval personnel.

Rebecca J. Rickly is an assistant professor at Texas Tech University, where she serves as codirector of composition and teaches undergraduate and graduate courses in rhetoric and writing. Her work focuses on rhetoric but includes such diverse applications as technology, feminisms, methods and methodologies, literacy study, contrastive rhetorics, and administration. She has served on the CCCC Committee on Computers and Composition and NCTE's Assembly on Computers in English, and she has chaired NCTE's Instructional Technology Committee. Her publications include *The Online Writing Classroom* (with Susanmarie Harrington and Michael Day), and her work has appeared in several edited collections as well as *Computers and Composition, CMC Magazine, The ACE Journal,* and *Kairos.*

Peter Sands is assistant professor of English at the University of Wisconsin-Milwaukee, where he teaches graduate and undergraduate courses in composition, rhetoric, computers and pedagogy, and American literary studies. He has been teaching in networked environments for more than a decade.

Donna N. Sewell is associate professor of English at Valdosta State University, where she directs the writing center, overseeing a wonderful staff of tutors, in addition to teaching writing and literature. She coedited *Taking Flight with OWLs: Examining Electronic Writing Center Work* (Lawrence Erlbaum Associates, 2000) with James A. Inman. She has also published chapters and articles on composition studies and writing center work.

Mary Corbin Sies is director of graduate studies and associate professor of American studies at the University of Maryland. Additionally, she is an affiliate faculty member of the Women's Studies Department and a member of the Historic Preservation faculty. Sies's research and teaching interests span material culture studies, planning history, architectural history, urban history, and cultural and social history of the United States in the 19th and 20th centuries. She is an authority on American suburbs from 1850 to the present, particularly planned, exclusive suburbs and the cultural landscapes, values, and lifeways established by their upper middle class residents. She is also working on a history of the American arts and crafts movement from the perspectives of consumers and is interested in theorizing and studying issues of gender, space, and the domestic-built environment.

Caroline Szylowicz is the Kolb-Proust Librarian and assistant professor of library administration at the University of Illinois at Urbana-Champaign Library. She is working on a bibliography of Proustian studies.

Stephen Tchudi is professor of English and chair of the English Department at the University of Nevada, Reno. He is a past president of the National Council of Teachers of English and former editor of *The English Journal.*

Alice Trupe directs the writing center at Bridgewater College in Virginia, where she also teaches writing courses, introductory linguistics, and young adult literature. She is currently working on a textbook for teaching writing. She can no longer remember how to teach in a room without computers.

Sherry Turkle is Abby Rockefeller Mauze Professor in the Program in Science, Technology, and Society at MIT. She has written numerous articles on psychoanalysis and culture and on the "subjective side" of people's relationships with technology, especially computers. Turkle is the author of *Psychoanalytic Politics: Jacques Lacan and Freud's French Revolution* (Basic Books, 1978; MIT Press paper, 1981; second revised edition, Guilford Press, 1992) and *The Second Self: Computers and the Human Spirit* (Simon and Schuster, 1984; Touchstone paper, 1985; second revised edition, MIT Press, forthcoming). Her most recent book, *Life on the Screen: Identity in the Age of the Internet* (Simon and Schuster, November 1995; Touchstone paperback, 1997) explored the psychology of computer-mediated communication on the Internet. Turkle's current research is about the psychological impact of computational objects as they become increasingly "relational" artifacts; she is studying a range of objects, from "affective computers" to robotic dolls and pets.

Myka Vielstimmig is the creation of Kathleen Blake Yancey (Pearce Professor of English at Clemson University) and Michael Spooner (director of Utah State University Press). These collaborators have written together under this pseudonym on issues of (electronic) genre, of multivocality, authorship, collaboration, and of textuality. *Vielstimmig* is German for "many voiced."

References

Adams, H. (1918). *The education of Henry Adams*. New York: Houghton-Mifflin.

Adler-Kassner, L., Crooks, R., & Watters, A. (Eds.). (1997). *Writing the community: Concepts and models for service-learning in composition*. Urbana, IL: NCTE.

Agre, P. (2000). Commodity and community: Institutional design for the networked university. *Planning for Higher Education, 29*(5–14), 2.

Alcorn, M. (1994). Self-structure as a rhetorical device: Modern ethos and the divisiveness of the self. In J. S. Baumlin & T. F. Baumlin (Eds.), *Ethos: New essays in rhetorical and critical theory* (pp. 3–36). Dallas, TX: Southern Methodist University Press.

Allen, N. (1996). Gaining electronic literacy: Workplace simulations in the classroom. In P. Sullivan & J. Dautermann (Eds.), *Electronic literacies in the workplace: Technologies of writing* (pp. 216–237). Urbana, IL: NCTE.

Allen, N., & Wickliff, G. A. (1997). Learning up close and at a distance. In S. A. Selber (Ed.), *Computers and technical communication: Pedagogical and programmatic perspectives* (pp. 201–218). Greenwich, CN: Ablex.

Alverman, D. E., Moon, J. S., & Hagood, M. (1999). *Popular culture in the classroom: Teaching and researching critical media literacy*. Newark, DE: International Reading Association.

Anderson, A., Mayes, J. T., & Kibby, M. R. (1991). Small group collaborative discovery learning from hypertext. In C. O'Malley (Ed.), *Computer-supported collaborative learning* (pp. 23–38). Berlin: Springer-Verlag.

Anderson, D., Benjamin, B., & Paredes-Holt, B. (1998). *Connections: A guide to online writing*. Boston: Allyn & Bacon.

Anderson, L. (Ed.). (1991). *Sisters of the earth: Women's prose and poetry about nature*. New York: Random House.

Aristotle. (1991). *Rhetoric* [G. A. Kennedy, Trans.]. Oxford, UK: Oxford University Press.

Aronowitz, S. (1992). Looking out: The impact of computers on the lives of professionals. In M. C. Tuman (Ed.), *Literacy online: The promise (and peril) of reading and writing with computers* (pp. 119–137). Pittsburgh, PA: University of Pittsburgh Press.

Ayers, E. L., & Thomas, W. (2001). *The difference slavery made: Two american communities and the coming of the civil war.* Retrieved September 20, 2001, from http://jefferson.village.virginia.edu/vcdh/AHR/article.html

Baker, N. (2001). *Double fold: Libraries and the assault on paper.* New York: Random House.

Bakhtin, M. M. (1981). *The dialogic imagination* [M. Holquist, Ed.; C. Emerson & M. Holquist, Trans.]. Austin: University of Texas Press.

Barker, T. T., & Kemp, F. O. (1990). Network theory: A postmodern pedagogy for the writing classroom. In C. Handa (Ed.), *Computers and community: Teaching composition in the twenty-first century* (pp. 1–27). Portsmouth, NH: Boynton/Cook.

Barlowe, J., & Hottell, R. (1998). Feminist theory and practice and the pedantic i/eye. In E. G. Peck & J. S. Mink (Eds.), *Common ground: Feminist collaboration in the academy* (pp. 269–281). Albany, NY: SUNY Press.

Barney, D. (2000). *Prometheus wired: The hope for democracy in the age of network technology.* Chicago: University of Chicago Press.

Barrett, E. (Ed.). (1989). *The society of text: Hypertext, hypermedia, and the social construction of information.* Cambridge, MA: MIT Press.

Bartholomae, D. (1985). Inventing the university. In M. Rose (Ed.), *When a writer can't write: Research on writer's block and other writing process problems* (pp. 134–165). New York: Guilford.

Bass, R., & Eynon, B. (1998). Teaching culture, learning culture, and new media: An introduction and framework. In R. Bass, T. Derricksen, B. Eynon, & M. Sample (Eds.), *Intentional media: The crossroads conversations on learning and technology in the American culture and history classroom. Works and Days, 16*(1–2), 11–96.

Batson, T. (1989). Teaching in networked classrooms. In C. L. Selfe, D. Rodrigues, & W. R. Oates, *Computers in english and the language arts: The challenge of teacher education* (pp. 247–255). Urbana, IL: NCTE.

Belenky, M. F., Clinchy, B. M., Goldberger, N. R., & Tarule, J. M. (1986). *Women's ways of knowing: The development of self, voice, and mind.* New York: Basic Books.

Berlin, J. A. (1984). *Writing instruction in nineteenth-century American colleges.* Carbondale, IL: Southern Illinois University Press.

Berlin, J. A. (1987). *Rhetoric and reality: Writing instruction in American colleges, 1900–1985.* Carbondale, IL: Southern Illinois University Press.

Berners-Lee, T., Hendler, J., & Lassila O. (2001). The semantic web. *Scientific American.* Retrieved Sept. 20, 2001, from http://www.sciam.com/2001/0501issue/0501berners-lee.html

Berthoff, A. E. (1981). *The making of meaning: Metaphors, models, and maxims for writing teachers.* Montclair, NJ: Boynton/Cook.

Berthoff, A. E. (1984). *Reclaiming the imagination: Philosophical perspectives for writers and teachers of writing.* Montclair, NJ: Boynton/Cook.

Besser, H. (2000). Digital longevity. In M. K. Sitts (Ed.), *Handbook of digital projects: A management tool for preservation and access.* Retrieved September 20, 2001, from http://www.nedcc.org/digital/TofC.htm

Bhabha, H. (1994). *The location of culture.* New York: Routledge.

Bigham, D. E. (1998). *Towns and villages of the lower Ohio.* Lexington: Kentucky University Press.

Biocca, F., & Levy, M. R. (Eds.). (1995). *Communication in the age of virtual reality.* Hillsdale, NJ: Lawrence Erlbaum Associates.

Birkerts, S. (Ed.). (1996). *Tolstoy's dictaphone: Technology and the muse.* St. Paul, MN: Graywolf Press.

Bishop, W. (1989). Qualitative evaluation and the conversational writing classroom. *Journal of Teaching English, 267–285.*

Bissex, G. (1980). *Gnys at wrk: A child learns to read and write.* Cambridge, MA: Harvard University Press.

Blackwell, E., Beeman, W. O., & Reese, C. M. (Eds.). (1988). *Object, image, inquiry: The art historian at work.* Santa Monica, CA: The Getty Art History Information Program.

Bloome, D., & Bailey, F. (1992). Studying language and literacy through events, particularity, and intertextuality. In R. Beach, J. Green, M. Kamil, & T. Shanahan (Eds.), *Multiple disciplinary approaches to researching language and literacy* (pp. 181–210). Urbana, IL: NCTE & NCRE.

Blyler, N. A. (1995). Pedagogy and social action: A role for narrative in professional communication. *Journal of Business and Technical Communication, 9*(3), 289–320.

Boiarsky, C. (1990). Computers in the classroom: The instruction, the mess, the noise, the writing. In C. Handa (Ed.), *Computers and community: Teaching composition in the twenty-first century* (pp. 47–67). Portsmouth, NH: Boynton/Cook.

Bolter, J. D. (1991). *Writing space: The computer, hypertext, and the history of writing.* Hillsdale, NJ: Lawrence Erlbaum Associates.

Bonk, C. J., & King, K. S. (Eds.). (1998). *Electronic collaborators: Learner-centered technologies for literacy, apprenticeship, and discourse.* Mahwah, NJ: Lawrence Erlbaum Associates.

Britton, J., Burgess, T., Martin, N., McLeod, A., & Rosen, H. (1975). *The development of writing abilities (11–18).* London: Macmillan.

Brockman, J. (1996). *Digerati: Encounters with the cyber elite.* New York: Hardwired.

Brooks, C., & Warren, R. P. (1938). *Understanding poetry.* New York: Harcourt, Brace.

Brown, D. (Ed.). (2000). *Interactive learning: Vignettes from America's most wired campuses.* Bolton, MA: Anchor.

Brown, J. S., & Duguid P. (2000). *The social life of information.* Cambridge, MA: Harvard Business School Press.

Brown, L. F. (1998, April 7). *Session H.17 at CCCC.* Retrieved August 16, 1999, from http://www.ttu.edu/wcenter/9804/msg00059.html

Bruffee, K. A. (1984). Collaborative learning and the "conversation of mankind." *College English, 46,* 635–652.

Bruffee, K. A. (1986). Social construction, language, and the authority of knowledge: A bibliographic essay. *College English, 48,* 773–790.

Bruffee, K. A. (1993). *Collaborative learning: Higher education, interdependence, and the authority of knowledge.* Baltimore: Johns Hopkins University Press.

Bruner, J. (1990). *Acts of mind.* Cambridge, MA: Harvard University Press.

The building blocks project (National Initiative for a Networked Cultural Heritage. Questionnaire and reports). (2000). Retrieved November 1st, 2000. http://www.ninch.org/bb/project/project.html

Bump, J. (1990). Radical changes in class discussion using networked computers. *Computers and the Humanities, 24,* 49–65.

Burke, K. (1950). *A rhetoric of motives.* New York: Prentice-Hall.

Burkhalter, B. (1999). Reading race online: Discovering racial identity in usenet discussions. In M. A. Smith & P. Kollock (Eds.), *Communities in cyberspace* (pp. 60–75). London: Routledge.

Burns, H. (1992). Teaching composition in tomorrow's multimedia, multinetworked classrooms. In G. E. Hawisher & P. LeBlanc (Eds.), *Re-imagining computers and composition: Teaching and research in the virtual age* (pp. 115–130). Portsmouth, NH: Boynton/Cook.

Bush, V. (1945). As we may think. *Atlantic Monthly.* Retrieved September 15, 1999, from http://www.isg.sfu.ca/~duchier/misc/vbush

Butler, H. J. (1995). Where does scholarly electronic publishing get you? *Journal of Scholarly Publishing, 26,* 174–186.

Campus Computing Project. (2000). *Struggling with it staffing* (K. Green, Ed.). Retrieved May 10, 2001, from http://www.campuscomputing.net/summaries/2000/index.html

Cherny, L. (1999). *Conversation and community: Chat in a virtual world.* Stanford, CA: Center for the Study of Language and Information.

Cherny, L., & Weise, E. R. (Eds.). (1996). *Wired women: Gender and new realities in cyberspace.* Seattle, WA: Seal Press.

Christensen, F., & Christensen, B. (1978). *Notes toward a new rhetoric: Nine essays for teachers* (2nd ed.). New York: Harper & Row.

Condit, C. M., & Lucaites, J. L. (1993). *Crafting equality: America's anglo-African word.* Chicago: University of Chicago Press.

Conference of College Composition and Communication. (1998). *CCCC promotion and tenure guidelines for work with technology.* Retrieved November 15, 1999, from http://www.ncte.org/positions/4c-tp-tech.html

Cooper, M. (1999). Postmodern pedagogy in electronic conversations. In G. E. Hawisher & C. L. Selfe (Eds.), *Passions, pedagogies, and 21st century technologies* (pp. 140–160). Logan: Utah State University Press.

Cooper, M. M., & Selfe, C. L. (1990). Computer conferences and learning: Authority, resistance, and internally persuasive discourse. *College English, 52,* 847–869.

Corbell, P. (1999). Learning from the children: Practical and theoretical reflections on playing and learning. *Simulation & Gaming, 30*(2), 163–181.

Corbett, E. P. J. (1971). *Classical rhetoric for the modern student.* New York: Oxford University Press.

Covino, W. (1999). Cyberpunk literacy; or, piety in the sky. In T. Taylor & I. Ward (Eds.), *Literacy theory in the age of the Internet* (pp. 34–46). New York: Columbia University Press.

Craig, T., Harris, L., & Smith, R. (1999). Rhetoric of the "contact zone": Composition on the front lines. In T. W. Taylor & I. Ward (Eds.), *Literacy theory in the age of the Internet* (pp. 122–145). New York: Columbia University Press.

Crane, G. (1998). The perseus project and beyond: How building a digital library challenges the humanities and technology. *D-Lib Magazine.* Retrieved July 20, 2000, from http://www.dlib.org/dlib/january98/01crane.html

Crane, G. (2000). Designing documents to enhance the performance of digital libraries. *D-Lib Magazine.* Retrieved July 20, 2001, from http://www.dlib.org/dlib/july00/crane/07crane.html

Crane, S. (1895). *Red badge of courage.* New York: Appleton.

Crook, C. (1991). Educational practice within two local computer networks. In C. O'Malley (Ed.), *Computer-supported collaborative learning* (pp. 165–182). Berlin: Springer-Verlag.

Crump, E. (1998). At home in the MUD: Writing centers learn to wallow. In C. Haynes & J. R. Holmevik (Eds.), *High wired: On the design, use, and theory of educational MOOs* (pp. 177–191). Ann Arbor: University of Michigan Press.

Daiute, C. (1985). *Writing and computers.* Reading, MA: Addison-Wesley.

Daly, J. A., & Miller, M. D. (1975). The empirical development of an instrument to measure writing apprehension. *Research in the Teaching of English,* 242–256.

Damrosch, D. (1995). *We scholars: Changing the culture of the university.* Cambridge, MA: Harvard University Press.

D'Angelo, F. J. (1975). *A conceptual theory of rhetoric.* Cambridge, MA: Winthrop.

D'Arms, J. H. (2000). *The electronic monograph in the 21st century.* Retrieved August 20, 2001, from http://www.acls.org/jhd-aha.htm

Davidson, N. (1999). *Cooperative/collaborative learning.* Retrieved November 15, 1999, from http://www2.emc.maricopa.edu/innovation/CCL/CCL.html

Davis, B. H., & Brewer, J. P. (1997). *Electronic discourse: Linguistic individuals in virtual space.* Albany, NY: State University of New York Press.

Day, M., & Batson, T. (1993). The network-based writing classroom: The ENFI idea. In M. Collins & Z. Berge (Eds.), *Computer-mediated communication and the online classroom* (Vol. 2, pp. 25–46). Cresskill, NJ: Hampton Press.

Day, M., Crump, E., & Rickly, R. (1996). Creating a virtual academic community: Scholarship and community in wide-area multiple-user synchronous discussions. In T. M. Harrison & T. Stephen (Eds.), *Computer networking and scholarly communication in the twenty-first-century university* (pp. 291–311). Albany: State University of New York Press.

Denley, P., & Hopkin, D. (1987). *History and computing.* Manchester, UK: Manchester University Press.

Derian, J. D. (1994). Lenin's war, Baudrillard's games. In G. Bender & T. Druckery (Eds.), *Cultures on the brink: Ideologies of technology* (pp. 267–276). Seattle, WA: Bay Press.

Desmet, C. (1998). Equivalent students, equitable classrooms. In S. C. Jarratt & L. Worsham (Eds.), *Feminism and composition studies: In other words* (pp. 153–171). New York: MLA.

DeWitt, S. L. (1996). The current nature of hypertext research in computers and composition studies: A historical perspective. *Computers and Composition, 13,* 69–84.

Dibbell, J. (1999). *My tiny life: Crime and passion in a virtual world.* New York: Owl.

Dirlik, A. (1998). Globalism and the politics of place. *Development, 41*(2), 7–13.

Dobler, B., & Bloomberg, H. (1998). How much web would a Webcourse weave if a Webcourse would weave webs? In J. Galin & J. Latchaw (Eds.), *The dialogic classroom: Teachers integrating computer technology, pedagogy, and research* (pp. 67–91). Urbana, IL: NCTE.

Doheny-Farina, S. (1996). *The wired neighborhood.* New Haven: Yale University Press.

Donaldson, M. (1978). *Children's minds.* New York: W. W. Norton.

Dorman, W., & Dorman, S. (1997). Service-learning: Bridging the gap between the real world and the composition classroom. In L. Adler-Kassner, R. Crooks, & A. Waters (Eds.), *Writing the community: Concepts and models for service-learning in composition* (pp. 119–132). Urbana, IL: NCTE.

Doyle, D. H. (1983). *The social order of a frontier community, Jacksonville, Illinois, 1825–1870.* Bloomington: University of Illinois Press.

DuBois, E. C., Kelly, G. P., Kennedy, E. L., Korsmeyer, C. W., & Robinson, L. S. (1985). *Feminist scholarship: Kindling in the groves of academe.* Urbana: University of Illinois Press.

Duffelmeyer, B. B. (2001). Critical computer literacy: Computers in first-year composition as topic and environment. *Computers and Composition, 18.* Retrieved August 20, 2001, from http://corax.cwrl.utexas.edu/cac

Duin, A. H., & Hansen, C. (1994). Reading and writing on computer networks as social construction and social interaction. In C. L. Selfe & S. Hilligoss (Eds.), *Literacy and computers: The complications of teaching and learning with technology* (pp. 89–112). New York: MLA.

Eddy, G., & Carducci, J. (1997). Service with a smile: Class and community in advanced composition. *The Writing Instructor, 16*(2), 78–91.

Ede, L., & Lunsford, A. (1983). Why write . . . together? *Rhetoric Review, 1,* 150–158.

Ede, L., & Lunsford, A. (1990). *Singular texts/plural authors: Perspectives on collaborative writing.* Carbondale: Southern Illinois University Press.

Edwards, A. D. N. (1995). *Extra-ordinary human-computer interaction: Interfaces for users with disabilities.* New York: Cambridge University Press.

Ehninger, D. (1972). *Contemporary rhetoric: A reader's coursebook.* Glenview, IL: Scott Foresman.

Eldred, J. M. (1989). Computers, composition pedagogy, and the social view. In G. E. Hawisher & C. L. Selfe (Eds.), *Critical perspectives on computers and composition theory* (pp. 201–218). New York: Teachers College Press.

English Department, University of Connecticut. (1998). *English 109 course description.* Retrieved August 16, 1998, from http://www.lib.uconn.edu/English/Undergraduate/109.html

Erikson, E. H. (1958). *Young man Luther: A study in psychoanalysis and history.* New York: Norton.

Erikson, E. H. (1963). *Childhood and society* (2nd ed.). New York: Norton. (Original work published 1950).

Erikson, E. H. (1969). *Gandhi's truth: On the origins of militant nonviolence.* New York: Norton.

Ermann, M. D., Williams, M. B., & Shauf, M. S. (1997). *Computers, ethics, and society.* New York: Oxford University Press.

Escobar, A. (1999). Gender, place and networks: A political ecology of cyberculture. In W. Harcourt (Ed.), *Women on the Internet: Creating new cultures in cyberspace* (pp. 31–54). London: Zed.

Evans, W. (1996). Computer-supported content analysis: Trends, tools, and techniques. *Social Science Computer Review, 14*(3), 269–279.

Faigley, L. (1990). Subverting the electronic workbook: Teaching writing using networked computers. In D. Daiker & M. Morenberg (Eds.), *The writing teacher as researcher: Essays in the theory of class-based research* (pp. 290–311). Portsmouth, NH: Heinemann.

Faigley, L. (1992). *Fragments of rationality: Postmodernity and the subject of composition*. Pittsburgh, PA: University of Pittsburgh Press.

Faigley, L. (1997). Literacy after the revolution. *College Composition and Communication, 48*, 30–43.

Fanderclai, T. L. (1996). Like magic, only real. In L. Cherny & E. R. Weise (Eds.), *Wired women: Gender and new realities in cyberspace* (pp. 224–241). Seattle, WA: Seal.

Feenberg, A. (1991). *Critical theory of technology*. New York: Oxford University Press.

Felder, R. M., & Silverman, L. K. (1988). Learning and teaching styles in engineering education. *Engineering Education, 78*, 674–681.

Fish, S. (1980). *Is there a text in this class? The authority of interpretive communities*. Cambridge, MA: Harvard University Press.

Fish, S. (1983). Short people got no reason to live: Reading irony. *Daedalus, 112*, 175–191.

Fisher, C. (1996). Learning to compute and computing to learn. In C. Fisher, D. C. Dwyer, & K. Yocam (Eds.), *Education and technology: Reflections on computing in classrooms* (pp. 109–127). San Francisco: Jossey-Bass.

Flores, M. J. (1990). Computer conferencing: Composing a feminist community of writers. In C. Handa (Ed.), *Computers and community: Teaching composition in the twenty-first century* (pp. 106–117). Portsmouth, NH.

Floriani, A. (1994). Negotiating what counts: Roles and relationships, texts and context, content and meaning. *Linguistics and Education, 5*, 241–274.

Flower, L. (1993). *Problem-solving strategies for writing* (4th ed.). Fort Worth, TX: Harcourt Brace Jovanovitch.

Flower, L. (1997). Observation-based theory building. In G. A. Olson & T. W. Taylor (Eds.), *Publishing in rhetoric and composition* (pp. 163–185). Albany: State University of New York Press.

Forman, J. (1991). Computing and collaborative writing. In G. E. Hawisher & C. L. Selfe (Eds.), *Evolving perspectives on computers and composition studies* (pp. 65–83). Urbana, IL: NCTE.

Forman, J. (Ed.). (1992). *New visions of collaborative writing*. Portsmouth, NH: Heinemann.

Forman, J. (1994). Literacy, collaboration, and technology: New connections and challenges. In C. L. Selfe & S. Hilligoss (Eds.), *Literacy and computers: The complications of teaching and learning with technology* (pp. 130–143). New York: MLA.

Formo, D. M., & Reed, C. (1999). *Job search in academe: Strategic rhetorics for faculty job candidates*. Baltimore: Stylus.

Foster, D. (1997). Community and identity in the electronic village. In D. Porter (Ed.), *Internet culture* (pp. 23–37). New York: Routledge.

Foucault, M. (1979). What is an author? In J. V. Harari (Ed.), *Textual strategies: Perspectives in poststructuralist criticism* (pp. 141–160). Ithaca, NY: Cornell University Press.

Freire, P. (1993). *Pedagogy of the city* (D. MacEdo, Trans.). New York: Continuum.

Freire, P. (1995). *Pedagogy of hope* (R. R. Barr, Trans.). New York: Continuum.

Freire, P. (1996). *Pedagogy of the oppressed* (M. B. Ramos, Trans.). New York: Continuum. (Original work published 1970)

Gajjala, R. (1998). *The sawnet refusal: An interrupted cyberethnography*. Unpublished doctoral dissertation, University of Pittsburgh.

Gajjala, R., & Mamidipudi, A. (1999). Cyberfeminism, technology and international "development." *Gender and Development, 7*(2), 8–16.

Galin, J. R., & Latchaw, J. (Eds.). (1998a). *The dialogic classroom: Teachers integrating computer technology, pedagogy, and research*. Urbana, IL: NCTE.

Galin, J. R., & Latchaw, J. (1998b). Voices that let us hear: The tale of the borges quest. In J. R. Galin & J. Latchaw (Eds.), *The dialogic classroom: Teachers integrating computer technology, pedagogy, and research* (pp. 43–66). Urbana, IL: NCTE.

Galland, C. (1991). Running lava falls. In Anderson (Ed.), (pp. 149–151).

Gardels, N. (1991, November 21). Two concepts of nationalism: An interview with Isaiah Berlin. *New York Review of Books*, 19–23.

Gere, A. R. (1987). *Writing groups: History, theory, and implications*. Carbondale: Southern Illinois University Press.

Gerrard, L. (1997). Thoughts on computers, gender, and the body electric. *Kairos: A Journal for Teaching Writing in Webbed Environments, 2*(2). Retrieved July 20, 2001, from http://Engish.ttu.edu/Kairos/2.2/coverweb/invited/lg.html

Gers, S. (1999, August 16). *Re: Session H.17 at CCCC*. Retrieved August 18, 1999, from http://www.ttu.edu/wcenter/9804/ msg00153.html

Gibson, W. (1984). *Neuromancer*. New York: Ace.

Gilbert, S., & Gubar, S. (1979). *The madwoman in the attic: The woman writer and the nineteenth-century literary imagination*. New Haven, CT: Yale University Press.

Gilster, P. (1997). *Digital literacy*. New York: Wiley.

Giroux, H. (1983). *Theory and resistance in education*. New York: Bergin and Garvey.

Gomez, M. L. (1991). The equitable teaching of composition with computers: A case for change. In G. E. Hawisher & C. L. Selfe (Eds.), *Evolving perspectives on computers and composition studies* (pp. 318–335). Urbana, IL: NCTE.

Goodman, K. S. (1973). *Miscue analysis: Applications to reading instruction*. Urbana, IL: ERIC.

Goodman, Y. M., & Marek, A. M. (1996). Revaluing readers and reading. In Y. M. Goodman & A. M. Marek (Eds.), *Retrospective miscue analysis* (pp. 203–207). Katonah, NY: Richard C. Owen Publishers.

Grasha, A. F. (1996). *Teaching with style: A practical guide to enhancing learning by understanding teaching and learning styles*. San Bernardino, CA: Alliance Publishers.

Green, D. (Ed.). (2000). *Copyright and fair use: Town meetings 2000*. Retrieved September 20, 2000, from http://www.ninch.org/copyright/townmeetings/report2000.pdf

Greenstein, D. I. (1994). *A historian's guide to computing*. New York: Oxford University Press.

Grigar, D., & Barber, J. F. (1998). Defending your life in MOOspace: A report from the electronic edge. In C. Haynes & J. R. Holmevik (Eds.), *High wired: On the design, use, and theory of educational moos* (pp. 192–231). Ann Arbor: University of Michigan Press.

Guha, R. (1988). Preface. In R. Guha & G. C. Spivak (Eds.), *Selected subaltern studies* (pp. 35–36). New York: Oxford University Press.

Gumbel, A. (1999). How e-mail puts us in a flaming bad temper. *Computers + Internet: Text 2*. Retrieved May 27, 2001, from http://www.projektlernen.de/pa_internet_email_language.htm

Haas, C., & Neuwirth, C. M. (1994). Writing the technology that writes us: Research on literacy and the shape of technology. In C. L. Selfe & S. Hilligoss (Eds.), *Literacy and computers: The complications of teaching and learning with technology* (pp. 319–335). New York: MLA.

Halberstam, J., & Livingston, I. (Eds.). (1995). *Posthuman bodies*. Bloomington: Indiana University Press.

Halio, M. P. (1996). Multimedia narration: Constructing possible worlds. *Computers and Composition, 13*, 343–352.

Hancock, J. (Ed.). (1999). *Teaching literacy using information technology: A collection of articles from the Australian Literacy Educators' Association*. Newark, DE: IRA.

Handa, C. (Ed.). (1990a). *Computers and community: Teaching composition in the twenty-first century*. Portsmouth, NH: Boynton.

Handa, C. (1990b). Politics, ideology, and the strange, slow death of the isolated composer or why we need community in the writing classroom. In C. Handa (Ed.), *Computers and community: Teaching composition in the twenty-first century* (pp. 160–184). Portsmouth, NH: Boynton.

Hanson, L. K. (2000). Advanced composition online: Pedagogical intersections of composition and literature. In S. Harrington, M. Day, & R. Rickly (Eds.), *The online writing classroom* (pp. 207–242). Cresskill, NJ: Hampton.

Haraway,. D. (1985). A manifesto for cyborgs: Science, technology, and socialist feminism in the 1980s. *Socialist Review, 80*, 65–105.

Haraway, D. (1991). *Simians, cyborgs, and women: The reinvention of nature*. London: Routledge.

Harcourt, W. (Ed.). (1999). *women@internet: Creating new cultures in cyberspace*. London: Zed.

Harrington, S., Day, M., & Rickly, R. (Eds.). (2000). *The online writing classroom*. New York: Hampton.

Harrison, S. (1998). *E-mail at work*. Retrieved May 27, 2001, from http://www.istc.org.uk/email.htm

Harste, J. C., Woodward, V. A., & Burke, C. L. (1984). *Language stories and literacy lessons*. Portsmouth, NH: Heinemann.

Hawisher, G. E. (1992). Electronic meetings of the minds: Research, electronic conferences, and composition studies. In G. E. Hawisher & P. LeBlanc (Eds.), *Re-imagining computers and composition: Teaching and research in the virtual age* (pp. 81–101). Portsmouth, NH: Boynton/Cook.

Hawisher, G. E. (1996, March). *Women on the net: Constructing gender in electronic discourses*. Paper presented at the annual meeting of the Conference on College Composition and Communication, Milwaukee, WI.

Hawisher, G. E. (2000). *Accessing the virtual worlds of cyberspace*. Keynote address to Furman University's conference on new information technologies and liberal education, Greenville, SC.

Hawisher, G. E., & LeBlanc, P. (Eds.). (1992). *Re-imagining computers and composition: Teaching and research in the virtual age*. Portsmouth, NH: Boynton.

Hawisher, G. E., & Selfe, C. L. (Eds.). (1989). *Critical perspectives on computers and composition instruction*. New York: Teachers College.

Hawisher, G. E., & Selfe, C. L. (Eds.). (1991a). *Evolving perspectives on computers and composition studies: Questions for the 1990s*. Urbana, IL: NCTE.

Hawisher, G. E., & Selfe, C. L. (1991b). The rhetoric of technology and the electronic writing class. *College Composition and Communication, 42*(1), 55–65.

Hawisher, G. E., & Selfe, C. L. (Eds.). (1996). *Literacy, technology, and society: Confronting the issues*. New York: Prentice Hall.

Hawisher, G. E., & Sullivan, P. (1998). Women on the networks: Searching for e-spaces of their own. In S. C. Jarratt & L. Worsham (Eds.), *Feminism and composition studies: In other words* (pp. 172–197). New York: MLA.

Hayles, N. K. (1999). *How we became posthuman: Virtual bodies in cybernetics, literature, and informatics*. Chicago: University of Chicago Press.

Haynes, C., & Holmevik, J. R. (Eds.). (1998). *High wired: On the design, use, and theory of educational MOOs*. Ann Arbor: University of Michigan Press.

Healy, D. (1997). Cyberspace and place: The Internet as middle landscape on the electronic frontier. In D. Porter (Ed.), *Internet culture* (pp. 55–68). New York: Routledge.

Heidegger, M. (1977). *The question concerning technology and other essays* (W. Lovitt, Trans.). New York: Harper.

Heim, M. (1993). *The metaphysics of virtual reality*. New York: Oxford University Press.

Hellekson, K. (1998). Leaving the academy. *The Journal of the Midwest Modern Language Association, 31*, 1–5.

Hennequin, M. W. (2001). *Annotated hypertext*. Retrieved May 26, 2001, from http://sp.uconn.edu/~mwh95001/249s/s01/annotated.html

Herring, S. C. (1994). *Gender differences in computer-mediated communication: Bringing familiar baggage to the new frontier*. Retrieved February 2, 1998, from http://www.cpsr.org/cpsr/gender/herring.txt

Herring, S. C. (1996a). Gender and democracy in computer-mediated communication. In R. Kling (Ed.), *Computerization and controversy: Value conflicts and social choices* (pp. 476–489). New York: Academic.

Herring, S. C. (1996b). Posting in a different voice: Gender and ethics in CMC. In C. Ess (Ed.), *Philosophical perspectives on computer-mediated communication* (pp. 115–145). Albany: State University of New York Press.

Herzberg, B. (1994). Community service and critical teaching. *College Composition and Communication, 45*(3), 307–319.

Hillocks, G., Jr. (1986). *Research on written composition: New directions for teaching*. Urbana, IL: NCTE.

Hofstetter, F. T. (1998). *Internet literacy*. New York: McGraw-Hill.

Hogan, J. P. (1997). *Mind matters: Exploring the world of artificial intelligence*. New York: Ballantine.

Holcomb, C. (1997). A class of clowns: Spontaneous joking in computer-assisted discussions. *Computers and Composition, 14*(1), 3–18.

Holdstein, D. H., & Selfe, C. L. (Eds.). (1990). *Computers and writing: Theory, research, practice*. New York: MLA.

Howard, T. (1997). *The rhetoric of electronic communities*. Norwood, NJ: Ablex.

Huckin, T. N. (1997). Technical writing and community service. *Journal of Business and Technical Communication, 11*(1), 49–59.

Hull, G., Rose, M., Fraser, K. L., & Castellano, M. (1991). Remediation as a social construct: Perspectives from an analysis of classroom discourse. *College Composition and Communication, 45*, 299–329.

Hunt, K. (1996). Establishing a presence on the World Wide Web: A rhetorical approach. *Technical Communication, 4*, 376–387.

Hunt, R. (1989). A horse named Hans, a boy named Shawn: The Herr von Osten theory of response to writing. In C. M. Anson (Ed.), *Writing and response: Theory, practice, and research* (pp. 80–100). Urbana: NCTE.

Huot, B. (1996). Computers and assessment: Understanding two technologies. *Computers and Composition, 13*(2), 231–243.

Hutchings, P. (2001, December). *Opening the classroom door: Lessons from a national initiative*. Paper presented at the Modern Language Association Conference, New Orleans, LA.

Imaging Florida Project, University of Florida. (2001). *Imaging Florida*. Retrieved November 15, 2001, from http://www.elf.ufl.edu/agency/imaging.html

Inayatullah, S., & Milajevic, I. (1999). Exclusion and communication in the information era: From silences to global conversation. In W. Harcourt (Ed.), *women@internet: Creating new cultures in cyberspace* (pp. 76–88). London: Zed.

Jackson, K. T. (1987). *Crabgrass frontier: The suburbanization of the United States*. New York: Oxford University Press.

Jarratt, S. J. (1991). *Rereading the sophists: Classical rhetoric refigured*. Carbondale: Southern Illinois University Press.

Jessup, E. (1991). Feminism and computers in composition instruction. In G. E. Hawisher & C. L. Selfe (Eds.), *Evolving perspectives on computers and composition studies: Questions for the 1990s* (pp. 336–355). Urbana, IL: NCTE.

Johnson, D. W. (1970). *The social psychology of education*. New York: Holt, Rinehart & Winston.

Johnson, D. W., & Johnson, R. T. (1995). Using structured academic controversy strategy to teach language arts. In R. J. Stahl (Ed.), *Cooperative learning in language arts* (pp. 357–384). Menlo Park, CA: Addison-Wesley.

Johnson, D. W., & Johnson, R. T. (1999). *What we know about collaborative learning at the college level*. Retrieved November 15, 1999, from http://www2.emc.maricopa.edu/innovation/CCL/whatweknow.html

Johnson, D. W., Johnson, R. T., & Holubec, E. J. (1995). Learning together in the language arts classroom: Practical applications. In R. J. Stahl (Ed.), *Cooperative learning in the language arts* (pp. 49–70). Menlo Park, CA: Addison-Wesley.

Johnson, E. (1993). Electronic Shakespeare: Making texts compute. *Computer-Assisted Research Forum, 1*(3), 1–3.

Johnson, E. (1998). The World Wide Web, computers, and teaching literature. *Teaching Literature with Computers*. Retrieved July 20, 2001, from http://www.dsu.edu/~laflinj/tlwc/webprof.html

Jones, S. G. (1995a). *Cybersociety: Computer-mediated communication and community*. London: Sage.

Jones, S. G. (1995b). Understanding community in the information age. In S. G. Jones (Ed.), *Cybersociety: Computer-mediated communication and community* (pp. 10–35). Thousand Oaks, CA: Sage.

Jones, S. G. (1998). The Internet and its social landscape. In S. G. Jones (Ed.), *Virtual culture: Identity and communication in cybersociety* (pp. 7–35). Thousand Oaks, CA: Sage.

Joyce, M. (1987). *Afternoon, a story.* Watertown, MA: Eastgate Systems.

Joyce, M. (1988). Siren shapes: Exploratory and constructive hypertexts. *Academic Computing, 3*(4), 10–4, 37–42.

JSTOR. (2001). *The history of JSTOR.* Retrieved August 20, 2001, from http://www.jstor.org/about/background.html

Kalmbach, J. (1997b). Computer-supported classrooms and curricular change in technical communication programs. In S. A. Selber (Ed.), *Computers and technical communication: Pedagogical and programmatic perspectives* (pp. 261–274). Greenwich, CT: Ablex.

Kaplan, C., & Rose, E. C. (1988). *The canon and the common reader.* Knoxville: University of Tennessee Press.

Kaplan, C., & Rose, E. C. (1989a). *Approaches to teaching Lessing's The Golden Notebook.* New York: MLA.

Kaplan, C., & Rose, E. C. (1989b). *Doris Lessing: The alchemy of survival.* Athens: Ohio University Press.

Kaplan, N. (1991). Ideology, technology, and the future of writing instruction. In G. E. Hawisher & C. L. Selfe (Eds.), *Evolving perspectives on computers and composition studies: Questions for the 1990s* (pp. 11–42). Urbana, IL: NCTE.

Kasson, J. F. (1976). *Civilizing the machine: Technology and Republican values in America, 1776–1900.* New York: Grossman.

Kelley, K. (1995). *Out of control: The new biology of machines, social systems and the economic world.* New York: Perseus.

Kemp, F. (1998). Computer-mediated communication: Making nets work for writing instruction. In J. R. Galin & J. Latchaw (Eds.), *The dialogic classroom: Teachers integrating computer technology, pedagogy, and research* (pp. 133–150). Urbana, IL: NCTE.

Kennedy, G. (1998). *Comparative rhetoric: An historical and cross-cultural introduction.* Oxford: Oxford University Press.

Kennedy, G. (1999). *Classical rhetoric and its Christian and secular tradition from ancient to modern times.* Chapel Hill: University of North Carolina Press.

Kiesler, S. (Ed.). (1997). *Culture of the Internet.* Mahwah, NJ: Lawrence Erlbaum Associates.

Kiesler, S., Siegel, J., & McGuire, T. W. (1984). Social psychological aspects of computer-mediated communication. *American Psychologist, 39*(10), 1123–1134.

Kim, A. J. (2000). *Community building on the Web: Secret strategies for successful online communities.* Berkeley, CA: Peachpit.

King, L., & Stovall, D. (1992). *Classroom publishing: A practical guide to enhancing student literacy.* Hillsboro, OR: Blue Heron.

Kinneavy, J. L. (1971). *A theory of discourse, the aims of discourse.* Englewood Cliffs, NJ: Prentice-Hall.

Kinney, K. (2001). Online communities, self-silencing, and lost rhetorical spaces. *Kairos: A Journal for Teaching Writing in Webbed Environments, 6*(1). Retrieved August 20, 2001, from http://english.ttu.edu/kairos/6.1/coverweb/kinney

Kline, M. J. (1998). *A guide to documentary editing* (2nd ed.). Baltimore: Johns Hopkins University Press.

Klonoski, E. (1994). Using the eyes of the PC to teach revision. *Computers and Composition, 11*, 71–78.

Kollock, P., & Smith, M. A. (1999). Communities in cyberspace. In M. A. Smith & P. Kollock (Eds.), *Communities in cyberspace* (pp. 3–25). New York: Routledge.

Kramarae, C. (Ed.). (1988). *Technology and women's voices: Keeping in touch*. New York: Routledge.

Kramer, P. E., & Lehman, S. (1990). Mismeasuring women: A critique of research on computer ability and avoidance. *Signs: Journal of Women and Culture in Society, 67*, 158–172.

Ladd, E. C., Jr., & Lipset, S. M. (1975). *The divided academy: Professors and politics*. New York: McGraw-Hill.

Lancashire, I. (1996). *Using TACT with electronic texts: A guide to text-analysis computing tools, Version 2.1 for MS-DOS and PC DOS*. New York: MLA.

Landow, G. P. (1992). *Hypertext: The convergence of contemporary critical theory and technology*. Baltimore: Johns Hopkins University Press.

Langer, J. A. (1992). Speaking of knowing: Conceptions of understanding in academic disciplines. In A Herrington & C. Moran (Eds.), *Writing, teaching, and learning in the disciplines* (pp. 69–85). New York: MLA.

Langer, S. (1957). *Philosophy in a new key*. Cambridge, MA: Harvard University Press.

Langston, D. M., & Batson, T. (1990). The social shifts invited by working collaboratively on computer networks: The ENFI project. In C. Handa (Ed.), *Computers and community: Teaching composition in the twenty-first century* (pp. 160–184). Upper Montclair, NJ: Boynton/Cook.

Langston, D. M., & Batson, T. (1991). The social shifts invited by working collaboratively on computer networks: The ENFI project. In C. Handa (Ed.), *Computers and society: Teaching composition in the twenty-first century* (pp. 140–159). Upper Montclair, NJ: Boynton/Cook.

Lanham, R. (1994). The economics of attention. *Proceedings of 124th annual meeting of the Association of Research Libraries*. Retrieved August 20, 2001, from http://sunsite.berkeley.edu/ARL/Proceedings/124/ps2econ.html

Lavazzi, T. (1998, April). *Communication on (the) line*. Presentation to the Northeast Modern Language Association, Baltimore, MD.

Leavis, F. R. (1962). *Two cultures? The significance of C. P. Snow*. London: Chatto and Windus.

LeFevre, K. B. (1987). *Invention as a social act*. Carbondale: Southern Illinois University Press.

Leiby, M. A., & Henson, L. J. (1998). Common ground, difficult terrain: Confronting difference through feminist collaboration. In E. G. Peck & J. S. Mink (Eds.), *Common ground: Feminist collaboration in the academy* (pp. 173–192). Albany: State University of New York Press.

Lemke, J. L. (1998). Metamedia literacy: Transforming meanings and media. In D. Reinking, L. Labbo, M. McKenna, & R. Kiefer (Eds.), *Handbook of literacy and technology: Transformations in a post-typographic world* (pp. 283–301). Mahwah, NJ: Lawrence Erlbaum Associates.

Leonard, D., & Sensiper, S. (2000). The role of tacit knowledge in group innovation. In D. E. Smith (Ed.), *Knowledge, groupware, and the Internet* (pp. 281–301). Boston: Heinemann.

Lewis, M. J., & Lloyd-Jones, R. (1996). *Using computers in history: A practical guide*. New York: Routledge.

Living Lab Collaborative, University of North Dakota. (1999). *Peer editing and writing as a process*. Retrieved September 20, 2001, from http://volcano.und.nodak.edu/vwdocs/msh/llc/is/pe.html

Lockyard, J. (1996). Progressive politics: Electronic individualism and the myth of virtual community. In D. Porter (Ed.), *Internet culture* (pp. 219–232). New York: Routledge.

Luke, C. (1996). *ekstasis@cyberia*. Retrieved May 1, 1998, from http://www.gseis.uscl.edu/courses/ed253a/Luke/CYBERDIS.html

Lunsford, A., & Connors, R. (1995). *The St. Martin's Handbook* (3rd ed.). New York: St. Martin's.

Lyotard, J. (1997). *The postmodern condition: A report on knowledge* (B. Massumi, Trans.). Minneapolis: University of Minnesota Press.

MacKinnon, R. C. (1995). Searching for the leviathan in usenet. In S. G. Jones (Ed.), *Cybersociety: Computer-mediated communication and community* (pp. 112–137). Thousand Oaks, CA: Sage.

MacNealy, M. S. (1999). *Strategies for empirical research in writing*. Needham Heights, MA: Allyn & Bacon.

Marcuse, H. (1964). *One-dimensional man: Studies in the ideology of advanced industrial society*. Boston: Beacon.

Markley, R. (1996). *Virtual realities and their discontents*. Baltimore: Johns Hopkins University Press.

Martin, L. E. M. (Ed.). (1997). *The challenge of Internet literacy: The instruction-web convergence*. London: Haworth.

Marx, L. (1967). *The machine in the garden: Technology and the pastoral ideal in America*. London: Oxford University Press.

Mawdsley, E., & Munk, T. (1993). *Computers for historians*. Manchester, UK: Manchester University Press.

McLaren, P. (1996). Paulo Freire and the academy: A challenge from the U.S. left. *Cultural Critique*, 151–184.

McLuhan, M. (1964). *Understanding media*. New York: McGraw-Hill.

Mechling, J. (1979). If they can build a square tomato: Notes towards a holistic approach to regional studies. *Prospects, 4*, 59–77.

Mengel, S. A., & Carter, L. M. (1999). Multidisciplinary education through software engineering. In *ASEE/IEEE Frontiers in Education conference proceedings* (pp. 13a3-12–13a3-17). Piscataway, NJ: IEEE.

Mezirow, J. (1998). On critical reflection. *Adult Education Quarterly, 48*(3), 185–199.

Miller, E. (1998, May). An introduction to the resource description framework *D-Lib magazine*. Retrieved August 20, 2001, from http://www.dlib.org/dlib/may98/miller/05miller.html

Minty, J. (1991). Why do you keep those cats? In J. Anderson (Ed.), (pp. 105–106).

Mitchell, W. J. (1996). *City of bits: Space, place, and the infobahn*. Cambridge, MA: MIT Press.

Modern Language Association Committee on Computers and Emerging Technologies in Teaching and Research. (1996). *Guidelines for evaluating computer-related work in the modern languages*. Retrieved November 15, 1999, from http://www.mla.org/reports/ccet/ccet guidelines.htm

Mohanty, C. (1991). Cartographies of struggle: Third world women and the politics of feminism. In C. Mohanty, A. Russo, & L. Torres (Eds.), *Third world women and the politics of feminism* (pp. 1–50). Bloomington: Indiana University Press.

Moore, M., & Potts, C. (1994). Learning by doing: Goals and experiences of two software engineering project courses. In J. L. Diaz-Herrara (Ed.), *Proceedings of seventh conference on software engineering education* (pp. 151–164). New York, NY: Springer-Verlag.

Morris, C. (1995). *Becoming southern: The evolution of a way of life, Vicksburg and Warren County, Mississippi, 1770–1860*. New York: Oxford University Press.

Morris, P. J., II, & Tchudi, S. (1996). *The new literacy: Moving beyond the 3Rs*. San Francisco: Jossey-Bass.

Mouffe, C. (1993). *The return of the political*. London: Verso.

Murphy, A. (1981). *Special children, special parents: Personal issues with handicapped children*. Englewood Cliffs, NJ: Prentice-Hall.

Nader, R. (1996). The body politic. *PC World, 14*, 193.

National Endowment for the Humanities. (1998, January 28). *Schools for a new millennium*. Retrieved February 26, 1998, from http://www.neh.gov

National Initiative for a Networked Heritage. *Building blocks* from http://www.ninch.org/bb/project/project.html. "Field areas" questionnaire reports from http://www.ninch.org/bb/field/field.html

National Science Council, Committee on Intellectual Property Rights and the Emerging Information Infrastructure. (2001). *The digital dilemma: Intellectual property in the digital age*. Retrieved September 20, 2001, from http://books.nap.edu/html/digital dilemma

Nesbitt, P. D., & Thomas, L. E. (1998). Beyond feminism: An intercultural challenge for transforming the academy. In E. G. Peck & J. S. Mink (Eds.), *Common ground: Feminist collaboration in the academy* (pp. 31–49). Albany: State University of New York Press.

Noble, D. F. (1977). *America by design: Science, technology and the rise of corporate capitalism*. New York: Knopf.

Noddings, N. (1991). Stories in dialogue: Caring and interpersonal reasoning. In C. Witherell & N. Noddings (Eds.), *Stories lives tell: Narrative and dialogue in education* (pp. 157–170). New York: Teachers College.

Nonaka, I. (2000). A dynamic theory of organizational knowledge creation. In D. E. Smith (Ed.), *Knowledge, groupware and the Internet* (pp. 3–42). Boston: Heinmann.

Norberg, A. L., & O'Neill, J. E. (1996). *Transforming computer technology: Information processing for the pentagon, 1962–1986.* Baltimore: Johns Hopkins University Press.

North, S. M. (1987). *The making of knowledge in composition: Portrait of an emerging field.* New York: Greenwood.

Novak, T. P., & Hoffman, D. L. (1998). *Bridging the digital divide: The impact of race on computer access and Internet use.* Retrieved April 24, 1998, from http://www2000.ogsm.vanderbilt.edu/papers/race/science.html

O'Brien, J. (1999). Writing in the body: Gender (re)production in online interaction. In M. Smith & P. Kollock (Eds.), *Communities in cyberspace* (pp. 76–104). London: Routledge.

Odell, L., & Prell, C. L. (1999). Rethinking research on composing: Arguments for a new research agenda. In M. Rosner, B. Boehm, & D. Journet (Eds.), *History, reflection, and narrative: The professionalization of composition 1963–1983.* Westport, CT: Ablex.

Ohmann, R. (1985). Literacy, technology, and monopoly capital. *College English, 47*(7), 675–689.

O' Malley, C. (1991). Designing computer support for collaborative learning. In C. O'Malley (Ed.), *Computer-supported collaborative learning* (pp. 283–297). Berlin: Springer-Verlag.

Ong, A. (1997). The gender and labor politics of postmodernity. In L. Lowe & D. Lloyd (Eds.), *The politics of culture in the shadow of capital* (pp. 61–97). Durham, NC: Duke University Press.

Ong, W. (1982). *Orality and literacy.* New York: Methuen.

Pajares, F., & Johnson, M. J. (1996). Self-efficacy beliefs and the writing performance of entering high school students. *Psychology in the Schools, 33,* 163–175.

Palmer, P. J. (1998). *The courage to teach: Exploring the inner landscape of a teacher's life.* San Francisco: Jossey-Bass.

Pastore, M. (2000a). Digital divide shows signs of narrowing. *CyberAtlas.* Retrieved October 18, 2000, from http://cyberatlas.internet.com/big picture/demographics.html

Pastore, M. (2000b). US schools all but wired. *CyberAtlas.* Retrieved May 10, 2000, from http://cyberatlas.internet.com/markets/education.html

Pastore, M. (2001a). Interest high, but Internet too costly for inner city. *CyberAtlas.* Retrieved May 10, 2001, from http://cyberatlas.internet.com/big_picture/demographics.html

Pastore, M. (2001b). Teachers say Internet improves quality of education. *CyberAtlas.* Retrieved May 10, 2001, from http://cyberatlas.internet.com/markets/education.html

Peck, E. G., & Mink, J. S. (Eds.). (1998). *Common ground: Feminist collaboration in the academy.* Albany: State University of New York Press.

Peterson, M. (1993). Life on the Internet: Portrait of collaboration. *The North American Review, 10*(1).

Piaget, J. (1926). *The child's conception of the world* (J. A. Thomlinson, Trans.). London: Kegan Paul.

Pike, D. (1998). 80 years of Frandsen. *Silver and Blue, 9.*

Pimentel, K. (1995). *Virtual reality: Through the new looking glass.* New York: Intel/McGraw-Hill.

Plato. (1993). *The republic* (G. M. A. Grube, Trans.). Indianapolis, IN: Hacket.

Postman, N. (1985). *Amusing ourselves to death.* New York: Viking.

Prigogine, I. (1999). A message from Ilya Prigogine. *First Monday, 4*(8). Retrieved August 20, 2001, from http://www.firstmonday.dk/issues/issue4_8/prigogine

Raymond, E. S. (1999a). *The cathedral and the bazaar.* Retrieved November 15, 1999, from http://www.tuxedo.org/~esr/writings/cathedral-bazaar/cathedral-bazaar.html

Raymond, E. S. (1999b). *How to become a hacker.* Retrieved November 15, 1999, from http://www.tuxedo.org/~esr/faqs/hacker-howto.html

Recchio, T. E. (1992). Parallel academic lives: Affinities of teaching assistants and freshman writers. *WPA: Writing Program Administration, 15*(3), 57–61.

Reich, R. B. (1991). *The work of nations: Preparing ourselves for 21st-century capitalism*. New York: Knopf.

Reinking, D., McKenna, M. C., Labbo, L. D., & Kieffer, R. D. (Eds.). (1998). *Handbook of literacy and technology: Transformations in a post-typographic world*. Mahwah, NJ: Lawrence Erlbaum Associates.

Reiss, D., Selfe, D., & Young, A. (Eds.). (1998). *Electronic communication across the curriculum*. Urbana, IL: NCTE.

Reither, J. A. (1993). Bridging the gap: Scenic motives for collaborative writing in the workplace and schools. In R. Spilka (Ed.), *Writing in the workplace* (pp. 195–206). Carbondale: Southern Illinois University Press.

Resnick, M. (1996). *Distributed constructionism*. Retrieved June 6, 2001, from http://lcs.www.media.mit.edu/groups/el/Papers/mres/Distrib-Construc.html

Resnick, M. (1998). Technologies for lifelong kindergarten. *Educational Technology Research and Development, 46*(4). Retrieved June 6, 2001, from http://lcs.www.media.mit.edu/groups/el/papers/mres/lifelongk

Rickly, R. J. (1995). *Exploring the dimensions of discourse: A multi-modal analysis of electronic and oral discussion in developmental writing*. Unpublished dissertation, Ball State University, Muncie, IN. Retrieved November 1, 1999, from http://labyrinth.daedalus.com/dissertations

Robbins, K. (1995). Cyberspace and the world we live in. In M. Featherstone & R. Burrows (Eds.), *Cyberspace, cyberbodies, cyberpunk: Cultures of technological embodiment*. London: Sage.

Robertson, D. L. (1996). Facilitating transformative learning: Attending to the dynamics. *Adult Education Quarterly, 47*(1), 41–54.

Rogers, E. (1995). *Diffusion of innovations* (4th ed.). New York: Free Press.

Ronald, K. (1988). On the outside looking in: Students' analyses of professional discourse communities. *Rhetoric Review, 7*, 130–159.

Rose, M. (1989). *Lives on the boundary*. New York: Penguin.

Roskelly, H. (1994). The risky business of group work. In G. Tate, E. P. J. Corbett, & N. Meyers (Eds.), *The writing teacher's sourcebook* (pp. 141–146). (3rd ed.). New York: Oxford University Press.

Roy, M. J. (1998). UConn highly ranked in "most wired" survey. *Advance, 16*(23), 1.

Roy, P. (1995). Cultivating cooperative group process skills within the language arts classroom. In R. J. Stahl (Ed.), *Cooperative learning in language arts* (pp. 17–48). Menlo Park, CA: Addison-Wesley.

Rugh, S. S. (2001). *Our common country: Family farming, culture, and community in the nineteenth century midwest*. Bloomington: Indiana University Press.

Ruis, B. (1999). Crossing borders: From crystal slippers to tennis shoes. In W. Harcourt (Ed.), *women@internet: Creating new cultures in cyberspace*. London: Zed.

Saldana-Portillo, M. J. (1997). Developmentalism's irresistible seduction—Rural subjectivity under Sandinista agricultural policy. In L. Lowe & D. Lloyd (Eds.), *The politics of culture in the shadow of capital* (pp. 132–172). Durham, NC: Duke University Press.

Sasken, S. (1994). *Cities in a world economy*. Thousand Oaks, CA: Pine Forge.

Schiller, D. (1999). *Digital capitalism: Networking the global market system*. Cambridge, MA: MIT Press.

Schiller, H. (1996). *Information inequality: The deepening social crisis in America*. New York: Routledge.

Schleifer, R. (1997). Disciplinarity and collaboration in the sciences and humanities. *College English, 59*, 438–452.

Schon, D. (1987). *Educating the reflective practitioner: Toward a new design for teaching and learning in the professions*. San Francisco: Jossey-Bass.

Schriner, D. K., & Rice, W. C. (1989). Computer conferencing and collaborative learning: A discourse community at work. *College Composition and Communication, 40*, 472–478.

Schroeder, R. (1996). *Possible worlds: The social dynamic of virtual reality technology*. Boulder, CO: Westview.

Schumpeter, J. A. (1950). *Capitalism, socialism, and democracy*. New York: Harper.

Schutz, A., & Gere, A. R. (1998). Service learning and English studies: Rethinking "public" service. *College English, 60*(2), 129–149.

Selfe, C. L. (1989). Redefining literacy: The multilayered grammars of computers. In G. E. Hawisher & C. L. Selfe (Eds.), *Critical perspectives on computers and composition instruction* (pp. 3–15). New York: Teachers College.

Selfe, C. L. (1992). Preparing teachers for the virtual age: The case for technology critics. In G. E. Hawisher & P. LeBlanc (Eds.), *Re-imagining computers and composition: Teaching and research in the virtual age* (pp. 24–42). Portsmouth: Boynton/Cook.

Selfe, C. L. (1998, April). Technology and literacy: A story about the perils of not paying attention. Paper presented at annual meeting of the conference on *College Composition and Communication, 50*(3), 411–436. Chicago, IL. Available online: http://www.ncte.org/forums/selfe/

Selfe, C. L. (1999). *Technology and literacy in the twenty-first century: The importance of paying attention*. Carbondale: Southern Illinois University Press.

Selfe, C. L., & Hilligoss, S. (Eds.). (1994). *Literacy and computers: The complications of teaching and learning with technology*. New York: MLA.

Selfe, C. L., & Meyer, P. R. (1991). Testing claims for on-line conferences. *Written Communication, 8*, 163–192.

Selfe, C. L., & Selfe, R. J., Jr. (1994). The politics of the interface: Power and its exercise in electronic contact zones. *College Composition and Communication, 45*, 480–504.

Shakespeare, W. (2000). *As you like it* (M. Hattaway, Ed.). London: Cambridge University Press. (Original work published 1623)

Shapiro, G., & Markoff, J. (1998). The coder as a black box instrument. In G. Shapiro & J. Markoff (Eds.), *Revolutionary demands: A content analysis of the cahiers de doléances of 1789* (pp. 60–72). Stanford, CA: Stanford University Press.

Sharples, M. (Ed.). (1993). *Computer-supported collaborative writing*. New York: Springer-Verlag.

Sheriff, C. (1997). *The artificial river: The Erie canal and the paradox of progress, 1817–1862*. New York: Hill and Wang.

Shor, I. (1987). *Critical teaching and everyday life*. Chicago: University of Chicago Press.

Shor, I. (1996). *When students have power*. Chicago: University of Chicago Press.

Shulman, L. S. (1999). Taking learning seriously. *Change, 31*(4), 10–17.

Simpson, J. H. (1998, April 8). *Re: Session H.17 at CCCC*. Retrieved August 16, 1999, from http://www.ttu.edu/wcenter/9804/msg00078.html

Singley, C. J., & Sweeney, S. E. (1998). In league with each other: The theory and practice of feminist collaboration. In E. G. Peck & J. S. Mink (Eds.), *Common ground: Feminist collaboration in the academy* (pp. 63–79). Albany: State University of New York Press.

Slattery, P., & Kowalski, R. (1998). On-screen: The composing processes of first-year and upper-level college students. *Computers and Composition, 15*(1), 61–81.

Smith, M. A. (1999). Invisible crowds in cyberspace: Mapping the social structure of the usenet. In M. A. Smith & P. Kollock (Eds.), *Communities in cyberspace* (pp. 195–219). New York: Routledge.

Snow, C. P. (1993). *The two cultures*. London: Cambridge University Press.

Snyder, G. (1990). *The practice of the wild: Essays*. San Francisco: North Point Press.

Snyder, I. (Ed.). (1998). *Page to screen: Taking literacy into the electronic era*. London: Routledge.

Spender, D. (1995). *Nattering on the net: Women, power, and cyberspace*. North Melbourne, AU: Spinifex.

Spilka, R. (1993). Influencing workplace practice: A challenge for professional writing specialists in academia. In R. Spilka (Ed.), *Writing in the workplace* (pp. 207–219). Carbondale: Southern Illinois University Press.

Spitzer, M. (1990). Local and global networking: Implications for the future. In D. Holdstein & C. Selfe (Eds.), *Computers and writing: Theory, research, and practice* (pp. 58–70). New York: MLA.

Spivak, G. (1999). *A critique of postcolonial reason: Toward a history of the vanishing present.* Cambridge, MA: Harvard University Press.

Stahl, R. J. (1995). Cooperative learning: A language arts context and an overview. In R. J. Stahl (Ed.), *Cooperative learning in language arts* (pp. 1–16). Menlo Park, CA: Addison-Wesley.

Standage, T. (1998). *The Victorian Internet: The remarkable story of the telegraph and the nineteenth century's on-line pioneers.* New York: Walker and Company.

Stillinger, J. (1991). *Multiple authorship and the myth of solitary genius.* New York: Oxford University Press.

Sullivan, P. (1996, March). *The changing faces of discourse: Women, e-spaces, and the World Wide Web.* Paper presented at the annual meeting of the Conference on Composition and Communication, Milwaukee, WI.

Sullivan, P., & Dautermann, J. (Eds.). (1996). *Electronic literacies in the workplace: Technologies of writing.* Urbana, IL: NCTE.

Sullivan, P., & Porter, J. (1997). *Opening spaces: Writing technologies and critical research practices.* Norwood, NJ: Ablex.

Swales, J. (1990). *Genre analysis: English in academic and research settings.* New York: Cambridge.

Tabbi, J. (1996). Reading, writing, hypertext: Democratic politics in the virtual classroom. In D. Porter (Ed.), *Internet culture* (pp. 233–252). New York: Routledge.

Takayoshi, P. (1994). Building new networks from the old: Women's experiences with electronic communications. *Computers and Composition, 11,* 21–35.

Taylor, E. W. (1998). *The theory and practice of transformative learning: A critical review.* Columbus, OH: ERIC Clearinghouse on Adult, Career, and Vocational Education. Retrieved August 13, 1998, from http://ericacve.org/mp_taylor_01.asp

Taylor, P. (1992). Social epistemic rhetoric and chaotic discourse. In G. E. Hawisher & P. LeBlanc (Eds.), *Re-imagining computers and composition: Teaching and research in the virtual age* (pp. 131–148). Portsmouth, NH: Boynton.

Taylor, T. W., & Ward, I. (Eds.). (1998). *Literacy theory in the age of the Internet.* New York: Columbia University Press.

Tebeaux, E. (1996). Nonacademic writing in the 21st century: Achieving and sustaining relevance in research and curricula. In A. H. Duin & C. J. Hansen (Eds.), *Nonacademic writing: Social theory and technology* (pp. 35–55). Mahwah, NJ: Lawrence Erlbaum Associates.

Teute, F. J. (2001, January). To publish and perish: Who are the dinosaurs in scholarly publishing? *Journal of Scholarly Publishing, 32*(2), 102–112.

The Tomás Rivera Policy Institute. (2001, July 20). Bridging the digital divide. Available online: http://www.trpi.org/dss/itstats.html

Tomlinson, J. (1997). Cultural globalisation and cultural imperialism. In A. Mohammadi (Ed.), *International communication and globalisation* (pp. 159–162). London: Sage.

Toulmin, S. (2001). *Return to reason.* Cambridge, MA: Harvard University Press.

Tovey, J. (1998). Organizing features of hypertext: Some rhetorical and practical elements. *Journal of Business and Technical Communication, 12*(3), 371–380.

Trimbur, J. (1991). Literacy and the discourse of crisis. In R. Bullock & J. Trimbur (Eds.), *The politics of writing instruction: Postsecondary* (pp. 277–296). Portsmouth: Boynton/Cook.

Trupe, A. (2002). Academic literacy in a wired world: What should a freshman essay look like? *Kairos: A journal for teachers of writing in webbed environments, 7*(2). Retrieved August 30, 2002, from http://english.ttu.edu/kairos/7.2

Tuman, M. (Ed.). (1992a). *Literacy online: The promise (and peril) of reading and writing with computers.* Pittsburgh, PA: University of Pittsburgh Press.

Tuman, M. (1992b). *Word perfect: Literacy in the computer age.* Pittsburgh, PA: University of Pittsburgh Press.

Turkle, S. (1984). *The second self: Computers and the human spirit.* New York: Simon & Schuster.

Turkle, S. (1995). *Life on the screen: Identity in the age of the Internet.* New York: Simon & Schuster.

Turkle, S. (1996). Virtuality and its discontents: Searching for community in cyberspace. *The American Prospect, 24*(7). Retrieved November 15, 1999, from http://epn.org/prospect/24/24turk.html

Twain, M. (1917). *A Connecticut yankee in King Arthur's court.* New York: Harper.

Udell, J. (1999). *Practical Internet groupware.* Sebastopol, CA: O'Reilly and Associates.

Udell, J. (2001). *Internet groupware for scientific collaboration.* Retrieved July 20, 2001, from http://software-carpentry.codesourcery.com/Groupware/report.html

Ulmer, G. (1997). *Electracy.* Retrieved November 15, 1999, from http://www.elf.ufl.edu

United States Department of Education. (1996). *GOALS 2000: Educate America Act.* Washington, DC: Government Printing Office.

Unsworth, J. (1996). Living inside the (operating) system: Community in virtual reality. In T. M. Harrison & T. Stephen (Eds.), *Computer networking and scholarly communication in the twenty-first century university* (pp. 137–150). Albany: State University of New York Press.

Vazquez, J. J. (1999). *The German historical school of economics: Theoretical and empirical considerations for the Lake George region in upstate New York.* Unpublished master's thesis, Rensselaer Polytechnic Institute, Troy, NY.

Vickery, R. L. (1983). *Sharing architecture.* Charlottesville: Virginia University Press.

Vygotsky, L. S. (1978). *Mind in society: The development of higher psychological processes* (M. Cole, J. John-Steiner, S. Scribner, & E. Souberman, Eds.). Cambridge, MA: Harvard University Press.

Vygtosky, L. S. (1986). *Thought and language* (A. Kozulin, Ed. & Trans.). Cambridge, MA: MIT Press.

Wahlstrom, B. J. (1994). Communication and technology: Defining a feminist presence in research and practice. In C. L. Selfe & S. Hilligoss (Eds.), *Literacy and computers: The complications of teaching and learning with technology* (pp. 171–185). New York: MLA.

Wahlstrom, B. J., & Scruton, C. (1997). Constructing texts/understanding texts: Lessons from antiquity and the middle ages. *Computers and Composition, 14*(3), 311–328.

Ward, I. (1994). *Literacy, ideology, and dialogue: Towards a dialogic pedagogy.* New York: State University of New York Press.

Warschauer, M. (1999). *Electronic literacies: Language, culture, and power in online education.* Mahwah, NJ: Lawrence Erlbaum Associates.

Weber, M. (1834). Power and bureaucracy. In S. M. Miller (Ed.), *Max Weber: Selections from his work* (pp. 59–82). New York: Thomas Y. Crowell Company.

Wiener, N. (1948). *Cybernetics; or, control and communication in the animal and the machine.* Cambridge, MA: Technology Press.

Wellman, B., & Gulia, M. (1999). Virtual communities as communities: Net surfers don't ride alone. In M. A. Smith & P. Kollock (Eds.), *Communities in cyberspace* (pp. 167–194). London: Routledge.

West, C. (1995). *Commencement address to Harvard University's John F. Kennedy School of Government.* Presentation to Harvard University School of Government, Cambridge, MA.

Whitman, W. (1881). There was a child went forth. In *Autumn rivulets.* Retrieved November 15, 1999, from http://www.liglobal.com/walt/childwentforth.html

Whitman, W. (1882). *Song of myself.* Retrieved November 15, 1999, from http://www.liglobal.com/walt/songofmyself/song51.html

Wilkenson, A. (1990). *Spoken English illuminated.* Milton Keynes, UK: Open University Press.

Winkelmann, C. L. (1995). Electronic literacy, critical pedagogy, and collaboration: A case for cyborg writing. *Computers and the Humanities, 29,* 431–448.

Winters, J. E. (1997). The effect of computers on college writing: A view from the field. In L. Lloyd (Ed.), *Technology and teaching* (pp. 13–21). Medford, NJ: Information Today.

Woodmansee, M. (1994). On the author effect: Recovering collectivity. In M. Woodmansee & P. Jaszi (Eds.), *The construction of authorship: Textual appropriation in law and literature* (pp. 15–28). Durham, NC: Duke University Press.

Woolsey, K. H. (1996). Hope and joy in a rational world: Kids, learning and computers. In C. Fisher, D. C. Dwyer, & K. Yocam (Eds.), *Education and technology: Reflections on computing in classrooms* (pp. 67–89). San Francisco: Jossey-Bass.

Wysocki, A., & Johnson-Eilola, J. (1999). Blinded by the letter: Why are we using literacy as a metaphor for everything else? In G. E. Hawisher & C. L. Selfe (Eds.), *Passions, pedagogies, and 21st century technologies* (pp. 349–368). Logan, UT: Utah State University Press.

Yagelski, R. P., & Gabrill, J. T. (1998). Computer-mediated communication in the undergraduate writing classroom: A study of the relationship of online discourse and classroom discourse in two writing classes. *Computers and Composition, 15*(1), 11–40.

Author Index

Subject Index

417

Printed in the United States
by Baker & Taylor Publisher Services